Electromagnetic Waves

in Space and History, for Amateurs and Professionals

Dipl.-Ing. Thomas Kaczmarek

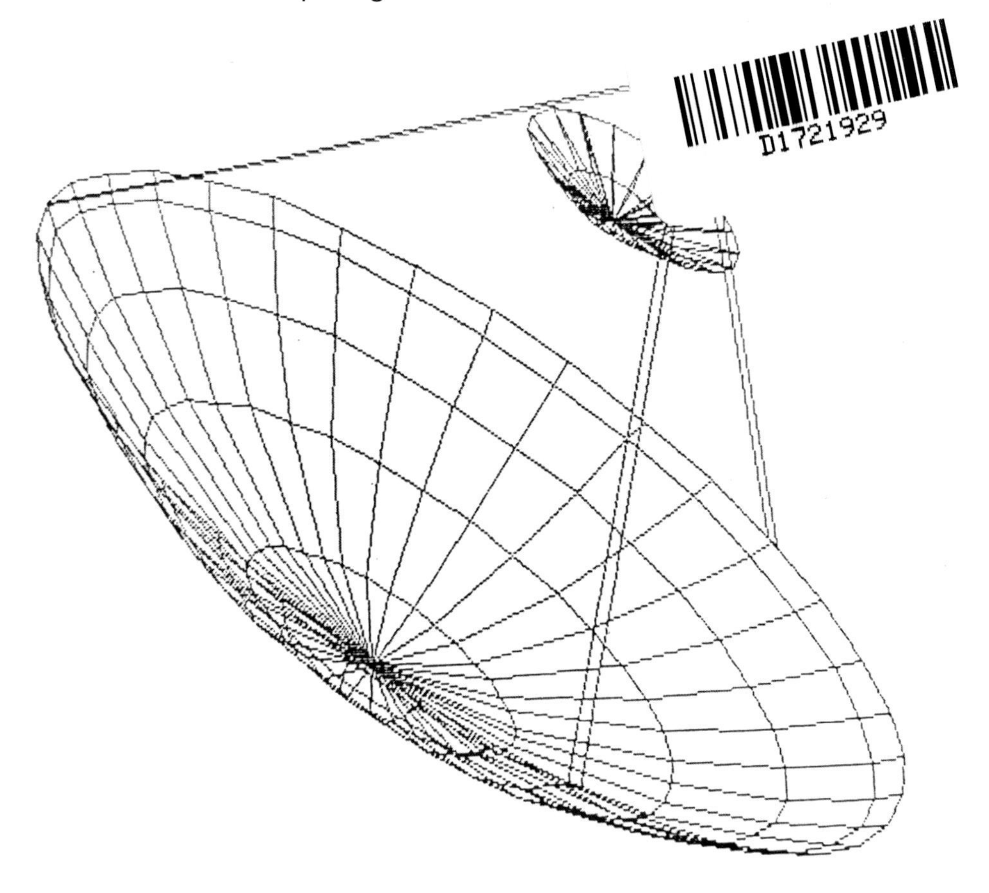

Wave propagation, antennas and frequencies,
analogue and digital modes,
digital systems and networks

About the author

Thomas Kaczmarek is a graduated engineer in RF- and communication technologies. He studied at the Ruhr-University in Bochum and was a member of the antenna and wave propagation group. He started as a development engineer in microwave technology. Then he continued as a team leader of an expert group for radio systems. He worked in telecommunication projects in London, Chicago, Berlin, Vienna, and Copenhagen.

In the field of automotive electronics he managed several projects. He was Lead Systems Engineer and responsible for an international team of engineers from fourteen nations who specified autonomous driving systems.

1st edition · June 2023

ISBN 978-3-910492-14-1

Legal notice

published by:	Thomas Kaczmarek
responsible:	Thomas Kaczmarek
layout:	Klaus-Georg Rump
overall production:	Blömeke Druck SRS GmbH, www.bloemeke-media.de

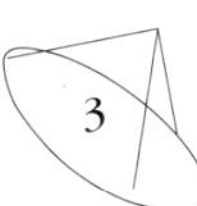

Before we start:

From the bottom of my heart, I thank my amazing wife. Without her, this book would not have been possible. With her positive energy and her tireless help, the book in its final shape and expression was created.

A big thank you to everyone who worked professionally in all fields, especially creating layout and realising production in all its details. Thank you, Klaus-Georg and Christian Rump and the phantastic team of Bloemeke Druck SRS GmbH.

Special thanks to Sabine Ewald for the competent editorial contributions.

Thomas

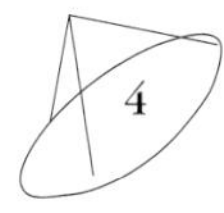

Table of Contents

Table of Contents

Table of Contents

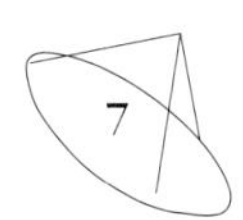

Table of Contents

Table of Contents

Table of Contents

Table of Contents

Table of Contents

Preface

This book was written for those who are interested in electromagnetic waves and communication systems, especially when they are interested in investigating some secrets about communication technology. We are dealing with electromagnetic waves in nature and technology without a huge load of mathematics. A lot of examples give a deeper insight into that. One focus is on the meaning of electromagnetic waves in systems engineering, considering what and how RF engineers are developing. Even radio amateurs, CB amateurs, and electronic enthusiasts will find new sources of inspiration and some well-known matters, too.

When you start reading, you can select three different ways to step into this book's content and to go through:

1. The classical start with the table of content and by reading straightforward or 2. The start by keywords from the glossary at the end of this book or 3. Simply to discover selected chapters that might be interesting. Those who are looking for their entry by keywords from the table of content at the beginning or the glossary at the end, will find a rather practical start in the content. Readers who like in-depth descriptions about their selected topics may refer to further literature beyond as listed at the end or as found on the World Wide Web. [1] is a good scientific book and a recommendation for in-depth analysis of radio waves with a mathematical approach.

"The air is literally full of radio signals." This simple statement is the right basis for all chapters in this book, which are about wave propagation and the nature of waves, their characteristics and behaviour, and also about radio systems and networks. Radio and communication systems have been developed and operated for 120 years by engineers and technical staff. Famous inventors from different professions and engineers all over the world from different faculties provide the base. We can look at electromagnetic waves from various sides with a closer look at the systems when we start to learn about details. Unfortunately we cannot deal with all aspects of each topic. Discussions about all the details would take too much space for a complete description. But there is literature for each topic available to study beyond the chapters in here. We will mention many popular topics and give an overview of electromagnetic waves. Examples: X-ray and ultraviolet radiation belong to the medical topics that need much more attention than we can give in this book. However, we will have a look at typical applications. Most of them belong to radio systems and applications from longwave frequencies with a range of 10 kHz up to microwave bands in a range of 1 to 30 GHz, also known as LW, MW, SW, VHF, UHF, SHF and beyond. Simply to cover frequencies beyond the visible light, it would take more literature.

The Gift – a Little Winter Story

It was a cold Christmas evening in the 1970s of the twentieth century, somewhere in the Ruhr area, one of the greatest industrial regions in the middle of Europe.

"The air is literally full of radio signals." This simple and fundamental phrase I read at the age of 14. It took my full attention and that did not stop for the following decades of my life. The magical phrase was part of the instruction manual for a shortwave radio kit. The kit was delivered by Radio Shack (Tandy Corporation) and it covered a frequency range from 3 to 30 MHz. The imagination of the air around us, which is full of radio signals, was like an ignition for my strong inspiration. The consequences have been similar to an infection with a kind of rf virus. It has been with me for a lifetime. The next step was to assemble the kit, a little shortwave radio. Somehow, I felt, like it was coming from my grandfather's heart when he gave me this meaningful gift. It was the same Christmas evening, when I started to assemble and install everything step by step according to the assembly instruction.

This was the way I got into contact with electromagnetic waves by receiving worldwide shortwave radio stations and some mysterious signals sounding like they are coming from another world. Some of them seemed to be melodic with different pitches, while others seemed to be bouncing and dancing through the atmosphere. The circuit filtered them from the air and provided an acoustic signal to the earphone, while the weather outside was unpleasant and cold. Some of the signals were noisy or had a certain rhythm, like the ticking of a clock, which was really strange. Turning the knob of the tuning capacitor, music appeared and turning in the other direction, I could listen to news from overseas in different languages. Are those really signals from all over the world? That was the point, my curiosity began to grow at double speed.

In the assembly instructions, I could also read that I should use a coil with 20 to 30 turns for monitoring at nighttime. But why? In the night, wave propagation is better at low frequencies, as we know today. So I used a larger coil. For daytime reception, the author recommended 2 to 20 turns. As I could hear, this was really a good idea. Without this knowledge, I would have been monitoring in the wrong frequency bands or without good results. I probably would have assumed a malfunction. Today, we know that it is our atmosphere which makes the difference. During nighttime the attenuation and refraction characteristics of the atmosphere are not the same as during daytime.

This little introduction shall be the frame for most of the topics in this book. We will read about wave propagation and the resulting various effects and applications of communication electronics within our daily lives. Simply looking at the various types of radio systems and their operational characteristics will take plenty of time. Some related fields will be mentioned briefly without a detailed background because the space here is limited. Examples: medical technologies, radio astronomy, biological interaction. However, space applications and radio astronomy have their own chapters in here.

We cannot see electromagnetic waves with our eyes, and we cannot feel them. In history, some intelligent and creative scientists had to discover them and to find a mathematical approach or experiments to describe them. Famous scientists as Maxwell, Tesla, Hertz, Marconi, and many others gave us the foundation. Early applications entered the stage hundreds of years ago. Nautical warnings could have been sent out, vessels could have contacted other vessels or coastal stations in case of an emergency. One of the most prominent examples is the Titanic, which could have send an emergency call in morse telegraphy before sinking in the disaster of 1912 when it rammed an iceberg. This could have really saved hundreds of lives in one of the greatest catastrophes in seafaring.

The use of electromagnetic waves and the operation of radio systems belong closely to communication technologies. Scientific depth is not the focus of this book, but one chapter is dedicated to a scientific and mathematical scenario. It reveals how scientists and mathematicians deal with numerical methods to solve electromagnetic problems while using fast calculators. In this example, we will face a special scenario about antenna structures and their reflection characteristics. We also discuss radio technology in general and focus on practical aspects too. Scientific aspects and technical components like antennas, waveguides and wave propagation are presented more detailed. Instead of mathematical perfection, we look at a set of topics that are almost described independently by different faculties. Customer devices, antennas, satellites, radio systems, frequency schedules, and the solar system are generally described in their own literature. Here, we have a mixture of them, and we explain how they interact. The link between all those topics are electromagnetic waves.

At the end of the book, the impact of social relations and social media are analysed within some examples. Today, we are still facing medical matters that we cannot explain. This happens again and again, and it makes nature and science exciting. Electric fields from transmitter power and their impact on biological organisms have been examined in depth for a long time by several authors. In 1990, the networks for digital cell phones were built. As a consequence, we saw a hype about new health issues when pseudoscientists entered the stage using evil methods. They scared people and switched off their clear minds by telling evil stories about diseases caused by electromagnetic waves, almost by cell phones and base stations.

Today, we face some new aspects, while we see a long time of digital transformation in the telecommunications sector. The consequences of 24/7 availability and the impact of apps on our individual health are not predictable. Even engineers who built up mobile networks in the 90s did not have any imagination of the use of those new technologies. Many lives could be saved by that, for example. But nobody had any idea about the upcoming negative consequences for the future. To hide behind a technology and stay anonymous, to mob, and attack other people, all this revealed the deeply bad nature and evil attitudes of some criminal users. Many of them are doing bad to others day by day over the networks. They hide in the shadow of anonymous accounts and act from the neighbourhood or from a long distance, so do psychopathic characters and scammers in distant countries, too. Meanwhile, we have many victims of threat and harassment.

Introduction

Electromagnetic waves cover a wide spectrum in the frequency domain. Wavelengths in the ELF and VLF bands (extremely low frequencies, very low frequencies) with some hundred metres wavelength can be found at the lower end. Towards the upper end, wherever it is, we will find mm waves with extremely high frequencies.

Between these extraordinary bands, we find the radio bands, which are of deeper interest in this book. We almost always consider wavelengths of 10 to 100 m (SW) as well as VHF, UHF, SHF, up to centimetres and mm waves (microwave), which are also used for radio communication. Those frequency bands of the electromagnetic spectrum are of special interest in this book. Even below the VLF-band we find some scientific applications. Also above the highest frequency bands we can find man made and natural electromagnetic transmissions such as visible light, ultraviolet radiation and x-rays.

In wireless communications electromagnetic waves are playing an important role. They carry energy and information from A to B. This can be transmitted from fixed stations or mobile stations, even from ships and planes. Of course, space communication and astronomy rely on wave propagation.

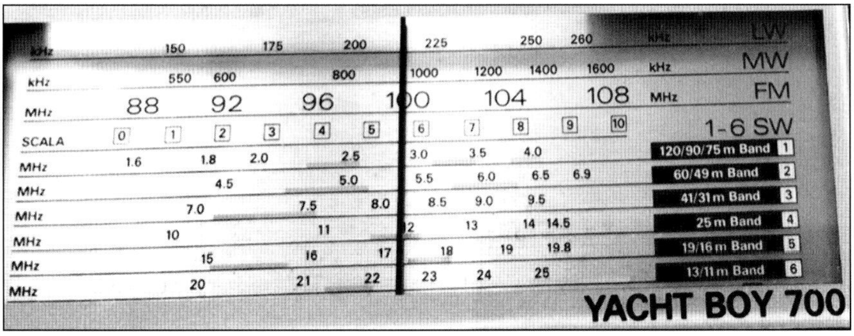

Figure 1: Tuning scale with long-, medium-, and shortwave bands plus FM. SW bands have been the preferred bands for worldwide listening to radio programmes.

The following chapters provide different aspects for a deeper look into functionality, technology and used components. Even networks are considered. The objective is to get a closer look at the involved systems and to learn more about wireless transmission by electromagnetic waves. Chapters deal with functionality, components, transmission coverage, noise, and network design. But we will also have a look at the people who use those technologies and those who create them. Finally, we discuss a possible wrong understanding of technical progress and some unexpected risks, as well as menacing consequences for personal health.

You will find a lot of practical experience in different areas like industrial development, construction, design, and electronics. Results from practical use cases and measurements will be discussed, too. We also try to demystify statements like "RF takes strange turns". Within each of the scenarios shown here, only a small amount of mathematics is used, because we do not want to bore or to stress the reader. It is allowed to skip the end of chapter one with a mathematical focus, because that scientific example uses numeric methods for calculating currents and fields. Here we get an impression of the way how numerical methods and computers will help engineers and designers in the field of research and development.

One major goal is to generate motivation for those readers who are already interested in technology and electromagnetics. Many chapters address the needs of young hobbyists and passionate radio enthusiasts. But the chapters also address professionals from other business areas, who initially read about topics around electromagnetic waves.

You will not find strictly scientific derivations, but a lot of basics in a good mixture of theoretical and practical knowledge. This collection you hold in your hand at this moment has different origins in related faculties. Electronics, nature, digital transformation, networks and media, and some more are presented and each description contributes different points of view. We appreciate giving motivation to many readers and supporting activities around science and technology. At last, we recognize overlapping areas with other professions like automotive electronics, biology, astronomy, and others. This book is a useful platform to get a look beyond the daily topics that we all have to deal with again and again. Everybody will find their own information and interesting ideas and probably new perspectives.

Beyond the technical aspects, we consider satellite monitoring and data transmission. This provides people predictive information for agricultural use, for example. Surface temperature, rain and weather, all this is a great benefit for us. Radio propagation also means information like news and entertainment programmes for all people in the world and a piece of freedom, even for those who may not having access to unfiltered news and to the World Wide Web. Through this little window, many people can nevertheless see what really happens in the world beyond the fake news of their own government. This means a blessing for all suppressed people in non democratic countries with restrictions that we can hardly imagine.

Chapter 1: Radio Waves and Radio Frequencies

The first chapter is a physical description of electromagnetic waves and their characteristics. This type of energy is everywhere, and it is also a little bit mysterious. We use RF-waves every day, and we are familiar with them. So it is a good idea to get a closer look at the applications we are using in our daily lives. In the second chapter we have a look more at components and systems. The third chapter includes a lot about networks, transmission lines and sites. We have just started to understand what the permanent use of some new technologies is doing to us and what the impact is on our society.

It is time to consider electromagnetic waves in the context of nature and history and how we enjoy all the helpful devices and products. Let us take some products we use every day. At home, we will find Bluetooth, wireless LAN, DECT and NFC technology. Nobody wants to do without them. Wireless phone calls, wireless access to printers, and giving other devices access to the World Wide Web is rather comfortable. Even special applications such as remote control of devices, listening to music by wireless earphones or even eavesdropping on rooms are part of a wide portfolio of wireless applications. Each of these technologies is worth being explained in an extra book or technical description. We will only have a brief look at most of them.

Big Inventors and Discoveries

To get an impression of the achievements that big inventors left for us, we will present some examples. The world of technology and inventions with a focus on communication technology is really fascinating, and it would take plenty books, to get a detailed overview of it. Nevertheless, let us have a look at some big inventors in history and their inventions in the context of electromagnetic waves.

Students of the scientific faculties are almost familiar with their names such as Gauss, Tesla, Ohm, Siemens and, a lot of others. Many names are used in international unit systems for quantities like voltage, current, resistance and many more.

James Clerk Maxwell is one of those very famous scientists whose name is still very present in mathematics and physics.This Scotsman has precisely described the characteristics of electromagnetic waves with his famous math equations for the use in the scientific and technological fields. It was in the middle of the 19th century when he discovered those fundamental relations. He found out some fundamental equations that combine electrical and magnetic fields with voltage and current. The results are the famous four equations, called Maxwell's equations. Any scenario in the field of electrodynamics and modern electrical system behaviour can be explained by Maxwell's equations. Nearly everything is based on those relations. They are even the basement of many computer simulations that are used for system design in electromagnetics. With modern software tools, engineers can use numeric methods which are based on Maxwell's equations and calculate quantities such as current on antennas and electric field strength in their vicinity. At the end of chapter one, you will find a sophisticated example of how to do this. But who else is in the line-up of famous inventors?

Michael Faraday is one of the inventors we all should know. Maxwell used some of his results for his own work. Faraday was born in London in 1791 and he was a scientist who liked to use experimental setups. Therefore, his reputation during his lifetime was not really high. Today, everybody knows at least one of his basic experiments. He held a magnet beneath a piece of paper with iron filings on it. Suddenly, the iron filings moved to a characteristic pattern. Faraday really made the magnetic field visible. Awesome and simple. We also heard about the famous Faraday cage, a device to protect people against high voltage and lightning. This simple cage makes the current run around the inner space of the cage, only within the metal lattice. By doing so, anybody inside the cage remains safe. Today, we see the same effect in a car while a thunderstorm with lightning strike the car. It was a fantastic discovery. Today, it is still used to protect people against lightning. Another famous phenomenon is the Faraday effect. It is also applied to technical components for satellite communication. Faraday found out how to change the angle of the polarisation plane of a transversal electromagnetic wave. This is basic knowledge for changing the polarisation plane to a different angle in a waveguide. The setup is rather simple. You only need a guided wave that can propagate in a waveguide with a circular profile. The waveguide is surrounded by a coil with a certain level of direct current. That produces a static magnetic field within the waveguide. With the help of that kind of magnetic field in the direction of propagation, it was easy to rotate the polarisation plane of the wave around its propagation axis. The rotation angle depends on the magnetic field strength of the static field inside the waveguide, consequently depending on the direct current through the coil. Satellite receiving equipment uses this effect for switching polarisation planes between horizontal and vertical. Faraday never could have imagined at all how this useful application could work 200 years later. In the nineties, thousands of satellite receiving systems for consumers used this effect.

Nicolas Tesla, born in Croatia in 1856, was one of the greatest inventors of all. Today, we can find his name in the unit system for the quantity of magnetic field strength. That's the way many students of the scientific and technical faculties become familiar with Tesla. Tesla applied for more than 700 patents in his career. He was obviously a very busy guy, and many famous trials and inventions came from his genius. One of them happened in the field of electrodynamics. It was the two-phase alternating current that is now used by everybody. He also worked in the field of wireless energy transport. One of the successful presentations was about making gas glow in a glass tube, just like the neon tubes we are using today. Tesla had an immense potential. We could write endlessly about his experimental setups and inventions and what they mean to us in a wireless world of technology.

Guglielmo Marconi was the one who could send energy over long distances using completely new methods and devices, just in the sense of modern communication technology. He built up a big transmission station in Europe and sent out radio waves over the Atlantic Ocean, which could be received on the American continent. His transmitters were designed for 25 kW power. Simple resonant circuits and huge antenna setups were typical for his equipment. He really was the prototype of later radio amateurs who had good theoretical knowledge but also worked with real technical devices, even if they were very big at his time. Marconi was an absolute professional, when we consider what he really achieved on both sides of the ocean. It was in the year 1899 that some people‘s lives could be saved by an emergency call with this new radio technology, which was also used on vessels later on. In 1900, Marconi owned the patent for wireless transmission of signals. Up to that time, this was only possible for short distances. Later, the range could be extended.The early types of transmitters needed a huge amount of energy. Marconi optimised the resonant circuits, which made the transmission system more effective. Marconi had to start without vacuum diodes. But a little bit later, he could also build up more efficient setups by using vacuum diodes.Today, many radio enthusiasts remember that technology, which was the basis for great technological progress in this century.

Heinrich Hertz was another very prominent inventor. When we talk about frequencies, his name is almost used as a unit name for cycles per second. In 1886, he was able to send radio waves from a transmitter to a receiver, even in the later FM band and even before Marconi‘s trials. He contributed the basics for Marconi‘s successful results, even if Hertz only achieved short distances of a few metres. But it was the idea that much more would be possible whenever someone could prove this.

Another famous scientist was **André-Marie Ampère,** an Italian inventor. His name is used as a unit name for electric current. He pub-

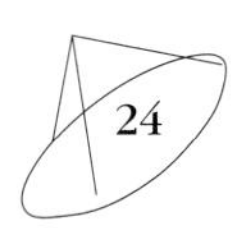

lished important papers since 1820, and he explained what electric current and voltage really are. He also found out Ampère's law. We will discuss that at the end of chapter one in the context of Maxwell's equations.

Looking for the inventor of the telephone as we know it today, we will find different developers. **Graham Bell** was one of the most famous telephone developers. 1922 is his year of death. His work brought transmission lines and telephone devices to the Americans. But he used knowledge and publications of other scientists and developers, who were also working hard in this field. By doing so, Bell could build up networks for telegraph transmission and phone calls all over the country. **Philipp Reis** also is one of the famous telephone inventors. In Germany, in 1861, he was the first one who successfully transmitted voice signals over a telephone line.

Figure 2: Ground stations are connected to the whole world by geostationary satellites. Phone calls, TV programmes, and data links are transmitted from several orbit positions.

Konrad Zuse was a pioneer, too. But he was working in another field of technology, which must be mentioned here. He constructed the first calculator in our civilization. Born in 1920, he built the famous Z1 computer and later, in 1943, the advanced Z3, which had an extra input and output unit. Zuse used telephone relays. These hardware components were already available, and he used them to execute operations and even programmes for calculation. To be honest, Zuse did not need electromagnetic waves for that and, he could not execute complex operations as later calculators did. But his work was the foundation for data transmission and calculators. Today data exchange and transmission are done on any digital wireless device.

There are still other inventors and scientists who have had an impact on technology in our modern world. **Alessandro Volta** is one of them. Born in 1745 in Italy, he worked as a developer and inventor who was busy in power supply. Without his work, we probably would have had trouble in supplying our mobile devices today. His knowledge created a system of tubes, that could be used for chemical processes to generate electrical voltage. There are still more inventors we could mention here. **Werner von Siemens** built up a network of transmission lines for telegraphs and telephones in Germany.

Ohm should be mentioned, too. In every electrical circuit, he reminds us of his work because his name became the unit name for resistors. Many others brought us new technologies in electrodynamics and related fields.

We could continue with further scientists and inventors, but now it is time to have a closer look at the characteristics of electromagnetic waves. In addition to the chapters in this book, you can find very sophisticated descriptions and animations concerning Maxwell's equations and electromagnetic waves on the World Wide Web or in scientific publications.

Waves in Communication Technology

In the first chapter, we talk about the nature and appearance of electromagnetic waves. Wireless communication can easily be explained by a model consisting of a transmitter, a receiver, and the space between, where waves will propagate through. That space can be a medium like air or water, but it can also be empty space or a vacuum as in the universe, where waves will propagate, too. They need no medium like acoustic waves that cannot propagate in space. In technical applications, you will almost always find modulated electromagnetic waves. That means waves carrying information require a stimulation of their amplitude or by phase or frequency changes coming from audio signals, speech or music, and data, too.

Origins of Electromagnetic Waves

Where do electromagnetic waves come from? And how are they produced? The biggest amount of radiation comes from natural phenomena. Waves are radiated by the sun in the visible spectrum, but also at infrared, ultraviolet and x-ray wavelengths. Other astronomical objects send out immense amounts of energy through x-rays. We can even measure this. Special satellites with different detectors and cameras carry instruments to measure electromagnetic waves in different frequency ranges. This type of radiation, coming from space, has not dedicated polarisation, neither linear nor circular. It is similar to light coming from a car headlight, which also has no defined polarisation. But we will see that it's completely different if we generate radio waves by our own means. These waves almost have a dedicated polarisation, when we send them out by an antenna.

So let's consider generated radio waves first. Man-made electromagnetic waves can easily be generated with simple resonant circuits, tuned to a selected frequency. With the right energy coupling, you can send out that energy as radio waves over an antenna. A battery for example feeds new energy permanently into the resonant circuit.

That's the way transmission works. Receiving goes just in the opposite direction. Your antenna is stimulated by a signal coming from a distant transmission site and feeds the energy into the resonant circuit of a radio. But radio communication is not the only application. There are fundamentally different applications for electromagnetic waves, like in medical areas. We use them for x-ray visualisation applications and Computed Tomography in the medical sector.

For a better understanding of electromagnetic waves, we have a closer look at a resonant circuit with a capacitor and coil as an inductance. You can imagine that the plates of the capacitor will be spread apart from each other. This is a way to imagine the transition from a resonant circuit with oscillating electrons to an antenna, also with oscillating electrons. In each cycle, they run from one plate of the capacitor through the inductance to the other plate. Then the direction changes and the second half of the cycle starts. You can see the visualisation of that process in Figure 11. The consequence

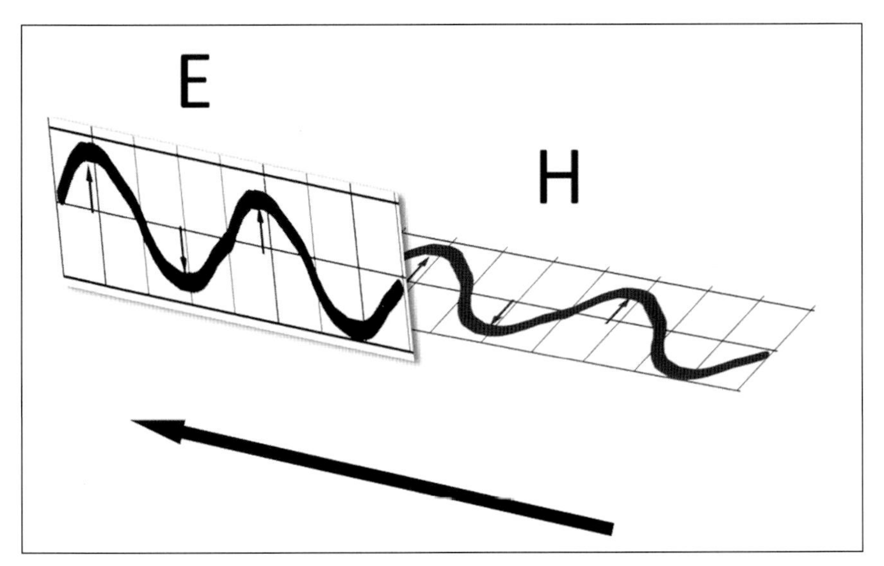

Figure 3: Electric and magnetic fields in their proper oscillation plane and their direction of propagation, a wave with linear polarisation. In the far field, both components are in phase, and in the near field, they have a phase difference of 90°.

of the moving electrons is that the electric field lines also move. During this process, the field lines suddenly detach and leave the electrons as their point of origin. They become a kind of closed circle and go out into free space (Figure 12). Now we have a propagating wave in free space with an electrical and a magnetic component. We will look at this later.

Linear Polarisation

The electrical component of a wave in free space has its direction in a plane, such as horizontal or vertical, depending on the type of transmission antenna. Generally, transmission is intended to be in one of both planes (antenna with horizontal or vertical polarisation). Both planes are orthogonal to each other and orthogonal to the direction of propagation. But there is a second pair of polarisation types, which are also orthogonal to each other. They are called circular, with a rotating E vector right hand or left hand. What is the secret and what is the purpose behind? We will see it in a later chapter.

We can describe waves with linear polarisation easily. For mathematical operations, we need an expression, that will characterise a wave by amplitude and phase. We use:

$$A(r) \cdot e^{-j\omega t} \cdot e^{jkr}$$

"A" means the amplitude, while the following terms express the phase. "A" can represent the quantity of electric or the magnetic field vector. The amplitude decreases by 1/r starting from the origin of the wave.

The expression $e^{-j\omega t}$ describes the oscillation with the angular frequency $\omega=2\pi f$ and $\omega=2\pi/T$. With f is the frequency of the wave and T is the cycle duration and j coming from the complex number calculation, while t represents any specific point on the time axis.

The expression e^{jkr} is the phase related term for propagation with a wave number of $k=2\pi/\lambda$ and $k=\omega/c$. λ is the wavelength and c is the speed of propagation of the wave, that means speed of light, in case of free space. Furthermore, r is the distance to the origin or the source of the wave.

This is more or less the easy part, and we will go on with only little math. One exception is the subchapter "Practical example of research and development" at the end of chapter one. There we will need some extensions for wave descriptions, which seems more complex than in the other chapters. The extensions come from dedicated relations within a coordinate system, in order to introduce a set of directions. That means that "A" is a vector quantity with x-, y-, and z components. Also the propagation term gets vector components, to enable waves to run into different directions. We can use cartesian coordinates as well as polar coordinates. The term would change to

$$\vec{A}(\vec{r}) \cdot e^{-j\omega t} \cdot e^{j\vec{k}\vec{r}}$$

The arrows indicate that it is a vector quantity. Let us consider linear polarisation in detail now. The electric field and the magnetic field are orthogonal components. They also have a phase relation that needs some attention. In the far field, both have the same phase. But in the near field of the wave, they have a phase difference of 90°. But what is the reason? When we look at the process when a wave detaches, the near field consists of some complex terms, which decrease by $1/r^2$ and $1/r^3$ (see Figure 12). That means, those terms have a greater impact in the near field, but only there.

In the far field it is a term which decreases by 1/r which dominates. This term has another phase relation and it dominates in the far field. Here, electric and magnetic fields have no phase difference. When we look at the quotient of the E-field quantity and the H-field quantity, we have the field impedance of free space or of vacuum as a result:

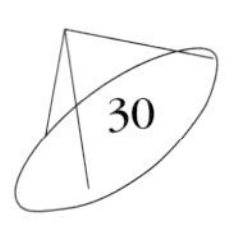

$$z_0 = \frac{E}{H}$$

The impedance of free space also corresponds to the product of magnetic field constant and the speed of light.

$$z_0 = \mu_0 \cdot c$$

A practical value as impedance of free space is the following approximation:

$$z_0 \sim 377\ \Omega$$

The meaning lies in the math description of the quotient of electric and magnetic field, when we are in the far field of the antenna.

Many antennas are derived from the basic form of the electric dipole. They almost work with linear polarisation, what is rather evident when charge carriers oscillating from one end to the other. This means a current which moves on one line. We know the monopole as an example and the λ/2-vertical. They radiate waves with a field vector in the same direction as the antenna. The magnetic field consists of concentric rings around the vertical radiator (Figure 13). In the far field, they are orthogonal to the electric field lines. Please find a detailed description in Figure 11 and Figure 12.

Dipoles can also have the horizontal orientation. In this case they radiate waves with electric field vectors, which are parallel to the earth's surface. Generally, we find applications in radio communication with horizontally and with vertically polarised waves. It depends on the type of system or application, which polarisation is the preferred one. Terrestrial TV in Germany used horizontal polarisation in the last century. Radio systems for mobile radio applications, i.e. for contacting taxis and police vehicles, use almost vertical polarisation. On short wave, we will find both for worldwide communication.

The reason is that radio waves can be rotated while being refracted by the ionosphere. In many cases, we will not receive radio waves with a pure polarisation. Pure polarisation may get lost, no matter which orientation the transmitted wave had, while detached from the transmission antenna.

Looking at the decoupling of radio waves horizontally and vertical polarisation, we will measure a typical isolation of 20 dB or more. When a transmitter sends out a signal with vertical polarisation, and we try to measure it with an antenna, that receives horizontally polarised waves, then we can find this difference. If we adjust the test setup 100%, maybe we will measure isolation better than 20 dB maybe 25 dB or 30 dB. But in some applications, it will typically be below 20 dB.

What does that mean for receiving radio waves? That means: Even if you receive a v-polarised signal with a h-polarised antenna, you will have a signal. But it will be about 3 S-steps lower in your measurement instrument, like an S-metre. If you send out signals with circular polarisation, you can receive them with a loss of 3 dB, no matter which orientation your linear receiving antenna really has.

This effect is valid, and vice versa. Mathematically, you can explain it as follows: any wave with a circular polarisation consists of two waves with linear polarisation, and a linear receiving antenna can only absorb one linear part. That means the half of the power, and consequently, minus 3 dB.

Circular Polarisation

Beyond linear polarisation in antennas for radio communication, we find circular polarisation, a special case of elliptic polarisation. To get an overview about their characteristics, it is a suitable model to imagine an E field line. In case of linear polarisation, their orientation remains constant. Circular polarisation, however, means, that the E-field vector rotates around the propagation axis, while the propagation speed is also the speed of light.The direction of rotation can be right-round or left-round.

The propagation direction is orthogonal to E-field and H-field and the tip of the E-field vector moves on a curve like a screw thread. The length of the rotating E-vector is constant (in circular polarisation, but not in elliptic). Only the phase is oscillating continuously by the rotation.

Mathematically, a circularly polarised wave can be characterised as follows: A circularly polarised wave is the result when two linearly polarised waves with x- and y-polarisation (orthogonal), same amplitude, same direction of propagation, and a phase difference of ±90° are added.

While receiving a circularly polarised wave with an antenna for linear polarisation, you will receive the same quantity of energy no matter which orientation (circular angle) the receiving antenna has. It is the same the other way round. When you use an antenna with circular polarisation while receiving a signal with linear polarisation, you can use any rotation angle around the propagation axis.

In both cases, however, you will lose 3 dB of receiving power. Helical- and spiral antennas can be used for circular polarisation as well as cross dipoles with a feed system with ±90° between both polarisation planes. Some antennas are working with left and right circular polarisation. They need two different feeds. We will see an example in a later chapter. The decoupling or isolation between both polarisations is approximately 20 dB.

Electromagnetic Waves and their Paths

There are different modes of propagation when looking at radio waves. First, we consider the basic categories: electromagnetic waves propagating in free space and waves, which propagate in different types of waveguides. It is an extraordinary case that waves in free space don't need any medium for propagation, that is very different in case of acoustic waves.

Electromagnetic energy can be transported through a vacuum (example: transmissions over communication satellites). In most applications, however, electromagnetic waves propagate in media like air, water, dielectric substrates like in cables or plasma (ionised gas).

Conducted Waves in Waveguides

Many types of cables belong to the well-known guides for electromagnetic waves, like coaxial cable and two wire lines. But there are further candidates, like rectangular waveguides, strip lines and optical fibres for extremely short wavelengths. The selection of the right waveguide depends on the application and the criteria for using it: low attenuation, light weight, mechanical robustness, bandwidth, and bending radius.

Waveguides with rectangular or circular profile for example can be used for 1-100 GHz, because they have very low attenuation for high frequencies, definitely lower than coaxial cables. But they are extremely large for lower frequencies. Microstrip structures are used for flat environments such as planar antennas, feed systems and printed circuit boards.

In waveguides, the mode of a wave has to be considered. Generally, the most relevant mode is the TEM mode (transverse electromagnetic wave). They are characterised by E-field lines and H-field lines, which are orthogonal to the direction of wave propagation. This mode is used in coaxial cables and in free space propagation.

Beyond this, we find TE modes and TM modes with only one wave component orthogonal to the propagation direction. For TE it is the electric field, and for TM, it is the magnetic field. They occur in waveguides, i.e. in the H_{10}-wave or higher modes in a rectangular waveguide. The H_{10}-wave is the basic mode in rectangular waveguides which carries the most part of energy.

Depending on the application, there are also modes with higher order, such as H_{20}- and H_{30}-mode. They can also propagate and carry a certain amount of the wave's energy. The higher the order, the more field minima and maxima are distributed inside the waveguide's resulting field.

Coaxial Cables

That type of waveguide is a well-known one occurring in many daily applications, i.e., antenna cable for TV sets. The outer appearance with an inner and an outer conductor is most characteristic. Between both is a dielectric layer, and the outer coverage is a robust isolating material. Electromagnetic waves are efficiently isolated by the world outside.

Waves will propagate at a speed of 230,000 km/s, approximately 79% of the speed of light. Coax diameters have different ranges of a few millimetres up to diametres like a garden hose or even a thick sausage. The inner conductor is hollow when the cable has a huge diametre in order to save weight and material.

Thin coaxial cables are flexible and have a lower bending radius, while thick cables can have a bending radius of some decimetres. Further characteristics are the cable's attenuation and field impedance. It is important to note:

This impedance cannot be measured easily with a multimeter. Most applications need cables with 50 Ω and 75 Ω, but there are different values for special parallel wired cables, for example, with 240 Ω.

When cables are used for long distances or high frequencies, the attenuation has to be considered before you decide which cable type is the right one. Aircell 7 cables will attenuate with 7 dB over 100 m length on 144 MHz. And it will have an attenuation of 20 dB at 1 GHz for the same length. The cheaper RG58 cable reveals disappointing values compared to Aircell 7. For 100 m, it shows 5 dB attenuation on 10 MHz and 55 dB at 1 GHz. For huge lengths, it cannot be operated in a reasonable way on higher frequencies, whereas a little length will have no negative impact.

Thick coaxial cables - Why should we use thick coaxial cables at all? And what does thick mean? The answer is simple. 1¼ inch cables have a low attenuation compared to cables with significantly lower diametres. Additionally, due to their mechanical robustness, they are good candidates for tall mobile radio towers with cable feeds.

Mobile radio applications need lengths of 30-50 m up to the top of the tower. The Attenuation values of thin cables would be too high for that.

The typical thickness is from ½ inch up to 1¼ inch. Good cables can achieve 6 dB transmission attenuation for 100 m at 3.3 GHz. For shorter lengths, a diameter of ½ inch will also work. The inner conductor will be almost hollow, like a tube for medium and thick cables.

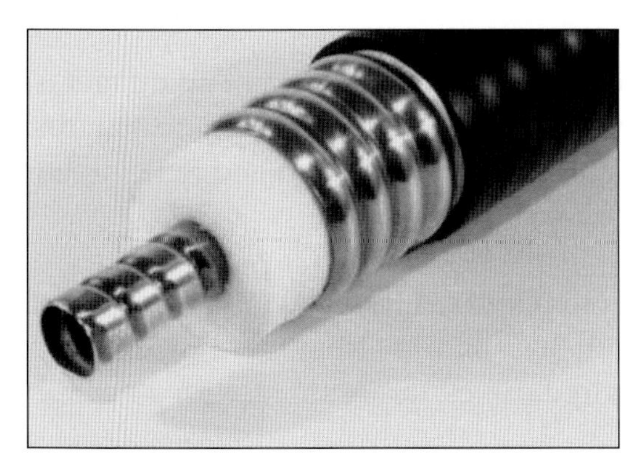

Figure 4: Coaxial cable for installations in cell phone networks with little loss up to GHz frequencies. The inner conductor is tubular. The bending radius should be greater than 200 mm. Typical diametres are 1/2, 7/8, and 1¼ inch.

Dielectric media as a substrate layer is typically made of polyethylene and includes a certain amount of air inside to keep the permittivity low. The outer conductor reminds one of the shape of a caterpillar, which makes the cable a little bit flexible and robust at the same time, with an acceptable bending radius of typically 200 mm.

Another important parameter is the isolation between the field inside the cable and the space outside. Good coaxial cables are specified for 100 dB and more. RG58 remains below that with 85 dB isolation, whereas Aircell 7 lies above 90 dB isolation.

This parameter is important when signals outside the cable with high signal levels may cause interference with the waves inside the cable. When you look at flat roofs with several antennas from several radio services, interference should be avoided, even if a coaxial cable lies in the vicinity of a transmission antenna, i.e., of mobile radio transmitters.

Trials in the 90s have revealed that there is a real potential for interference from GSM antennas for the 1800 MHz band and the intermediate frequency range of analogue satellite TV, going down in a cable from the SAT LNB to the receiver indoors. When it is installed near the GSM transmission antenna, there was an interference (of moiré type) on one or more SAT channels in the range of 950-2050 MHz. In a coaxial cable, the field lines go radially between the inner and the outer conductors, while the magnetic field lines run like concentric rings between the inner and outer conductors. Due to the orthogonality, we can classify them as TEM waves propagating in a coaxial cable.

In highly sophisticated coaxial cables, their characteristics are based on a good choice of material. For some applications, it is a key feature that the cable is resistant to flames and explosions. That‘s needed for use in tunnels or areas with gasoline or gas.

Two-Wire Lines

Another type of cable are parallel two-wire lines. In the 60s of the last century, they were used for TV between the roof antenna and the TV set inside. Even today, they are still very popular in amateur radio as symmetric feed lines. Their wave impedance is almost 100 Ω or 240 Ω, but also 300 Ω, 400 Ω or more. They are perfect for building antennas like Slim Jim or J-pole.

As they are not isolated by an outer conductor, they are less resistant to the field from outside the cable. In some applications, however, not in amateur radio, two-wire lines are twisted, i.e., in embedded systems for automotive applications as the CAN bus or in wiring for computer LAN. The propagation speed is approximately 60 % of the speed of light. But in the fields mentioned this doesn't cause any problems in general.

Optical Fibre

Fibre optic cables have been popular since the late 70s and found their place in electronics. Today, they are the backbone of broadband wired communications, and they stand for high-capacity transmission lines in modern broad-band networks. Mechanical issues and splicing have no meaning any more than in the early years. The profiles of fibre optics are meanwhile optimised in such a way that interfering wave modes are suppressed by a clever design of the cables' gradient in diametre (optimised profile). The speed of propagation in fibre optical lines and their shortening factor also depends on the relative permittivity ε_r. The speed is lower than the speed of light, roughly speeking, similar to coaxial cables.

Microstrip lines

Waveguides in microstrip technology and strip lines are flat and don't need much space. They can be used on pcb as rf line with

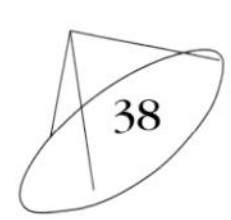

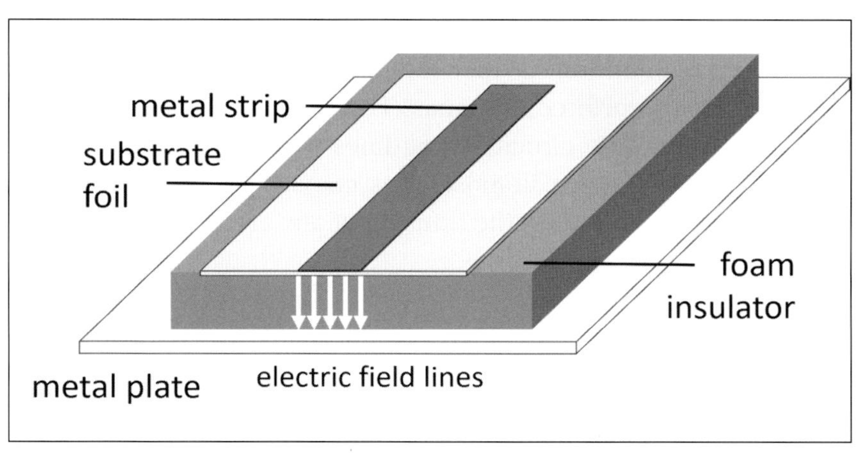

Figure 5: Microstrip line over a metal plate. In applications, wave impedances can be designed by the size of the line and the distance of the plate. 50-100 Ω are typical.

50 Ω, for example and in flat antennas as a feed system. Their geometry defines the exact wave impedance of the line. So they can also be designed for 100 Ω or any other impedance. The design of a stripline is defined by almost two parameters, the width of the line and the distance from the ground plate below or both ground plates below and above. The taller the line is, the higher its wave impedance.

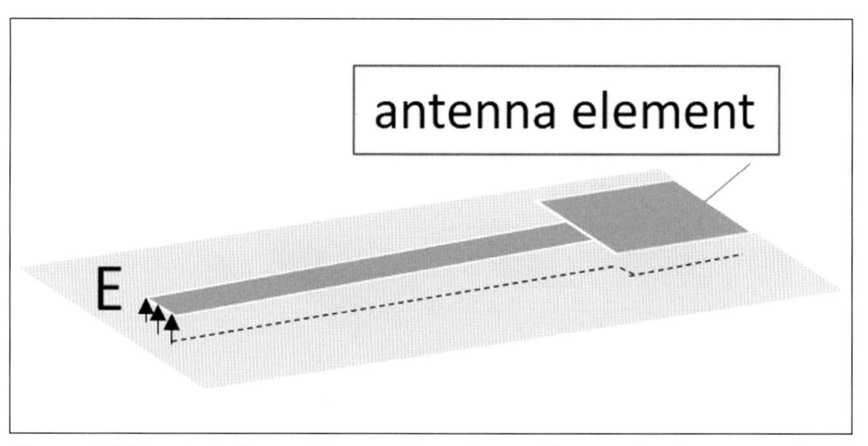

Figure 6: Microstrip line with a patch antenna. Two variants are usual: galvanic coupling or capacitive coupling.

Microstrip lines can be used on PCBs (printed circuit boards), where the bottom is a continuous metallic layer. They can feed electronic components and printed antennas (Figure 6). They are a practical choice for electronic circuits using lines with a defined wave impedance and little space. In most cases, coaxial cables would not be a practical solution within the housing of any electronic circuit.

Stripline technology can also be used in setups with two parallel plates. In this case, the metal layer of the stripline is part of a polyethylene film. The dielectric layer with the metal lines lies between the parallel plates, just in the middle. To keep a constant distance from the plates the dielectric layer is supported by one or two foam layers as a substrate (Figure 5).

We can see how the metallic structure and the polyethylene layer are supported by the substrate. Cheap foam layers are available from packaging materials. They often have a low attenuation and a practical height for being used in flat parallel plate setups. Layer systems like this are almost always used in flat satellite antennas for TV-reception.

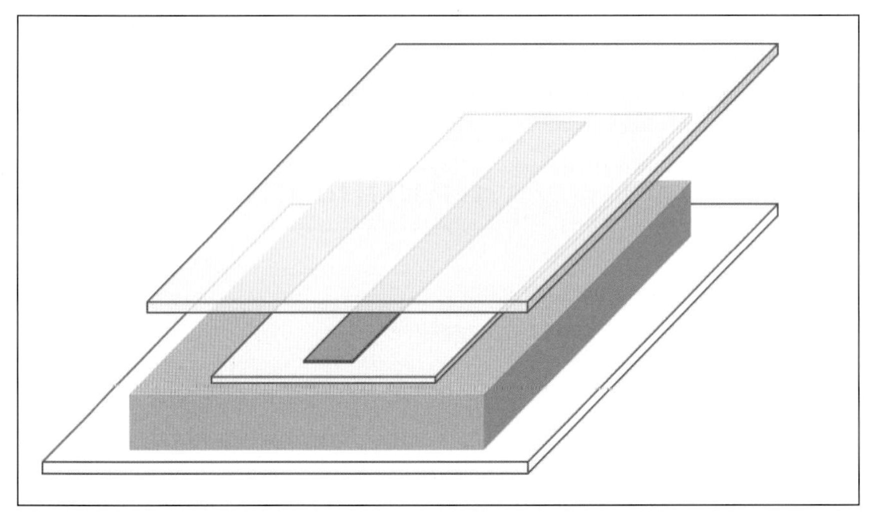

Figure 7: Microstrip line on a foil between two metal plates, lying on a substrate made of foam.

Waveguides

Waveguides are well suited for high-frequency applications because they can transport RF energy from A to B with extremely low transmission loss.

There are profiles with a rectangular shape and with a circular shape, which can even deal with two orthogonal polarisations. In the frequency range of 10 to 15 GHz, the diametre reminds one of a garden hose.

They are used in the feed systems of satellite dishes and for diplexers with filters (Figure 8). For low frequencies of 1 GHz, they become huge and less practical. In these cases, coaxial cables might be the better choice.

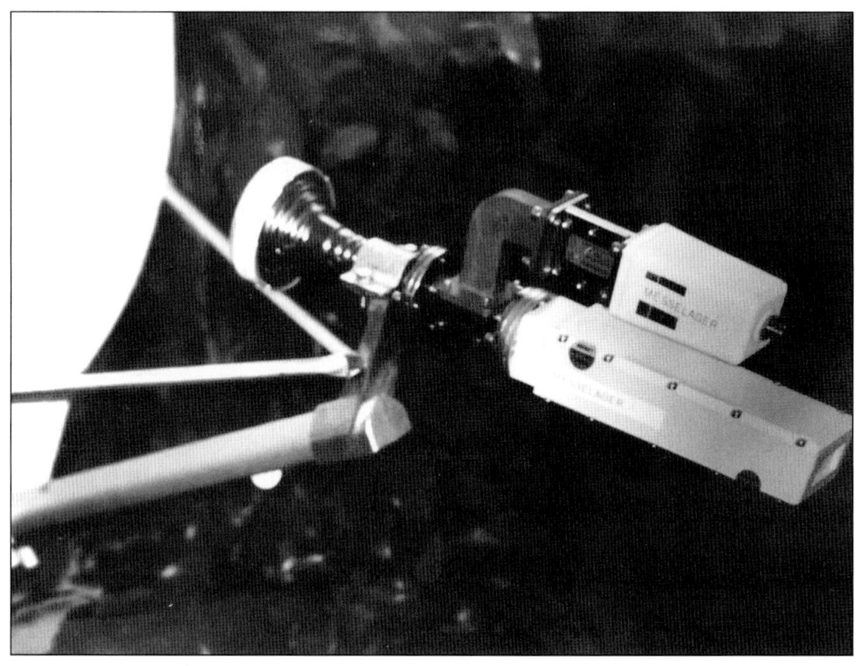

Figure 8: Feed system with a horn antenna, OMT, microwave filter and 2 LNBs (transmission lower one/receiving upper one).

Looking closer at the feed system in Figure 8, we can see an OMT, or orthomode transducer, a component with three gates: two for rectangular waveguides and one for a circular waveguide. In the focus of the parabolic dish, we can see the feedhorn, a type of antenna which can illuminate the dish or vice versa can collect all the focused energy coming from the parabolic reflector. The feedhorn is connected with a flange to the OMT. With its symmetric shape, the feedhorn couples energy into the OMT with two orthogonal planes of the E-field vector (h + v).

The OMT is constructed with two output flanges, one for vertical and one for horizontal polarisation. In Figure 8, you can see the feedhorn on the left side. During transmission, the feedhorn will illuminate the whole parabolic reflector, that increases the gain of the whole antenna system significantly. The fields of the horizontal and vertical polarisation planes are available, each at one gate with a rectangular flange. We remember that rectangular waveguides can only be used for one polarisation.

In this case, the electric field vector is parallel to the narrow side of the waveguide. At the top, we can see the LNB for the receiving mode. The other bigger one is the module for transmission mode. During transmission, the system needs an extremely huge amount of isolation between both rectangular ports h+v in order to protect the receiver LNB from signal levels beyond the specification.

The whole transmission energy goes through the gate of the feed horn. During reception, the OMT makes sure that only received signals from the dish come to the receiving port, where they can be detected in the LNB. The separation is done by polarisation separation with a special construction inside the OMT. The receiving path has an additional filter with a reduced bandwidth. Signals beyond the usable frequency range cannot interfere with the receiver. By this protection, only allowed rf signals can pass, and a better large signal behaviour is ensured.

Consequently, the most important function of the OMT is the separation of the h- and the v-port (rectangular flanges). In Figure 9, we can see the direction of the electric field lines in the waveguide. They are parallel to the narrow side of the waveguide's profile. With this knowledge, we can assume which port is for horizontal and which is for vertical polarisation. While the circular waveguide is working for both, h and v, we can see that the upper, smaller LNB is for reception and that horizontal polarisation is the plane for received signals. Consequently, transmitted signals coming from the bigger module are vertically polarised.

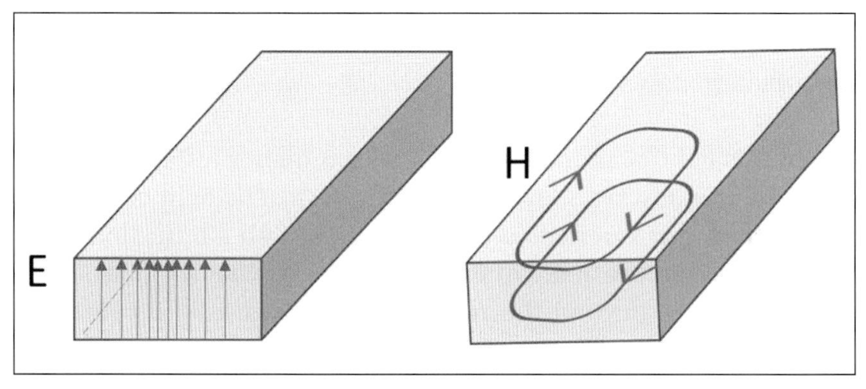

Figure 9: Rectangular waveguide with the electric field of the H_{10}-mode (left). The electrical field strength in the middle is higher than towards the edges. Right: magnetic field lines.

Figure 9 gives us a first impression of the field distribution in a rectangular waveguide. Electromagnetic waves can be fed into the waveguide by a kind of small monopole inside the waveguide in the middle of the wide side and a defined distance from the end of the waveguide. Generally, in the waveguide, there are different modes that may occur. Each of them carries energy. But it is almost the H_{10}-mode which is a TEM mode and is used for transmission as a basic mode. Other modes may occur and produce a complex field distribution. For special applications, they might be used as well, but in most cases, they are unwanted. Waveguides are specified for very low transmission losses. They are manufactured with an extremely smooth surface inside the waveguide. Attenuations of a few hundredths can be achieved for distances of several wavelengths. In microwave laboratories, they can be designed with equipment like the HP 8510 or similar network analysers.

Electromagnetic Waves at Boundary Layers

When media with different characteristics, one with a higher density and one with a lower density, have a common boundary, there will be some extraordinary effects to be observed in the transition area. Remember some trials in school days when the teacher explained the prism or the refraction effects of a light beam going through an aquarium.

When this light beam meets media with a higher refraction coefficient, we can observe some interesting effects: The light beam splits into a beam that is reflected on the surface, with a reflection angle that is similar to the angle of incidence. That's what happens at the boundary layer. Beyond the reflected beam, there is also a beam going through the surface into the water, but with a slightly different angle. This is the result of the different densities of the media, air and water when the beam is passing. With a prism made of acrylic, it would be the same. It is not only the angle of the beam in the prism that differs from the angle of incidence, but the light waves also propagate at a lower speed in the medium with a higher density and the higher refractive index. The relation is:

$$n = \frac{c_0}{c_m}$$

with c_0 as the speed of light in free space and c_m according to propagation speed in the denser medium as water.

The strength of the refraction effect depends on the characteristics of the media. The suitable physical description can be done by the parameters ε_r and μ_r They depend on the used material, and they have an impact on wave propagation, referring to the behaviour in a vacuum. The resulting refractive index is n with:

$$n = \sqrt{\varepsilon_r \mu_r}$$

ε_r means the permittivity, which depends on the material, and μ_r means the relative permeability. This example can also be used for electromagnetic waves, because light also has the characteristics of electromagnetic waves. Light is also a wave, even if it has some additional particle properties, according to another model of characterisation. When a wave goes through material of higher density, the propagation speed will decrease. Coaxial cables with a dielectric layer (i.e., polyethylene) are a good example. The density of polyethylene is higher than in air, which results in a shortening factor depending on the refractive index of polyethylene in the cable and a reduced speed of the wave inside as a result.

In systems with ferrite antennas or coils with a ferrite core, a similar effect is applicable. The magnetic field is literally pulled inside the ferrite material as a consequence of the higher inductance of the coil with a ferrite core. This makes ferrite antennas a good solution for small radios and the medium-wave reception gets unexpectedly good quality from a very small antenna.

To come back to the prism: When the light leaves the acrylic medium, we will observe that the refraction just goes the other way around and the light beam will be refracted again at a slightly different angle. We also remember the famous rainbow colours as a result of different wavelengths of different colours in the light.

Each wavelength propagates at a slightly different speed, and the original white light beam is fanned out into different colours. Look at Pink Floyd's "The Dark Side of the Moon" album cover. Considering metallic boundary layers, electromagnetic waves show another special behaviour. They are reflected completely. But what does that mean exactly? The metal surface is a good conductor, and as the wave is reflected at 100 %, the field vector of the incident wave and the field vector of the reflected wave will add to zero on metal or on other good conductors. By doing so the reflected wave gets a phase difference of 180° after the reflection compared to the incident wave's phase relation. This behaviour is also valid for circularly polarised waves. As the rotation direction is the same for incident and reflected waves, the type of circular polarisation must change for the reflected wave because of its opposite direction of propagation. That means, a left circularly polarised wave will change to a right circularly polarisation after reflection, and vice versa.

Water is no obstacle to radio waves. They can penetrate water and propagate. However, the penetration depth is generally poor, except for links at extremely low frequencies. Radio waves at 15 kHz and 76 Hz can penetrate 10 m or even 300 m for the lowest frequencies. Applications can be found in the military field for submarine links or scientific research. The penetration at higher frequencies cannot be used. Submarines have to surface for radio contacts, or they need special communication devices to get radio waves sent out or received.

Windows are made of materials like glass or acrylic. They are transparent, at least we think they are. But in some cases, they have a metallic coating with an extremely thin layer. When GSM providers wanted to feed radio signals into trains and waggons, communication was often not really possible due to poor radio signal levels. Most issues were observed in modern ICE trains. Finally, technicians found out that the metal-coated panes caused an unexpected loss of radio signals. Many trains got additional equipment to amplify radio signals in both directions and compensate for losses while penetrating into the train. One solution was the use of broadband repeaters, which could be installed easily and provided better radio coverage inside the trains. They picked up radio signals outside the train with a roof antenna and radiated the amplified signal in the train with another

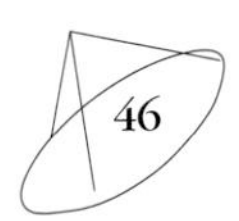

small patch antennas. This process worked in both directions. Phone signals from inside the train could be amplified and radiated from a roof antenna to the next mobile radio tower.

Radio Waves Detaching from a Linear Antenna

At the beginning, we learned that electromagnetic energy can be generated by a resonant circuit. Antennas are working like elements in a transition process to radiate the energy coming from the circuit. Guided waves can propagate into free space through an antenna just like radio waves. But they can also be radiated unintendedly as interfering signals. Most of those cases can be found in poor waveguides with changing wave impedances and sudden transitions. They leave the waveguide system in a damaged socket or with the wrong type of cable or adapter. We almost always detect this type of radiation as interference, narrowband, or broadband. That depends on the signal diffusing outward. Intended radiation is simply the use of a radio set with a good match between the transmitter and antenna. This can be achieved when the wave impedance of the cable meets the impedance of the antenna input. The antenna base impedance is what you measure at the antenna input. Generally, the value is a complex mathematical term like the impedance of a stretched dipole with $z = (73+j42)\ \Omega$ in theory. To achieve a correct coupling of the feedline to the antenna input impedance, the impedance of the feed system needs to be conjugate complex to the antenna impedance like, $z_{cc} = (73-j42)\ \Omega$ in the case of a dipole. This can be achieved by a transformation circuit (i.e., gamma match) or by using an antenna matching device.

The antenna input impedance must not be mixed up with the radiation impedance of the antenna. The radiation impedance is well described in [1]. In some cases, misunderstandings lead to wrong interpretations. A brief description is the following: P is the active power, which is fed to the antenna by a transmitter. The active power is divided into two parts: power loss P_V and radiated power

P_S. The active component R of the complex input impedance of the antenna can consequently be divided into R_S and R_V, with R_S as the radiation resistance of the antenna and R_V as the loss resistance of the active component of the input resistance. At this point, we always have to take care, because the radiation resistance leaves room for misunderstandings and misinterpretations.

Now, it's time to have a look at the process by which waves detach around an electric linear antenna (Figure 12).

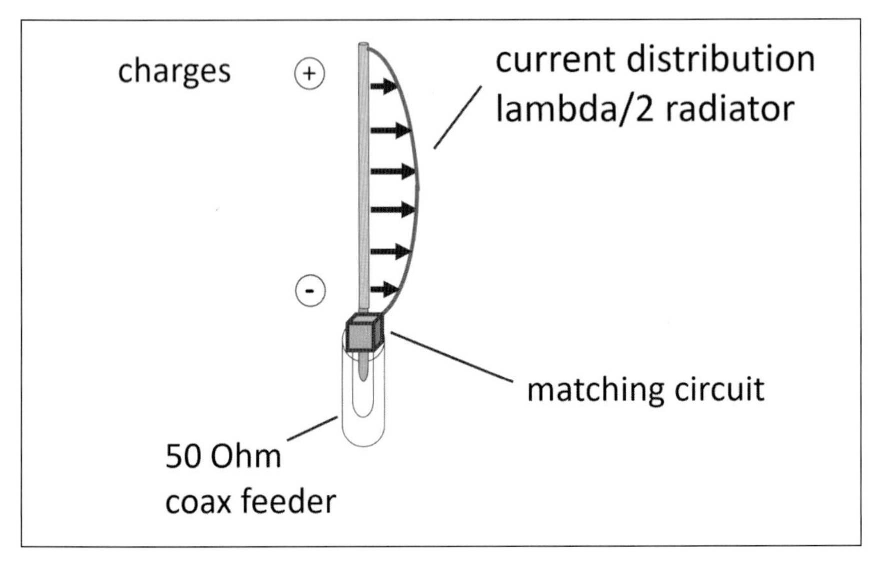

Figure 10: λ/2-vertical radiator with a typical distribution of current strength (arrows with a current maximum in the middle). The electric field has a parallel orientation to the rod. The magnetic field consists of concentric circles around the rod.

First, we look at the current distribution on the antenna. In the case of the resonant λ/2–radiator, it shows a current maximum just in the middle according to Figure 10. This is real current which can even warm up the antenna with high power on a thin radiator. Both ends will have a voltage maximum and a current minimum, and there are indeed charge carriers at both ends. Like in a resonant circuit, they move to the other end and back again in a swing-back-and-forth process.

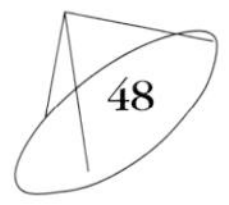

By this, the electric field lines begin to detach at a certain point. First, the electric field lines lie between the electric charges and during the movement of the charges they begin to form closed field lines.

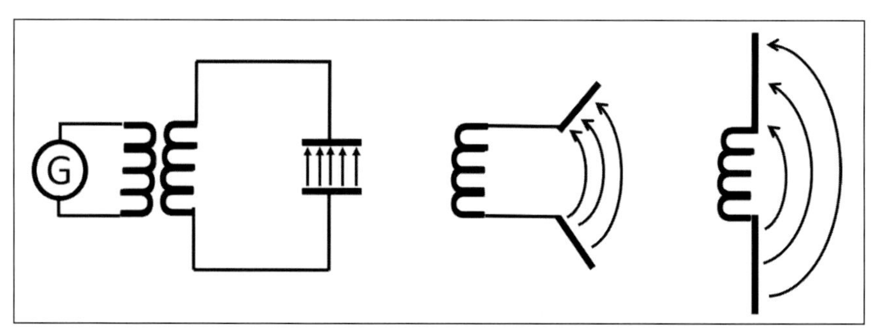

Figure 11: Imagination of the transition from a resonant circle to a radiating antenna, when the circuit is more and more spread. The electric field lines can be seen between the plates of the capacitor (left). The energy is fed by a generator.

Figure 12 shows a λ/2-antenna with charges on it, which get their energy from a coaxial cable and a transmitter. The antenna has a good match without any step or sudden change in the impedance at the antenna input. The figures show how electric field lines start on positive charges and end up on negative charges. We can also see how the movement of the charges bends and deforms the field lines. The sequence starts with a and b. The charges move towards the middle of the radiator. In c and d we can see the transition when a field line goes into free space, where the charges are passing the current maximum. Just in the moment, they pass the mid and go to the other end of the antenna, one half wave will detach, and the field line goes out into space. At this time, a new field line starts to expand, according to the field line in a, but with reverse polarity or direction.

It is the same field line that will initiate the second half wave in the process, which makes a new field line with a reverse direction detach and go out into free space. The difference in polarity results from the reverse direction of the charges on the radiator. Finally, a complete wave has been generated that will run through space

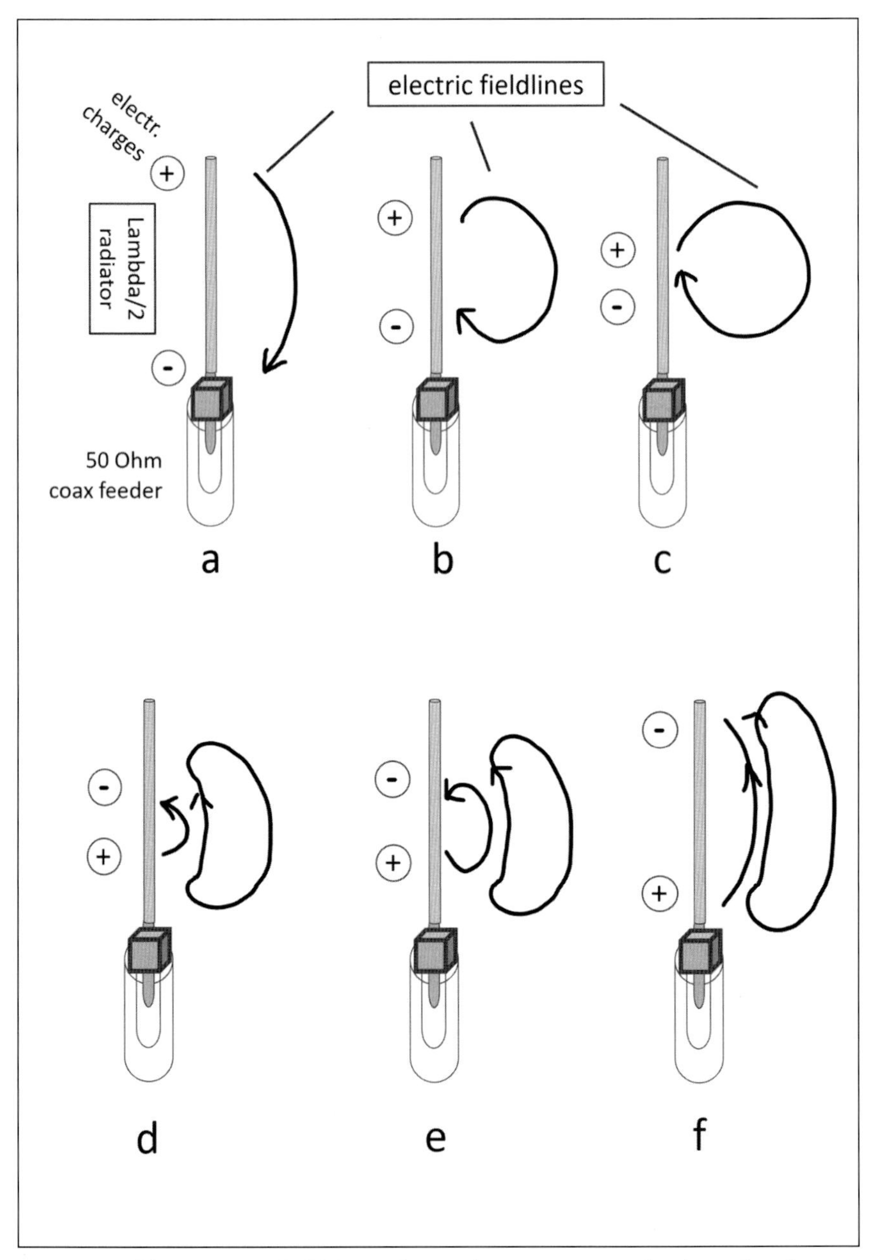

Figure 12: The electric field lines leave the antenna rod when the charges oscillate. Then the antenna radiates energy, which we can receive as a radio signal.

with the speed of light. The corresponding magnetic field lines are concentric rings around the radiator. In the far field, some wavelengths away from the antenna, the E- and H-field will have no phase difference. But in the near field, the phase difference is 90°.

And how will that process work in the case of receiving an electromagnetic wave? Which are the frequencies produced by the antenna? These are valid questions to be answered in this context. The second question can easily be answered if you imagine a waterfall diagram of a sdr (software defined radio). The charges in the antenna's material oscillate at all frequencies of stimulation at the same time, a kind of superposition of movements.

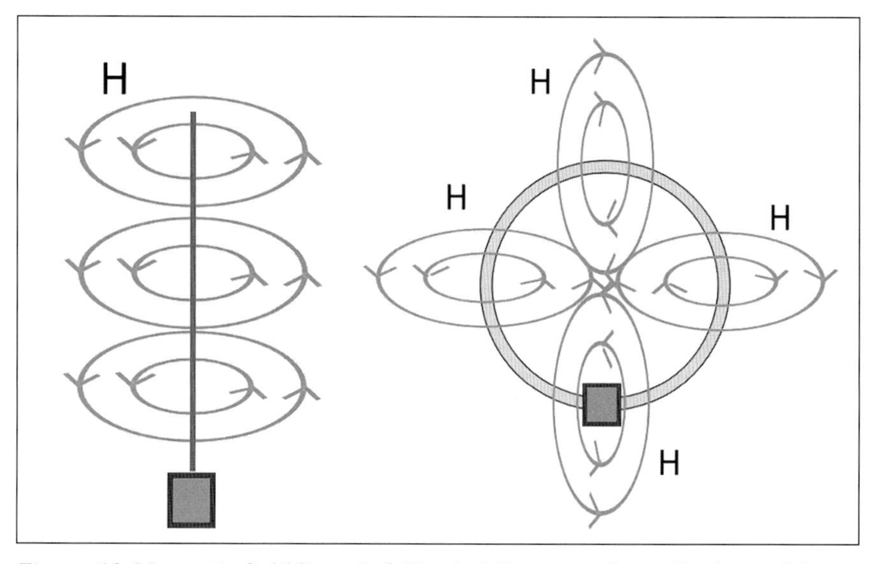

Figure 13: Magnetic field lines. Left: Vertical linear radiator. Horizontal beam characteristic: omnidirectional. Right: Loop antenna with magnetic fieldlines in the center of the loop antenna. Beam: Maxima to the right and to the left.

They are stimulated by different transmitter signals and their propagating field lines on all frequencies, the antenna is specified for. The selection by channels is done by the receiver's signal processing, after the antenna has coupled its energy to the resonant circuit's first stage. By that, the first question is almost answered, too. The use of a resonant antenna increases the level of received signals at certain frequencies. Any short antenna, however, will attenuate the received signal more or less, i.e., while using one metre of wire as an electric linear antenna for MW reception. The wavelength is far beyond 200 m and a short antenna will catch only little signal power.

But how do magnetic antennas, i.e., magnetic loop, work when we compare them to electric antennas? In this type of antenna, a current generates a magnetic field around the conductor while transmitting, just like in the basic lessons at school with current through a wire. The field lines are concentric around the loop, with a field concentration in the loop as a result (Figure 13).

It is a strong coupling between the antenna's current and the magnetic field component of the radiated field. In the case of reception the antenna works vice versa, and we can say that the antenna is almost stimulated by the magnetic component of an incident wave. The result can be observed inside buildings, where electric antennas almost suffer from man-made noise with a lot of interference. Magnetic antennas offer a more silent background for reception, especially at low frequencies (LW, MW, lower SW bands). Electric antennas for in-house reception have one weakness: they are also sensitive to the electrical component of unwanted interfering fields like man-made noise coming from several sources inside buildings. The accumulated interfering noise level can be very high and make reception impossible in some locations.

Propagation in Free Space

The following imaginary model can be used for a better understanding of how electromagnetic waves start to propagate in free space, even without being guided through any medium. Starting from a central point, wave minima and wave maxima move away in radial directions. Electric and magnetic field lines are orthogonal to the wave's propagation direction and form a wavefront.The relative angle between E- and H vectors is 90° at the beginning. In free space, it is a TEM wave (transversal electric and magnetic), when we consider a range of some wavelengths away from the antenna. The closer we approach to the antenna, the field structure changes to a complex near-field distribution of field components.

E- and H-fields have a phase difference, and the amplitudes of the fields consist of additional components with an amplitude decrease of $1/r^2$ and $1/r^3$ where r is the distance of a field point from the antenna. Even the orthogonality of the field components gets lost in the near field area.

For short antennas the far field begins at a distance r with

$$r = 2\,\lambda$$

For long antennas with a length of L, the far field begins at

$$r = \frac{2\,L^2}{\lambda}$$

With the following simple model, we will have a better understanding of the propagation. When we throw a stone into water, we can observe how waves go radially away from the centre. The wave fronts with minima and maxima seem like circles. When the distance increases more and more, the amplitude of the wave will decrease. This can be considered the amplitude reduction in free space, simply by dissipating.

This two-dimensional model can even be extended to three dimensions. The result is a model with propagating waves that will have decreasing power density when they move away from the centre of stimulation. This practical approach can easily explain the propagation loss in free space and the vacuum. It is also suitable to provide the mathematical base.

Looking at the surface of a sphere, it will become more and more a flat surface if the radius or the distance from the centre of radiation is large enough. Many mathematical approaches become easier by using this simple presumption without having significant deviations compared to the correct analytical calculation. We can find different definitions of the far field in the literature. Generally, the far field begins some wavelengths from the transmitting antenna. It depends on the antenna gain and the dimensions of the antenna. In many cases, we can assume a distance of 2 wavelengths where the far field begins. The near field has a more complex field distribution, which is not needed for most considerations. Near field considerations are valid in the vicinity of the radiator, where electromagnetic waves detach from it. Anything in this area will have a significant impact on the antenna's field distribution and may change radiation characteristics, including the far field radiation diagram. Any calculation related to far field considerations can use the following equation where P_S is the transmission power and P_e is the received power and r is the distance between the transmitter and receiver:

$$P_e = P_s \frac{\lambda^2}{(4\pi r)^2}$$

When larger antennas with a certain gain are used, we have to consider the gain of the transmitting antenna G_S and receiving antenna G_e as follows:

$$P_e = G_e G_s P_s \frac{\lambda^2}{(4\pi r)^2}$$

D is the path loss expressed in dB in this case:

$$D = 10\,log \frac{(4\pi r)^2}{\lambda^2}$$

The path loss is strongly dependent on the distance between the transmitter and receiver and also on the wavelength λ. Attenuation values such as path loss are always positive figures, almost always using the logarithmic measure dB.

Extraordinary Propagation Effects

Generally, electromagnetic waves propagate linearly, straight forward. The higher the frequency, the more we can recognise propagation characteristics similar to those of optical light. The behaviour of electromagnetic waves in the GHz range as an example is very similar to that of light. Considering a wave propagating through the air, we observe a linear propagation with a constant wavelength and frequency. But this is not the complete truth. Real propagation is much more complex, as we will see later on, and there are indeed some propagation characteristics we should discuss in details, due to their extraordinary causes.

Diffraction and refraction effects. We can observe this when light is passing an edge or a wall. Behind the wall, we can expect shadows. But there is a certain amount of refracted light, too, even in the shadow. We can also observe this during dusk and dawn. When the sun goes down, there will be a transition phase until it is completely dark. Radio waves show a similar effect. Even in a dense urban environment, there is always a path for the radio waves, even when there is no more line of sight. Sure, in the end, any radio contact will be disrupted, when the field strength gets too low. Diffraction effects often occur while listening in the medium wave band. At night, we can often hear stations from far behind the horizon.

Rotation of Polarisation. When a radio wave with linear polarisation meets a metal plate, it will be reflected. The orientation of the polarisation will not be changed except for a phase step of 180°. The plane of the polarisation is not changed, and it will be a linearly polarised wave after being reflected.

In the case of circular polarisation, things will change. When the circularly polarised wave meets the metal plane, the E-field of the reflected wave will have a phase difference of 180°. That's because the metal is a shortcut for any voltage on it. As the rotation direction of the reflected wave's E-field vector, we can observe that left-hand circular polarisation changes to a right-hand one and vice versa.

When radio waves propagate as **ground waves,** they can be reflected or diffracted several times at different angles, i.e. by buildings. This has a significant impact on the wave's polarisation plane and amplitude. Any type of interference of the radio wave on its way will have a mixture of effects. One of them is the rotation of the polarisation plane. Those effects can be observed for both polarisation types, linear and circular. Ground reflection can also have an impact on wave propagation, which is different for horizontal and vertical polarisation. Terrestrial TV-transmission almost always uses horizontal polarisation in the VHF- and UHF-bands. When the line of sight conditions were acceptable, no unexpected interference occurred and the polarisation plane remained free from rotation.

By that, any additional losses could be avoided. In most scenarios, even a quasi optical line of sight was sufficient for good TV reception.

Ionosphere – Refraction and Reflection. That's of big interest in the case of sky wave propagation on short waves. Reflections by ionised atmospheric layers can produce one or more "hops". This can lead radio waves over thousands of kilometres to other continents.

One variant of ionospheric propagation effects occurs when a wave enters the higher layers. The wave will be reflected at the top and at the bottom like in a waveguide. This will also produce immense ranges of propagation. Radio communication with stations on other continents will then be possible. While the radio waves propagate anywhere in the atmosphere, there will be no signal detectable on the ground in a certain zone between the transmitter and receiver, the so-called dead zone. There is still another application of propagation effects related to the ionosphere, which was always very important for radio coverage on shortwave. The effect works in such a way that it fills up the dead zone with radio signals.

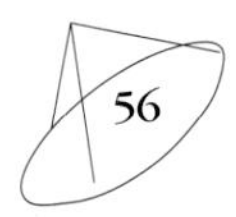

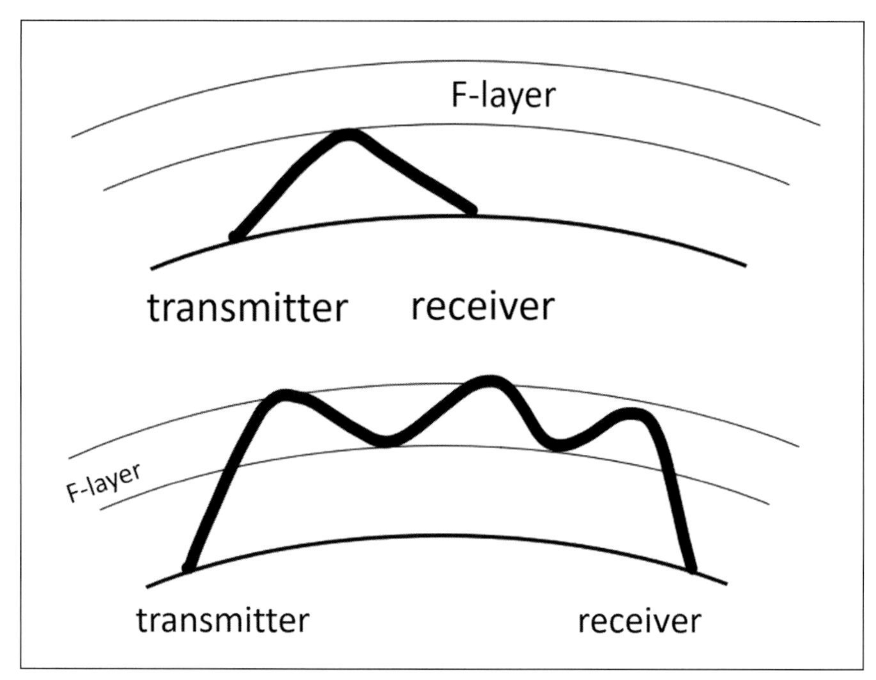

Figure 14: The ionosphere changes the direction of the wave propagation by refracting and reflecting waves. They can propagate over a distance of thousands of kilometres. The ionosphere is located at a height of 150 - 600 km in the atmosphere.

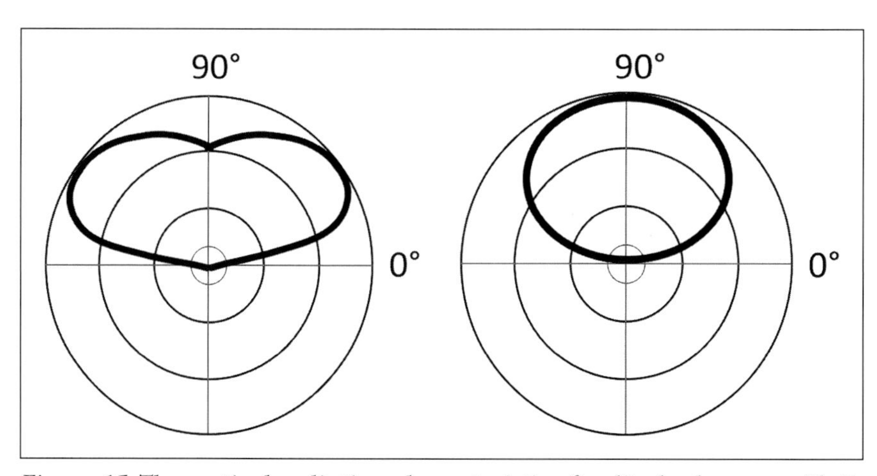

Figure 15: The vertical radiation characteristic of a dipole changes with its height over ground. Left: Height over the ground is 5/8 λ. Right: below 1/4 λ.

In many cases of transmitting radio waves from a short wave station over long distances, we observe a relatively large zone without radio coverage. By using NVIS (near vertical incidence skywave) this can be avoided. When radio signals are transmitted at steep angles or even straight up into the sky, a large range around the transmitter will have good coverage for radio reception. Antennas used for this effect can be dipoles with a short distance over the ground. The inverted-V is a good candidate for this. See Figure 15 right hand. The negative effect of this type of coverage is poor flat radiation towards the horizon. Unfortunately, dx-propagation over very long distances is significantly reduced by that.

Good results can be expected by using the method of greyline-dx, when the path of propagation lies in the dawn or dusk zone. Then the E-layer has not completely disappeared. So the signal can be refracted in such a way that it propagates with a shallow angle up to the higher layers in the atmosphere (F-layer). The path reminds one of the refraction experiment with the prism, where the light beam changes its direction. The subsequent mirroring at a shallow angle increases the propagation range significantly. That's typical for twilight zones. But what is the origin of all those effects in the ionosphere? Where does sky wave propagation with reflections come from? The reason behind this is a high concentration of free charge carriers in the F-layer of the earth's atmosphere.

We can say that large regions of the F-layer have become a plasma now, which means ionised gas. With a number of 10^{12} electrons per m^3 it is rather high. Free electrons are the origin for a good electrical conductivity, just as in metal. Even if their concentration is far below that of metal (10^{28} electrons per m^3), their density in the plasma is high enough to have a significant impact on the reflection characteristics of that layer. Now we know the reason for the atmosphere's behaviour like a good reflector. The lower E-layer also has similar properties. But with a lower electron concentration of 10^{10} electrons per m^3 the effect remains less clear and becomes more intense during the summer. In those months, increasing ranges of propagation can be expected, even in the 50 MHz band and in the 144 Mhz band.

Doppler effect. When we observe the Doppler effect, the received frequency seems to change slightly. That is caused by the movement of the transmitter in most cases. The received frequency drifts up or down, that‘s the so-called Doppler shift f.

$$f = f_0 \frac{1}{\sqrt{1 \pm v/c}}$$

In the above equation, we have f_0 as the transmitted frequency and v as the speed of a moving object (transmitter or receiver) and c as the speed of light. We know this effect from the siren of a passing ambulance car, when the siren‘s frequency suddenly seems to change down to a slightly lower frequency just in the moment of passing by.

The received frequency is higher when the transmitter moves towards the receiver. The wave fronts are compressed in this case. When the transmitter moves away from the receiver, we have just the opposite effect. The frequency is slightly lower and the wave fronts are extended. It is the relative speed between the transmitter and receiver that causes the effect. When a receiver moves, we will observe the same. This very special behaviour of radio waves (observed for the first time by Christian Doppler) enables scientists on earth to draw conclusions while watching and analysing astronomical constellations in the universe.

The Doppler shift is a very important and valuable instrument for remote sensing and observations in astronomy. It is used when galactical radiation comes from objects like quasars or x-ray sources, is analysed on earth. Another application is a special type of direction-finding antennas. These systems also use the Doppler effect to calculate the direction of incidence for electromagnetic waves. Round direction-finder systems with 10 or more vertical dipoles in a circle can easily estimate the radio source position by using the Doppler shift. They are used for finding transmission sites. By switching from one antenna to the next, the receiving system acts like one single virtual moving antenna running on the circle line. This effect is a valuable instrument in all scenarios with a relative movement between a source of transmission and a receiver. Conclusions about the relative speed are possible as is the direction of movement, towards or away from the receiver.

Fast Fading. Sometimes we receive radio signals, and we would expect a stable and constant signal level. But the level isn't constant and changes sometimes faster and sometimes slower. But why? There are different causes for that. Let's have a look at the fast-fading first. When a vehicle is moving between buildings in a metropolitan area and transmits while driving, the received signal will fluctuate more or less according to the speed of the vehicle and the numerous obstructions in the city.

When the vehicle is far away, the received signal can even have dropouts when the path loss increases by buildings and trees, until the contact gets lost completely. Fading can also come up during multiple reflections and multipath propagation, even if the transmitter and receiver do not move. Planes on a low flight level above the transmission link can produce an additional reflected signal which will interfere with the ground wave signal and produce some kind of periodic fading for a while.

Slow Fading. When we receive radio signals the amplitude can also change unexpectedly and slowly. It can be a periodic change in amplitude, but it can also be a kind of aperiodic change. What can be behind this? In some special radio bands for communication and broadcast, we can observe a strange kind of unexpectedly slower change in signal level. Some of the changes are even very slow, which makes the observation difficult. Various speculations about this can be heard.

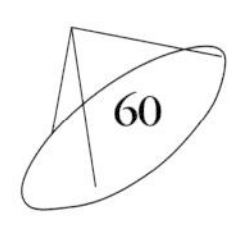

Simple causes such as a moving transmitter or the movement of antennas will not give us a satisfactory explanation of the changes in signal level in most cases. But in addition to the mentioned movements, which might cause a significant decrease in signal strength, other causes should be discussed as well.

Some origins can be found in the atmosphere. Weather phenomena like temperature inversion are one of the reasons why VHF- and UHF-signals suffer changes in signal strength. This can happen through reflections at inversion layers. On short wave bands, signal attenuation almost changes slowly when a radio wave propagates through the atmosphere. We should also take into account that temperature changes also have an impact on materials, for example. Sudden and clear temperature changes may cause shifts in resonant elements simply by changing dimensions and geometry. We also know about humidity as a hidden issue. When humid air condenses inside elements such as wave guides, cables, and antennas we sometimes face dramatical problems. Those causes are hard to determine, and they will often lead to misinterpretations and plenty of wasted time searching for the wrong cause.

Driving through a forest during a transmission, you will find out that trees will also cause signal fluctuation, in some cases until the link is disrupted. Even a single tree can interfere, and the signal decreases until complete loss.

On high frequencies above 1 GHz these diffraction and shadowing effects are more clear. On medium waves, we also observe a type of slow fading. This may result from cross-modulation effects in the atmosphere or even changes within the atmosphere itself. Fluctuation almost always happens in the range of 20 dB, in some cases even more. For professional radio systems engineers, they should always provide a sufficiently large safety reserve. Forests need, for example, a reserve of 3.5 dB on 2 GHz to compensate for the path loss difference in summer and winter, that is with and without foliage.

Ground Wave Propagation

The ground wave in FM radio broadcast systems has provided good propagation conditions for decades. Other applications also use the advantage of safe ground wave propagation by using the right radio bands and antennas. Professional radio communication in taxis, bus and police are good examples of the use of VHF- and UHF bands over ground waves. Another prominent example is the DCF77 transmitter on 77.5 kHz near Frankfurt, which controls some million clocks, I would estimate. Day and night, it sends out time signals for Europe on very low frequencies in order to provide a safe link. The ground wave, however, is not as stable in every case as we might assume. While propagation over ground waves seems to ensure stable signal levels below 100 MHz, fluctuations will increase slowly in the UHF bands around 430 MHz and even more above 1 GHz. Exceptions: Links with a free line of sight, as in urban areas with a high density of 4G and 5G sites. In this case, even at high frequencies of some GHz we can expect stable links.

Unstable signal levels can be observed near the windows inside buildings when we receive weak radio signals in the 70 cm band. We will observe a strong signal fluctuation with deep valleys in field strength. The changes seem to depend on the exact location inside the building. By moving the receiving antenna inside the room, there will be positions with strong but also weak signal levels. Those minima and maxima can also be observed on the terrace. Where do those fluctuations come from, and why do they fluctuate over time and by changing the position of the receiver antenna? Many of these scenarios are the result of superposition. Different paths of the incident wave will cause a complex field distribution. We can assume one dominant path but also one or more refracted field components that will interfere. They are not as stable as the path over the line of sight. Edges and reflections can produce interfering field components. This can happen near the window, near other buildings and walls, and these effects are more clear at higher frequencies. The long way of a ground wave passes a lot of obstructions. That will change a stable signal into a composition of several field components. Even when the signal on the direct path gets attenuated, its polarisation plane can be rotated, and the sky wave may provide additional field components. All this can have an impact on the ground wave. Last but not least, the positions of minima and maxima of complex field distributions will be even more dependent on the frequency of an incident wave. Consequently, we might assume that the stability in the 70 cm band is reduced compared to the 2 m band. When those interfering effects become too strong, we have only one choice: finding the antenna position with line of sight, or at least finding a position with one strong dominant path of the wave and low levels of the interfering wave components. This can be achieved by trying different positions near the window or on the terrace. The probability of increasing the signal levels is rather high and we can expect 10 dB or more of an increase in field strength while finding this out with patience.

The propagation behaviour of radio waves cannot easily be predicted due to the different impacts that have to be considered. One of the parameters is the wave length of the RF signal. Long waves and medium waves can propagate along the earth‘s surface over a long distance of some 100 km, even if they should propagate straight ahead. Waves above 1 GHz, however, will propagate almost straight ahead. They would not move along the earth‘s spherical surface.

We can conclude: the higher the frequency, the more important it is to have a free line of sight to the transmitter. Visible light and laser beams are good examples. For radio signals from 3 MHz to some hundred MHz propagation characteristics change significantly. We will look at some examples.

The ground wave is dependent of the ground conductivity. With better ground conductivity, the wave will be less attenuated and the range will increase. Over a long distance, the ground wave is more and more braked and compensated. This effect is more clear on dry ground (Figure 16). Topography, buildings, forests, all these obstructions will cause a decrease in radio signal level over a long distance of propagation.

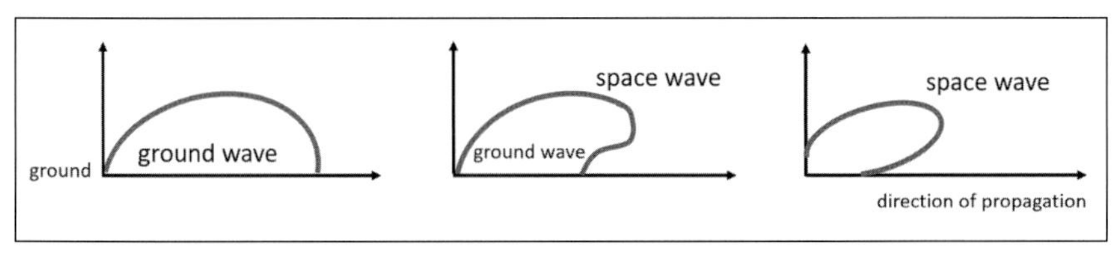

Figure 16: The conductivity of the ground can change the propagation radius of the ground wave. a) very good conductivity; b) medium conductivity; c) poor conductivity. The vertical axis represents the elevation angle of the beam.

But how can we work with better ground conductivity? Especially near the transmitting antenna, we have some options. When the antenna is placed in the garden, we can use a kind of artificial increase by using metal strips on the ground or slightly beneath the ground. This will provide better radiation conditions, especially in the shortwave bands or even below. The metal strips have a similar effect as the radials of a roof top antenna. To be honest, this is not a full metal ground which would not be compatible with the environment or with the family. But the measurement with the strips or even wires will show good results.

On VHF and UHF radio amateurs can work with repeater stations from a distance of 100 km or more if they have a proper location with a certain height over ground. The best antenna positioning is on the roof, or at least on a balcony. Most effective is a good height over the ground, as on hills, for example. Under these conditions, combined with a good antenna, there may even be more than one repeater on the channel, sometimes up to three or four. This is the benefit of a good site for transmitting and receiving, and you can expect even more than 100 km of range on VHF and UHF. Unfortunately this can also cause interference on your radio, such as co-channel interference. Generally, those overrange scenarios are not appreciated. Some of them can be avoided by the regulation of the telecommunication authorities, as well as with neighbouring countries. But there will always be channels that will cause problems.

Meteorscatter and inversion layers are natural phenomena. They can temporarily cause interference on channels that are sufficiently separated by a large distance. Those interfering signals from distant stations will disappear again when the meteor scatter or inversion layer disappears.

When frequency usage is badly regulated, then worst case scenarios can happen, like the following: one country uses the uplink band for repeaters, which is used by another neighbouring country as the downlink band, and vice versa. In the case of overrange propagation this might cause really bad coupling effects.

A good range for amateur radio stations can also be achieved on a balcony with some metres of height over ground. With a good antenna and without severe obstructions in the close vicinity, several repeaters from 50-100 km distance can be received. In rural areas without any infrastructure or repeaters, reception of other stations might be rare, and signal levels will be poor. This can really be disappointing. One last option for any successful link can be the use of special operating modes like FT8. This will give unexpectedly optimistic radio reports, and the hobby can be enjoyed again.

Generally, the best condition is a good height of the antenna over ground. When the antenna is able to look far into the country, many successful links and contacts can be expected.

Experts who plan microwave radio links in a backbone network have to consider additional rules beyond the line of sight criterium. The first Fresnel zone has to remain free of any object. It is a kind of tubular or elliptical zone between the transmitter and receiver. Its geometry depends on the wavelength used and the distance. No building, no trees or vehicles are allowed in this zone. Then the link will be free from any interfering signal reflections (Figure 17).

$$b = \sqrt{\frac{\lambda \cdot D}{2}}$$

The least allowed distance b is valid in the middle of the link, with D as the length between transmitter and receiver. Looking at the ground from the link axis, there should be no obstacle that might cause any additional or interfering field components.

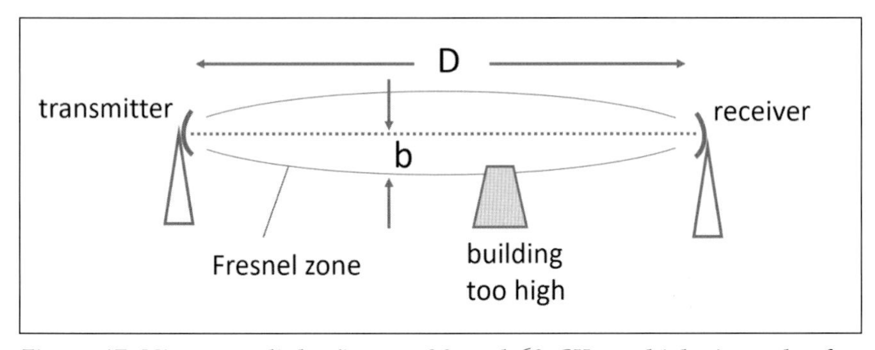

Figure 17: Microwave links (i.e., on 30 and 60 GHz or higher) need a free Fresnel zone without obstacles.

When the link has a free line of sight and the first Fresnel zone remains free too, the link is safe and multipath interference is avoided. As the digital modulation in backbone networks uses high-speed data transmission, the links are sensitive and need this type of precondition. By this, bit error rates will be low and the link will have sufficient quality.

When the telecommunications companies build up and maintain their networks, they need radio experts who are able to care about the link conditions and who can make measurements and document this. LOS-tester (Line Of Sight) is one of the jobs. They will check and estimate the planned links to see, if any surprises can be expected and if the line of sight is free. The radio link planner defines

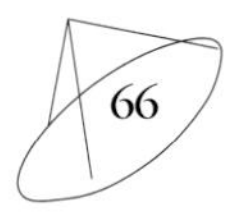

the links of his transport network. Radio planner is another high-qualification job in telecommunications companies. When both have done a good job, the links in a whole region will work properly, and mobile communication goes on without issues.

Superposition of Electromagnetic Waves

When a signal is transmitted to your receiver, it is not only the dominant or direct path on which the wave propagates. There are almost always different paths for the wave, which are caused by reflections.

This will produce different incoming signals with the same information, but with different power levels and timing. Coming in at the receiver, all those shares will superpose and form the complete received signal. In some cases, the amplitude of reflection is nearly as high as the level of the dominant direct path. The phases, however, will differ from the phase of the wave on the main path. This can produce strange effects. The amplitude of the received signal might oscillate, the modulation seems noisy, and bit error rates in digital modulation can increase dramatically. Sometimes superposition can be observed while waiting at a red traffic light, slowly moving the car, and listening to the analogue FM radio. The signal level might fluctuate. Moving a few metres ahead, the amplitude can increase again, or the signal disappears completely. In this case, the car is moving through minima and maxima, caused by the superposition of one signal on different wave paths. The distance between minima and maxima depends on the signal's frequency or wavelength.

Interference also occurs when a signal is transmitted from two different locations, i.e., when radio planners intend to broadcast from two or more different transmitter sites to achieve better signal coverage (single wave network). When different coverage areas from adjacent transmitters with the same signal frequency overlap, the result will be a superposition zone. Here, interference can cause trouble and reception becomes difficult or impossible.

This zone is critical because signal amplitudes mix up and change over time. The most critical are phase differences and time delay effects. Digital receivers can lose synchronisation or the signal amplitude is getting oscillating and unstable. Interference-free reception can be completely impossible in certain areas.

But there is a method to resolve that: the critical overlapping zone shall remain small. In the best case, both illuminations do not overlap at all. This can be achieved by clearly dominating only one signal in each assigned area. A spatial separation is a good solution. Valleys, hills, and shadowing can be used to separate the signal shares coming from different sites in a single-wave network.

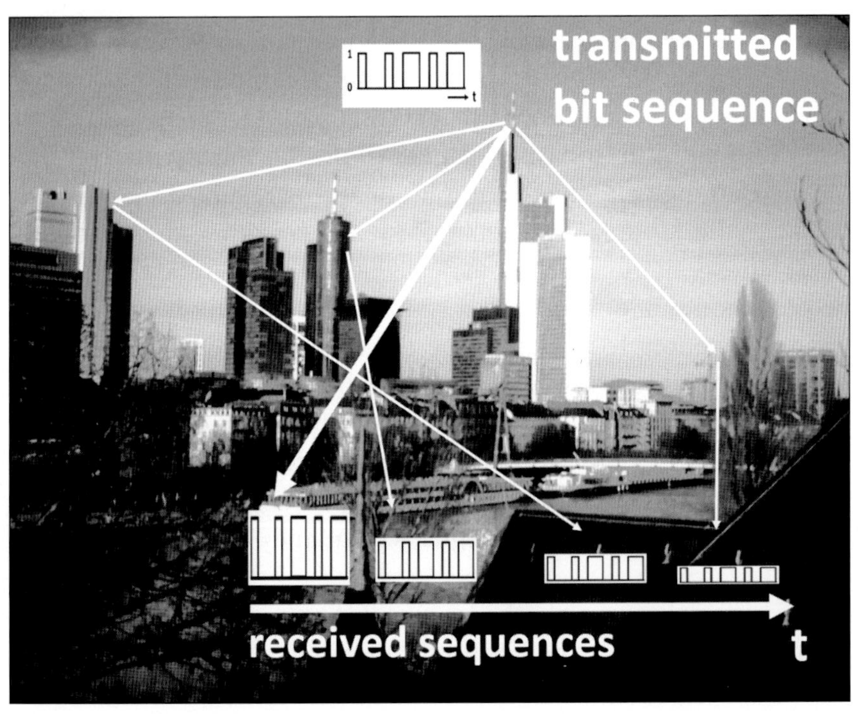

Figure 18: Multipath propagation. With the right digital signal processing, the receiver will combine different signals from the same source into one resulting signal. The reflections are added by eliminating the time difference caused by different path lengths of reflected signals. The interference is now eliminated. In GSM systems, the Rake-Receiver applies an algorithm to do this.

Digital operation modes have generally better options to handle interference by multipath reception than analogue modes. DSP (digital signal processing) is the technology that can separate and combine signal components by fast calculation. DSP can sum up the usable signal components and suppress the unwanted shares. Every digital receiving mode and every receiver hardware have their own concepts to handle different incoming signal fragments on the time axis.

In GSM technology, it is the RAKE receiver, which uses a special algorithm to handle multipath fragments while receiving phone signals. See Figure 18. This special algorithm can sum up different signals into one result by eliminating their phase differences or time shifts. Even in interfering zones close to a repeater station, on both sides and behind the repeater, any interfering overlapping is calculated away. Clear signals on receiver side are the result.

In single-wave networks, some digital operation modes are not prepared for such a challenge. They will have severe issues in some of their coverage zones. Single-wave operationd can produce severe interference and disruptions. In areas with overlapping zones with different illuminations with the same signal from different transmitting sites, the following issues can be observed: Any receiving station in the critical zone may have a good signal level, but the signal amplitude can oscillate, and the digital synchronisation will be lost within seconds or milliseconds.

Often, there is no signal acquisition possible, even if the signal level seems sufficient. How can this be avoided? In single-wave networks, a slight detuning or frequency deviation between the transmitting sites can help. Synchronisation to one share within the signal mixture is supported by this measure.

Multipath propagation can also be observed with analogue signals. We can see a kind of flutter fading with an obvious oscillation of the amplitude. When two signals come along on different paths, then the phase relation will have an impact on the superposition. Especially if the length of the paths changes, i.e., by moving, this effect is clearly visible. Even aeroplanes above will cause this type of changing phase relation and amplitude oscillation. Nils Schiffhauer has analysed this phenomenon in depth in [4]. In analogue single-wave networks the same strategy with a slight frequency deviation between the different sites of the network can help suppress unwanted interference effects.

Analogue and Digital Transmission

In this chapter, we will discuss the huge variety of analogue and digital radio signals, that are "in the air" continuously. We can see numerous types of signals that are used by the different services of countries all over the world. To avoid chaos and permanent interference, a certain regulation is needed. Radio transmissions, as well as voice and data transmission services, all need guidelines and allocations of frequency bands, maximum power, and other parameters. When we look at the radio bands in Figure 19, we know that AM is used and that a clear channel spacing must be applied. This is important for the coexistence of all transmitters in the world with only a minimum of interference.

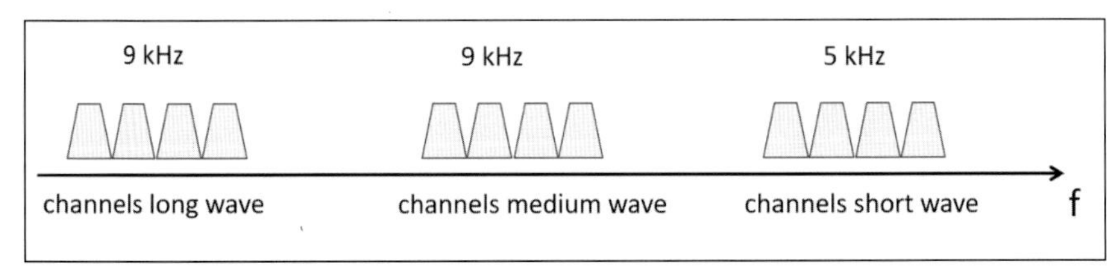

Figure 19: Channel spacing for LW, MW, and SW ranges.

Channel spacings as shown in Figure 19 exist for all radio services in every band. This is a good base for a worldwide regulation. In the long and medium wave bands, the spacing is 9 kHz in Europe; on shortwave broadcast bands it is 5 kHz.

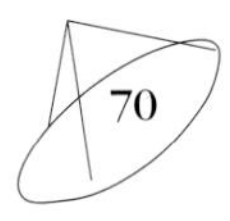

There are plenty of additional guidelines for channel spacing and transmission parameters for all applications in the field of professional radio services. We will look at this later. For public mobile radio, the allocation and channel spacing regulations are also very important. The GSM system works with a grid of 200 kHz. Each channel has a bandwidth of 200 kHz available for voice and data transmission.The regulated grids use specifications that ensure that no neighbour channel interference can occur. Standardisations by the telecommunication authorities are based on manufacturer specifications and other scientific sources. In some cases, interference cannot be avoided. Shortwave receivers without a sharp selection may produce a 5 kHz tone resulting from two neighbouring broadcast stations. Modern receivers can handle this, and even the good old Yachtboy even had a 5 kHz filter against this type of interference.

When rules and specifications regulate the operation, it is quite natural that exceptions are allowed and sometimes even accepted. On short-wave, it is the digital mode DRM (digital radio mondiale), which needs more than 5 kHz of bandwidth. DRM is a mode that gives a radio broadcast on shortwave the sound of an FM station with a 100 kHz bandwidth. DRM can achieve this unexpectedly good sound with a bandwidth of 4,5-30 kHz in different variants. The sound is extremely good. Generally, short-wave stations have almost always a rather poor sound compared to FM radio, especially when we listen to overseas broadcast stations such as once HCJB Quito or others. Sure, we need a signal with a certain level and stability, but the results are extraordinary in every case.

FM radio uses in Germany a bandwidth of 100 kHz for mono transmissions. Stereo needs 250 kHz because additional carriers have to be added to the channel bandwidth. One stereo channel uses L+R and L-R and RDS identification.This makes frequency planning more complex, because the existing 100 kHz grid shall also be used for stereo transmissions with its own grid.The only way to solve this, was to allow a clever channel re-use and the use of direct neighbours only at a big distance. By doing so, the coexistence of all transmitters for FM radio (88-108 MHz) without too much interference can be ensured.

Digital Signals and Digital Transmission

We can imagine audio signals and analogue voltage from a microphone, for example. Sinusoidal waves from an RF transmission can also be easily understood and visualised. But there are still two aspects to consider for a better understanding of the digital transition. The first is the transition from analogue to digital on the audio layer. The other is to find out what it makes on the RF layer and how it is used. In the chapter "Modulation Types" we can find some explanations concerning the RF-layer. This will point out the differences compared to the analogue modes of AM, FM, SSB and CW. But let's look deeper into the audio layer first, and see what happens with the analogue signal of microphone in the process of digitalisation.

Digital Voice and Music

Today, the consumption of music and voice almost runs on digital channels. Everybody can prove this for himself by checking his personal channels as CD, VOIP, DAB+, DVB T, DVB S, LTE, 5G. But a CD will store data instead of analogue signals originating from music records. The analogue representation of music is a superposition of a variety of sinusoidal signals over a huge frequency range. Somehow, this mixture of waves and oscillations in the audio band has to be transformed into a digital representation in a standardised format.

Resolution. The transformation from analogue to digital representations almost starts with sampling. This means that the resulting audio curve is sampled on discrete equidistant points in time. Thus, the level of the curve is detected just at one exact point on the time axis. The system will then allocate a digital data word to each sample which includes the amplitude information.When we choose a resolution of 8 steps on the level axis, we can work with data words of 3 bits in length. Meanwhile, the calculation speed of DSP systems (digital signal processing) is so high, that transformations of this type will run so fast that it seems like a real-time process.

Sampling rate. But how fast shall this sampling be? The answer comes from the Nyquist theorem, which says that the sampling rate has to be twice as high as the highest frequency in the audio fragment, which has to be digitalised.When the FM broadcast system of a studio uses signals with approximately 15 kHz for the daily music and news as an example, the Nyquist requirement necessitates a sampling rate of at least 30 kHz for that source.A data stream with one data word for each sample enables a reasonable reconstruction of the signal in a receiver where it is changed back to an analogue signal. Voice quality in older phone systems only needed an audio bandwidth of 3 kHz.The use of sampling rates better than 6 kHz is suitable for that. When the sampling rate reaches a certain speed, the original sinusoidal spectrum can be reproduced completely. See also Figure 25.

Codecs. Digital voice needs good intelligibility at the end of the channel where users are listening.An appropriate codec is the precondition for good voice quality in radio, telephone, or music in HiFi systems. Voice quality in telephone systems in the 90s was poor.The available codecs used the given bandwidth almost without extensive optimisation.

For higher demands, like music on CDs some appropriate digital methods have been used. Puls Code Modulation (PCM) was the basic instrument for digitalisation and this hasn‘t changed over the years. Once developed by Sony, PCM uses a bit stream of 64 kbit/s and a resolution with a sampling depth of 16 Bit. Even if digital technologies came from the last century, they were continuously optimised, and some amazing results are the benefit of our century.

Voice-over-phone links sounds like voice on a high-quality FM radio channel. Another really big achievement is the MP3 format, which uses much less space on a data carrier than PCM with a similar quality. But how is the MP3 codec is able to reduce the needed memory space so dramatically? It is done by using some technical tricks during digitalisation where parts with lower relevance in the audio spectrum are cut off. This psycho-acoustic approach is based on the fact that listeners do not need frequencies above 22 kHz or extremely quiet passages, either. They can be eliminated, and this will save memory space. This type of compression causes little loss, irreversible indeed, but for mp3 applications, there is no use case or intention to restore the original format in all details and with full quality.

For mobile radio codecs it became urgent to find a similar way because the channel resources and bandwidth were restricted and the number of users grew steadily. In the first half of the nineties GSM used a Full Rate Codec. It was a good choice for the early years of digital mobile radio. A few years later, in the middle 1990s the Half Rate Codec provided a similar voice quality but with less space in channel capacity. This provided additional channels for the network. Code snippets with unneeded frequencies were simply eliminated. So radio channels could handle more voice and data from more users.

The spectral properties of voice in different languages have been analysed more deeply since then. Codecs were even adjusted to the special characteristics of domestic languages. When you compare the melodic Mandarin, the official Chinese language, with the rather hard sounding German, there are differences in the audio spectrum that can be used for optimisation that depend on the language. Appropriate codecs focus available data capacity on the relevant voice characteristics in the audio spectrum, and codecs filter out those parts that are not needed. Today, our transmission systems are equipped with advanced codecs for voice and music.

MP3 is one of the most famous compression methods. It is used all over the world, and it is very effective. CD quality is not far away from MP3 quality. Saying there is no difference at all would cause some trouble in the community of HiFi experts. But what is the truth? Indeed, the ears of experts will find slight differences in music quality. But within the daily use of MP3 in almost noisy environments, nobody will say that it has poor quality.

When we look at methods and file formats for the highest audio quality and music enjoyment, we should consider the music quality that is provided by streaming services. There we will find amazing music and sound coming from original sources and data streams produced with extremely high quality. Some of them, i.e., the TIDAL streaming service, use their quality even as a unique selling point, which others cannot achieve with their codecs. Compared to SPOTIFY it is TIDAL that offers that very high-quality streaming service. To enjoy this, you need the appropriate equipment, like a high-end headphone like the legendary Omega combined with a high-end amplifier. Without that, high-end audio streams will be reduced to average music enjoyment. But once your high-quality audio equipment is plugged in, you should enjoy at least the FLAC format (free lossless audio codec), which provides CD quality at 1411 Kbps. For high demands, you should try the AAC-Code with 320 Kbps as a basic option. High resolution streaming products like this can even reach 2304-9216 Kbps. To be honest, these are options for real audio experts. AAC (advanced audio coding) is a further development of MPEG 2 and it can provide the very best streaming quality.

Back in the world of codecs for professional radio and public mobile phones, we recognise a lot of effort in development to provide better and more effective standards. TETRA, for example, uses ACELP. This is more a family of codecs and means Algebraic Code Excited Linear Predictive. In digital voice modes of amateur radio, it is an IMBE codec which is used by the APCO P25. D-STAR uses AMBE 2020 and DMR uses AMBE 2+ from the family of AMBE Codecs.

Today public mobile radio uses much better codecs than the former Half Rate Codec which was once used in GSM mobile networks. The sound in modern phone networks is much clearer and has more volume, and this high-quality is of course important for modern users. For LTE it is HD-Voice, a codec, which provides with 7 kHz rather good results. But even here, we will have an increase in voice quality due to the coding standard EVS (enhanced voice services), which provides 20 kHz, which is similar to FM radio. This means top quality in the telephone device, even in phone calls.

Comparing audio quality in digital and analogue voice transmissions of professional radio, there seems to be no big difference at first glance, especially when we optimise the voice audio by using the treble and bass tuning and the microphone amplifier adjustment. But we can't deny that there is a sharper and clearer characteristic in the digital voice system. It is typical that any background noise is missing, except when the transmission itself contains noisy fragments.The digital sound seems more powerful, which is also caused by an impression of high dynamic while speaking. On the other hand, the digital system shows sudden drop outs as soon as a certain signal level gets lost. It is rather strict behaviour. The transition of the analogue system is quite smooth when the signal level decreases, and we can listen until the noise overlaps the voice signal completely. This is rather good natured behaviour in the case of weak signals.

Modulation Types

Electromagnetic waves in their pure state with a constant frequency and a constant amplitude generally don't include any information without being modulated. When modulation is applied to the wave, it can change the amplitude, the frequency, or the phase. This means that the wave will now carry relevant information and propagate it. In addition to the good old modulation types such as AM and FM numerous modulation types have been developed. Each of them has its purpose. Some are used for digital voice, others are used for

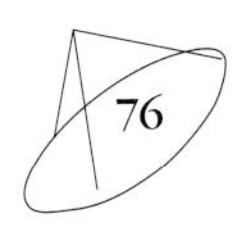

data communication under certain conditions. FSK (frequency shift keying) for example, is a mature mode that is still used in many applications. It is used with two and sometimes with four or even more frequencies. In the last century, FSK became popular and was almost called RTTY (RadioTeleTYpe or simply Telex). RTTY was used for weather report transmissions, telegrams, and news. Modulation in general is applied by imprinting the information as letters and figures to the wave. In the case of digital modulation, the bit streams with the bits' states are imprinted to the carrier. This will give a rhythm to one of the wave's parametres as amplitude, frequency, or phase, depending on the chosen modulation type.

Analogue signals like voice from a microphone can also modulate the silent carrier of a wave. Analogue voice signals consist of a mixture of sinusoidal voltages and are used to change to the amplitude or frequency of the carrier, as in AM- and FM radio and narrow band FM, too. Digital signals which give information to the wave can originate from voice, too, but only after the digitalisation of the analogue voice signal. Any data stream coming from music, measurement data, authentication data or graphical data, can modulate the wave and use the wave as a carrier.

Today, systems use almost QPSK, OFDM, QAM or GMSK (Gaussian minimum shift keying) for modulation. In addition, special applications use their own modulation, such as FT8, which is used by amateur radio and which can be detected even with very low levels even below the noise level. So the range of propagation can be extended dramatically. Noise and interfering effects of bad locations can be overcome, and a new option to use the radio set really promises many unexpected contacts to other stations.

While using links to space vehicles, such as space probes, some very special operation modes are needed. When probes leave our solar system, the received power levels are so poor that modulation types should be able to detect signals beneath noise. This can be

achieved by spreading spectrum modulation types with low bit rates, for example. But in space, FM is also popular for standard voice communication. We can check this at 145.980 MHz with a simple FM receiver and a telescope antenna. When the ISS is above Europe and the footprint of the ISS-transmission covers our own location, we can sometimes listen to voice traffic from astronauts on the ISS when they spend some time for amateur radio contacts. Another option for listening to space vehicles is the AMSAT Fox 1B operated by radio amateurs. Almost every day, signals can be heard on 145.780 MHz when the spacecraft comes to a closer point above us in its orbit.

AM and FM

In the early decades of radio broadcasting, the transmitters used amplitude modulation (AM). The rf carrier is modulated with an audio signal in a way that voice or music makes the amplitude follow the signal audio. In the frequency domain, we will measure this like in Figure 20. In the middle, just on the nominal frequency the carrier can be seen, while on the left and right hand we find the sidebands.

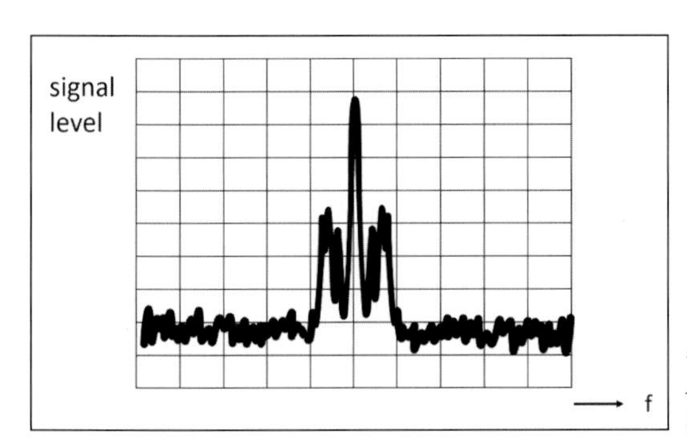

*Figure 20:
AM signal: carrier with two side bands*

The horizontal axis is the frequency axis, and the vertical axis is a grid for signal levels. Both sidebands LSB and USB on each side of the carrier contain the full amount of signal information. The carrier itself without any sideband carries no information at all. The level of the sidebands in relation to the level of the carrier means the modulation degree of the AM signal. When the relation is approximately 0.5 or 50 %, the audio sounds almost natural. When it is too low, the radio signal will have a weak sound, whereas at a degree of modulation of far over 50 % the sound will be noisy with interference, and the bandwidth of the RF signal will extend to unspecified values. But when the carrier doesn't have any information and consumes most of the transmission power, why do we need it? In voice transmission systems, it is the single-side band (SSB) operation mode, which provides advantages for use on short waves. The complete transmission power can be put into one sideband, which makes it more effective. It is a question of standards and regulation if LSB or USB shall be used in a frequency band. But from a technical point of view, the choice doesn't really matter if a system uses USB (upper side band) or LSB (lower side band). In the receiver, the carrier can be reproduced and added, as its frequency is known by the transmission. The whole audio signal can be restored. The only thing that is needed is a little tuning for the correct frequency relation between the carrier and sideband. Cheap radio sets for the consumer segment will contain only AM and FM because SSB always means some effort in circuit technology, filters and costs. Using the carrier of AM for reception will reduce that effort and makes simple receiving circuits possible.

On shortwave radio amateurs use LSB below 10 MHz, while USB is used above 10 MHz. Professional services on shortwave prefer USB on all channels. We have seen that in the case of SSB transmissions, the carrier can be added on the receiver side before demodulation starts. On the transmitter side, the carrier is suppressed and is not transmitted at all. One type of circuit for adding a carrier to an SSB transmission in the receiver is the use of a BFO (beat frequency oscillator). Grundig's Yacht Boy 700 from the eighties uses that feature, which needs some sensitivity in the fingertips for the right adjustment. But it is just this feature that enables a good reception of SSB and FSK signals with good results even while receiving data transmissions. Later communication receivers were equipped with a product detector, which allowed for better handling and more stability for listening. From a technical point of view, the product detector also adds the missing carrier signal to the receiver, which is missing in SSB transmissions.

On shortwave AM and SSB are the dominant voice modes, while FM is used on the VHF bands, or UKW in German. By the way, in the frequency domain, a narrow band FM signal is rather similar to an AM signal.

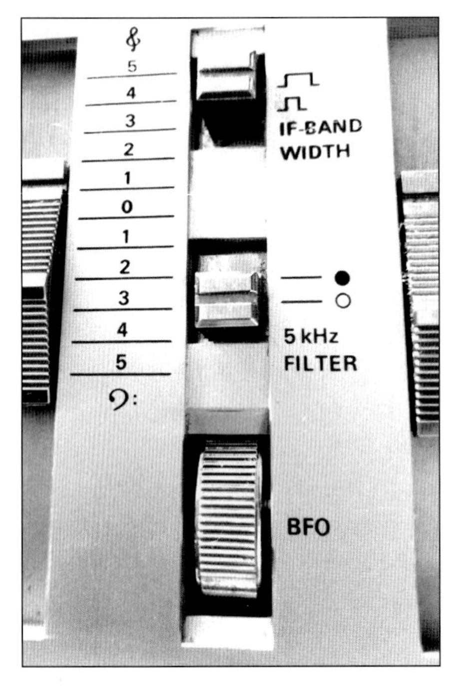

Figure 21:
Features of a portable radio.
The 5-kHz interference tone on SW could be suppressed and even SSB mode could be used (BFO).

Narrow band FM is used by professional radio services as taxi and maritime radio. They use typical bandwidths and grids of 10, 12.5, 20 and 25 kHz, while broadband FM with 100 MHz is almost always used in VHF broadcasts for public radio. The wide FM mode provides much more quality in voice and music transmissions. The stereo variant on UKW (VHF) uses even 250-300 MHz and it is a very popular mode. As the grids for mono and stereo have to be used in the same frequency band, there is a certain risk of interference. This can only be solved by the good frequency planning of the radio experts who are responsible for the network operation.

FM-Threshold. When comparing AM and FM concerning transmission quality and background noise without a signal, some differences become obvious. First is the FM voice quality, which seems to be better, and FM background noise without-signal seems stronger. The difference comes from the demodulator circuits used for FM and for AM. FM belongs to the angle modulation types and the sig-

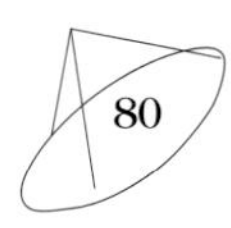

nal-to- noise ratio before and after demodulation is not linear compared to AM (see Figure 22). Before the FM signal can be clearly heard (behind the demodulator), its input needs a higher signal to noise ratio in the rf domain (C/N, before demodulation) compared to AM. For amplitude modulation, this is represented by the straight line in Figure 22. In this case, a voice signal is still detectable with a C/N even below 15 dB (in the diagram left beside the FM threshold) and voice will be almost understood for values down to 10 dB C/N. This means more range, which is a real advantage for communication in aviation. On the other side, right beside the FM threshold we can see that the S/N shows better values, which means better intelligibility for FM. For aircraft radio, this would be of no great benefit because the S/N is high even for AM due to free line of sight in most cases. The only limitation is the spherical surface of the earth. When the line of sight gets lost, the signal level will decrease within a short time because wave propagation goes strictly straight forward above 100 MHz.

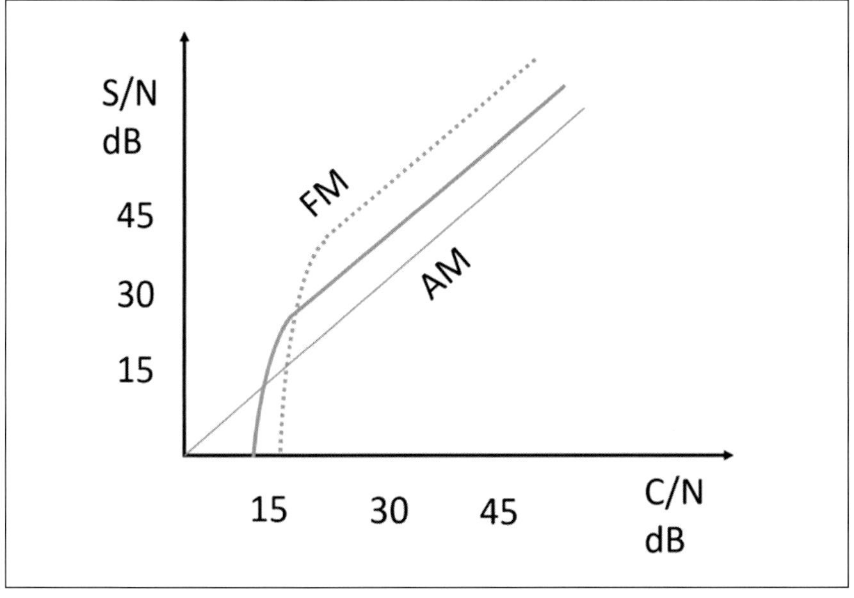

Figure 22: FM threshold and the behaviour of the S/N after demodulation. Modulation index of η =10 (dashed curve) and η=5 (solid curve).

In Figure 22 the dashed curve means a higher modulation index of $\eta=10$. The solid curve means a modulation index of $\eta=5$. Broadband FM in broadcast networks uses an audio bandwidth of 15 kHz or more and a modulation index of $\eta=5$. Narrowband FM as in taxi and maritime radio uses $\eta=1.5$. The typical bandwidth of the low frequency signal (audio) lies between 300 Hz and 3 kHz. In amateur radio the frequency deviation remains around 1.7 kHz while professional radio is allowed to use 2 kHz. Typical channel grids for amateur and professional radio are 10/12.5/20/25 kHz and some of them can work with a little higher frequency deviation. This will give a stronger sound to voice modulation.

The relation between frequency deviation, one important parameter of FM, modulation index η and the highest frequency of audio for transmission is as follows:

$$\eta \cdot f_S = \Delta f$$

In other words, the higher the product of modulation index and highest audio frequency, the higher the frequency deviation will be.

In the age of digital operation modes for voice we might assume that AM and FM will have no meaning any more. And it is a fact that the use of digital modes is steadily growing. But there are some areas where AM and FM still have great importance for communication: in professional radio and in aircraft radio, civil and military, too. In the VHF and UHF aero bands AM dominates. In professional voice radio, FM dominates.

But in the last decades, we have observed that a switch to digital professional radio has been pushed in some areas, even if analogue professional voice services are running successfully for more than 50 years. On shortwave SSB mode is quite common i.e., for off-route flight operations, but also in areas without infrastructure, such as oceans and wide areas with deserts in Africa and Asia. As ground services are missing, SSB on shortwave is the only way to keep in contact with aircrafts because the spheric surface of the earth averts contacts on VHF. SSB is efficient and focuses transmission power into one side band of 3 kHz, which increases the propagation range and ensures reliable contacts.

Figure 23: When this airliner left Chicago O'Hare 1995 for Europe, the crew was connected to the ground stations via VHF in AM mode. Later, while crossing the Atlantic, SSB channels on shortwave were used for reports and waypoint checks.

For decades, FM has also been successfully in many applications where robust radio links have to be implemented without too much configuration effort.

FM devices are easy to handle and to maintain, and the operation doesn't need any backbone networks. Most professional users work in public transportation, bus and taxi companies, and maritime and inland shipping.

In some areas, digital followers have started or are ready to substitute the running systems. Radio operation for taxis and transportation can be substituted by digital applications which use existing networks with good coverage, such as LTE. They only need a suitable software application for managing calls, customers, and orders. For taxi drivers customer requests come to the car's display after being released from the coordinator. The dense coverage of the public phone networks is really a great benefit. They can be used for the transmission of data such as addresses, timing, and availability. It's only the data interface and application software that need to be purchased and implemented. Nevertheless, there are still many cars that are called by the 4 or 5 tone selective call of the ancient individual radio systems of some taxi companies. These good old radio systems were always used for small talk between drivers on boring nights and in emergency cases. There was a time in the 60s when, in Germany, many taxi drivers were robbed. Their VHF radio was a good device for emergency calls, and many colleagues came immediately to help drivers in distress, even at night in dark areas. We remember that nobody had a mobile phone to call the police in the 1960s.

CW

CW Morse is one of the most famous modes of operation. The associated modulation type is almost A1A with a silent carrier which is switched on and off in the rhythm of dots and dashes. Very few stations used CW (continuous wave) in another mode by switching a tone with dots and dashes and a constant frequency as modulation of a silent rf carrier, a kind of CW in AM mode. This type of CW needed more bandwidth and was less economical in power consumption. On the other hand, this type could be heard on any small and cheap shortwave radio, even without a BFO or product detector. It was a big advantage for the so called "services" who wanted to send out secret messages over hundreds of kilometres to a receiver who wanted to remain unobtrusive or to stay hidden.

A1A is one of the narrow band modes, and it really has some characteristics of a data transmission mode. It was used long before the age of digitalisation with a simple method of switching a voltage on and off and producing letters by dots and dashes, defined in a look up table (morse alphabet). Samuel Morse, who died in 1872, recognised the potential for applying this new technology for transmitting telegrams over wires and lines throughout the country. Even for ships, morse signalling became very important. For a long time, CW was the standard for data transmission until microprocessor-controlled systems provided new data transmission modes.

Even in the 80s and 90s, CW survived, and it was used for a long time in Soviet and Russian maritime systems, civil and military. It was a time when maritime transmitters used CW with such immense speed that it seemed to be generated by computers even if CW experts were also able to send and listen to transmissions with very high speed. But this extreme type of high-speed CW could not be identified by ear as a chain of dots and dashes. It was simply too fast. No human ear could distinguish or separate any letters or identify a structure. Later, it was identified to be CW indeed, but it was only possible with a high-end decoder like the W4010 from Wavecom. CW was really competing with FSK for a short time. But FSK won the race.

All in all, the format with dots and dashes is robust and good-natured while using a normal speed. And it is a good candidate for using a computer to read out the signals with a small circuit and a little program, both self-made.

CW has survived until today. And maybe there are electronic enthusiasts for whom it is the right project to start with a self-made decoder. All you need is a signal converter that changes the audio of a tone to an on-off-signal which can be checked by the input of a computer. What comes behind is only a little program with an algorithm to identify if the length of a tone is a dot or a dash and if the pauses between are short or long. Then the results can be analysed and checked in a look up table and sent to a display like a monitor. Letters can be "short long" for "a" or "long long long" for "o".

FSK

Frequency shift keying or FSK changes the frequency of a carrier wave in the rhythm of the bit sequence, i.e., by an offset or shift of 250 Hz or 1700 Hz. Different offsets exist. Almost two different frequencies are used, one for "low" and one for "high" for each bit. There are also special modes using 8 or 16 different modulation frequencies. The difference between both states in the frequency domain of a RTTY transmission is called the shift. Still today, some services use FSK combined with the CCITT-No.2-Code which became popular as the Baudot-Code.

Emile Baudot has invented this code, which is used as an operation mode for transmitting signals with 5 bits per letter or figure. In addition, the code almost needs a start bit and a stop bit for each letter because it is in asynchronous mode. This can be advantageous because the receiver does not need any synchronisation. The score

letter	figure	encode	letter	figure	encode
A	-	11000	N	,	00110
B	?	10011	O	9	00011
C	:	01110	P	0	01101
D	+	10010	Q	1	11101
E	3	10000	R	4	01010
F		10110	S	'	10100
G		01011	T	5	00001
H		00101	U	7	00011
I	8	01100	V	=	01111
J		01101	W	2	11001
K	(	11110	X	/	10111
L	)	01001	Y	6	10101
M	.	00111	Z		10001

Table 1: Alphabet of the Baudot code with two levels, one for letters and one for numbers, according to CCITT No.2.

is simply derived automatically by the knowledge of the structure with a start and stop bit for each letter. Figure 24 shows the table. Many authorities and governmental institutions used the Baudot code for broadcasting their reports and news over 24 hours a day, almost at a speed of 50 Baud. One governmental station in the eighties was the MFS (Ministerium für Staatssicherheit der DDR the former GDR) who sent out propaganda all day. Other agencies in Western countries, such as DIPLO Paris, did the same. Weather reports from all over the world could be received in this simple format because the short wave was a worldwide medium for electromagnetic wave propagation, even if it wasn't reliable for 100 % within 24h. For ships on all oceans in the world, it was the only way to get information about thunderstorms and other navigational warnings in their area.

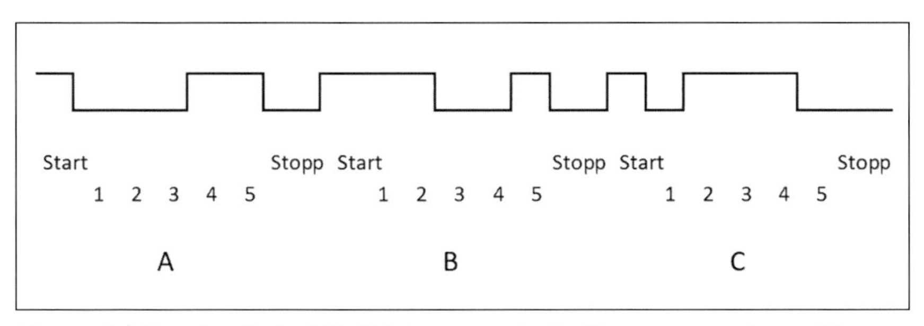

Figure 24: Baudot-Code "ABC": bit stream including start- and stop-bits.

FSK is a basic mode for many transmissions. ARQ and FEC systems often use this modulation type in different variants. Even the old modems that we needed for internet access in the 90s had this typical sound of a fast FSK bitstream. Once they used a speed of 256 kbit/s it gave the modems the screaming sound of a circular saw. Many of us still know that sound. Today, the dimensions of transfer rates are completely different and much higher, and the number of modes has increased. Phone connections use rates of 150 Mbit/s and they are able to carry a high number of TV channels and a whole family's data traffic for gaming and the internet, due to broadband technologies.

PSK

Phase shift keying (PSK) means that the phase relation within a sinusoidal signal can change between two states of phase with a difference of 180°. Bitstreams are the origin of the signal excitation of an rf-carrier and when the state changes, from 0 to 1 or from 1 to 0, it's only the phase that changes and other parameters such as the amplitude and frequency of the signal remain constant.

This type of modulation provides some interesting options and modes of operation such as PSK31 and its derivatives with higher data rates. PSK31 is used in amateur radio for the exchange of text messages and conversations and needs a very low bandwidth. PSK31 signals sound like a continuous tone with a constant frequency. Disruptions caused by the phase change cannot be heard. PSK63 and PSK125 are derivatives and work more economically related to occupied bandwidth.

To be honest, even 70 years ago, PSK was already used. Once, there were standards for TV reception based on PSK in a special way. In the 50s millions of households in the United States used NTSC, which was also based on an analogue variant of phase modulation for the picture in a TV signal. In Germany, it was PAL, which was used for TV-sets. They used analogue phase modulation for the picture and FM for sound transmission. Today, TV systems use digital modes for the whole transmission.

BPSK

Binary phase shift keying (BPSK) uses two different phases in a signal, while the amplitude is constant. The phase relation changes with the state of the transmitted bits, with a phase difference of 180°. For BPSK, there are only a few applications. One of them is PSK31 on amateur radio. But BPSK is the base for advanced modes such as QPSK and QAM which are working more efficiently.

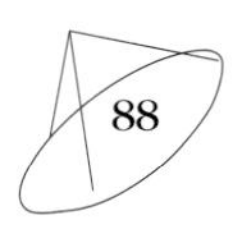

QPSK

Quadrature phase shift keying (QPSK) uses four different phase angles of a signal, and it contains 100 % more information for transmission compared to BPSK. To get a visual impression, we can imagine a cartesian coordinate system: Each of the four phase angles lies in one quadrant of the coordinate system, and they are named 00, 01, 10 and 11. Every time the phase changes, the system jumps to another quadrant, and so the signal has four options for its phase. That means that the four options can be represented by 2 bits. Any state has two bits. That's indeed more efficient than BPSK. In addition to this, some QPSK modes avoid zero crossing in the coordinate system because this would be a larger phase shift, which might cause amplitude changes and negative impacts on the spectral domain of the signal. Example: By using $\eta/4$ QPSK those negative effects can be avoided.

PAM

This type of modulation is used in Local Area Networks (LAN). Pulse amplitude modulation is based on the method of sampling in the time domain, while the sampling rate needs to be high enough with respect to the signal changes. PAM provides analogue values for the sample and for transmission. Figure 25 shows the relations of sampling on one side and digital or analogue representation on the other side. PAM is used in LAN standards for lower speed, working with 5PAM and with 5 steps for a voltage between –1 V and +1 V. 10 Base T uses a no return to zero code with a voltage of ±2..3 V. LANs with higher speeds use modes with ternary coding.

PCM

Puls Code Modulation (PCM) plays a role in the world of the compact disc (CD). PCM is based on sampling in the time domain, too. The difference compared to PAM is a digital representation (data word) instead of an analogue representation like voltage values, for example. Any sampled amplitude gets its appropriate bit sequence for transmission. By using a resolution of 64 bits for example, the system works with the quality of a CD player.

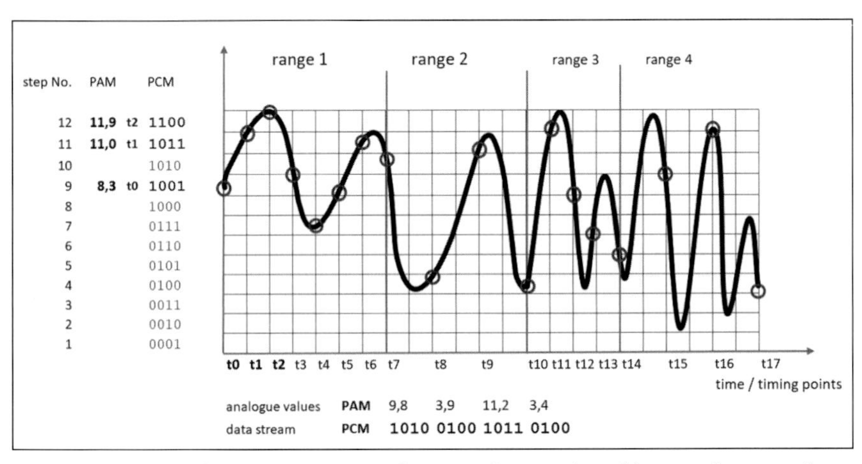

Figure 25: Digital representation of an analogue signal by analogue values, (PAM) and digital values or data words (PCM). The signal needs a minimum sampling rate, as in range 2, to get an unbiased reproduction of the signal. In range 4, it is not possible due to the poor sampling rate.

ASK and QAM

Amplitude shift keying (ASK) is a modulation type with changes in the amplitude of a carrier. A simple variant uses only "on" and "off". But generally, it is possible to divide up the amplitude into more steps. Combining ASK and QPSK, produces the QAM (quadrature amplitude modulation). To get a better understanding of QAM, we should consider QAM as two coexisting carriers with the same frequency, but with a phase difference of 90°. They are called I-signal and Q-signal or I/Q-signal and they can be seen as a sine- and a cosine-component with the same frequency. Each of the signals (I and Q) is modulated independently from each other, which means a multiplication of both. After that, they are added. The resulting signal is a QAM signal. When a QAM signal is demodulated, the described process runs in the opposite direction. QAM signals can have even more states if the amplitude has more different steps than 4 x 4 within a matrix. A mode with an increased number of steps in amplitude can be very efficient, but it is also less robust against interference.

Advanced modulation types (8QAM, 16QAM, …, 1024QAM) can be created by adding more carriers with phase shifts of 180° and 270° and with additional steps in amplitude. ASK and QAM are used in modems, and even DSL and WLAN use those modulation types.

Special Modulation Types

GMSK is a very famous modulation type, that is used in many systems for many years. The well-known FSK modulation simply gets an additional low-pass filter as an extension. That's the way to create that new modulation type called GMSK. What is the reason for a low pass filter just here? The low-pass eliminates phase discontinuities in the FSK and provides better signal shaping in the frequency domain. This modulation type has good transmission characteristics for use within a 200 kHz grid, like in GSM systems, which is a very successful and robust implementation.

DRM (Digital Radio Mondiale) is a modulation type that makes shortwave transmissions sound like high-quality FM broadcast transmissions. DRM uses a hybrid type of modulation. COFDM (coded orthogonal FDM) is combined with QAM Modulation. As a result, DRM gets its properties and robustness. This is an operation mode for wide band transmissions over thousands of kilometres through a medium with fluctuating characteristics. Here, DRM can show its unexpectedly good performance.

LTE and 5G, DAB and DVB-T. The public mobile networks of the new millennium need standards with the highest data rates and the highest capacities for transmission lines. The limit is just moving from 50…500 Mbit/s in the 4G networks up to 10 Gbit/s in the new 5G networks. To achieve the needed performance, telecommunication companies use efficient modulation types and more bandwidth at higher frequencies. Many of the new technologies are based on OFDM, one of the most effective multiplexing technologies, combined with QPSK and QAM as modulation types. Suitable QAM variants are 16-, 64- and 256QAM. For extremely fast 5G, 1024QAM is also used. Also, DAB, DAB+, DVB-T, and DVB-T2 use COFDM and various modulation types of the same family: QPSK, 16-, 64- and 256QAM.

For some extraordinary applications, there are other modulation types available that are suitable for them, such as 8 PSK for EDGE when it once enhanced GSM data transmission capacities.

When we look at the table, 9 we can suspect a high number of modes for special applications, that have been developed over time and that new types will be developed in the future, too.

Bits and Symbols

In this chapter, we look at some transmission details, like letters, numbers, and information, generally spoken. In digital modulation like FSK there are almost two frequencies that are alternately switched on and off. Each bit is represented by one tone or state or by the other. They may be called "0" and "1" or "high" and "low" or "mark" and "space".

QPSK as an example, shows that there can also be four states per step (per symbol) in a signal. In this case, the different states are represented by two bits. One symbol is represented by two bits. This can be extended with even more bits per symbol for more complex modulation types. In this case, a symbol is represented by even more bits, depending on how many states a signal can have. The symbol rate is almost called the baud rate, and it shows how many symbols per second can be transmitted.

Multiplexing, Organisation and Access to Signals

It is typical that in digital modes, a certain organisation of channels is needed. Before sending out a bitstream, data is nested, stacked, and concatenated according to certain rules. While receiving a digital message, it works just the other way around. When a bitstream is received on a GSM channel, for example, it is divided up into sub-streams to extract the different voice channels (of each user), which are packed into time slots. That is typical in TDMA systems.

Additional signalling channels have to be filtered out and reconstructed to get important data with information about the system status and the transmission parameters. When all these information streams from one GSM channel at 200 kHz are separated, voice can be demodulated, and signalling data can trigger certain commands, reducing the output power of a cell phone because the received signal level is high enough.

Many other functions are integrated and can be addressed by parameters in a transmission while a phone call is running. One of them is the bit error rate, which provides a kind of quality level. The channel id is another one. Commands are also transmitted. They can be used to trigger functions like a hand over to a neighbouring cell. One frequency channel in GSM carries 6 slots for voice communication and a seventh one for switching to a second frequency channel, if it is available.

In this case, there are several voice channels that increase the capacity of the cell. All this seems to be happening at the same time because the data rate is rather high. But the truth is, bits and bytes are packed according to a strict set of rules and standards. Data are assigned to voice channels in different time slots, data channels for different analyses and commands to be executed. GSM is only one example. Other digital modes, such as TETRA use similar procedures within similar logical streams and a similar physical infrastructure.

FDMA

FDMA can easily be imagined when you think of a radio scale, like the old radio sets used on their front ends. Radio stations and voice channels are placed side by side on a frequency axis, almost within a defined grid. The grid and its spacing are important standards, like a guideline for each radio band. This will protect against interference by neighbouring channels.

This is extremely important in every operation mode. The standards used belong to governmental regulations, which have almost to be coordinated with international instances. That affects a lot, frequency bands, channel spacing, filter specifications, modulation types, RF power etc. When interference occurs, it can happen as shown in Figure 26:

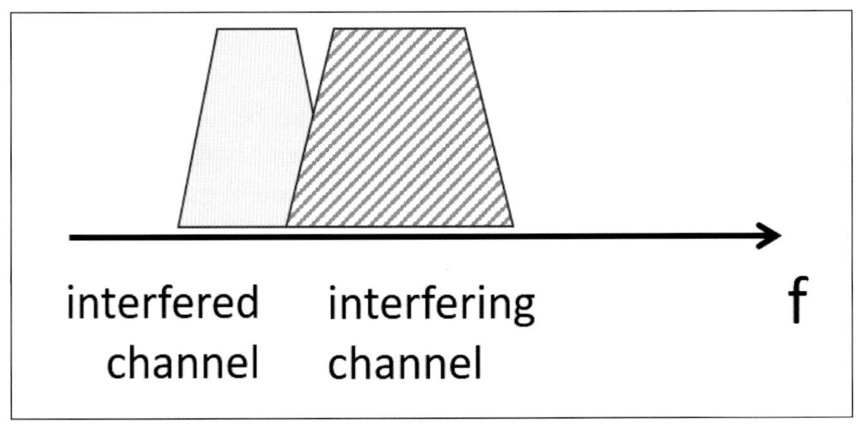

Figure 26: Adjacent channel interference. When the band width of a signal gets too wide in the frequency domain. It is almost the dominant signal that interferes with the weaker signal, and sometimes both are disturbed.

On the channel that is interfered with, voice turns into a noisy mixture. It sounds scratchy with disruptions, and no usable information comes through. In the case of interference on a digital channel, the bit error rate will increase dramatically, and the channel cannot be decoded any more.

TDMA

Digital systems allow new concepts for the use of one physical channel for more than one user. By TDMA (time division multiple access) several signals from different users can be put into timeslots of a single radio channel on one frequency with an adequate bandwidth. The slots are transmitted so fast, one after the other, that everything seems to happen in real-time.

Here is a simplified model: 5 voice channels, each of which is represented by its own bitstream. When those 5 voice channels are added to the time axis, one common data stream will be the result. The rule is that each of the data streams fills alternately one of five timeslots with a defined length again and again, always cyclical. In the end, there is a final data stream that contains five voice channels. On the receiver side, a voice channel (ch1) can be restored by using the same rule, just focusing on one timeslot that was addressed to it. After that, four other timeslots pass, and the next belongs again to the stream of ch1.

Finally, it is the organisation scheme, that puts the bytes of the different voice channels together in such a way that the resulting data stream can be resolved in order to provide any originating voice data stream after the transmission. This is a simplified description of a process that is more complex in real systems because additional signaling data has to be transmitted too, and the system with the slots needs a proper structure with slots and frames. It is important to understand that the number of channels for multiplexing is limited. The reason is that the capacity of the RF channel increases with an increasing data rate. And the specification of 200 kHz per channel in GSM is a hard limit.

2G or GSM is a classical and very successful application that uses the idea of TDMA. The summarising of all data channels as described above is only a simplified description. Real systems are much more complicated because they use a lot of additional data in logical and physical channels. When a system like GSM or TETRA is running in a country-wide network, the system handles parameters like bit error rate, neighbouring channel id, and a lot of other parameters, i.e., for controlling the link and disrupting it when needed. The closer we look at the used methods and at the details of communication procedures, the more we will conclude that voice data is only a small part (some say voice is a waste product) in a big and complex data transmission system. In such a system, the real protagonists are the machines talking to each other. And this is not exaggerated. All those channels running in parallel to phone calls are necessary to provide any type of data for each instance in the network. Technical instances need status information on network elements, and the accounting department needs billing data.

CDMA

In 3G-networks CDMA (code division multiple access) has been introduced to overcome the limitations of 2G like the speed of data transmission and generally the limited network capacity. Looking at the basic principle of CDMA, we will recognise a broadband approach that follows new methods and uses a new structure in signal processing. Broadband systems such as CDMA use a spread spectrum technology. Digital data streams with a narrow bandwidth, i.e., a voice data stream of a user, are mathematically combined with a spreading code by multiplication. Before this step, spreading codes (sometimes called chips) include no information. The result is a spectral mapping of the voice signal to a bandwidth of 5 MHz for example. Data streams of other users are also combined with their own spreading code and mapped to the same 5 MHz of bandwidth. But why can different information sources be mixed up in the same frequency band? Isn't it a kind of broadband noise? How can the originating data streams be restored after transmission? It is really a kind of noisy signal, but it is like noise with a certain intelligence. The method is: each user data stream gets their own spreading code

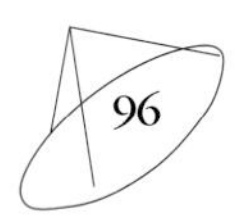

assigned (by multiplication), and the used spreading codes are orthogonal to each other. This is the key property of the set of spreading codes: orthogonality. When these spreading codes are applied to the user data streams, they will not interfere with each other, and on receiver side they can be restored when the individual spreading code is known there. That means that the use of orthogonal spreading codes avoids interference and even ensures multiplexing and demultiplexing of all user data streams on one mobile radio site with several users, for example.

By using this spread spectrum technology in a mobile network, voice channels, signaling data, and other logical channels are combined and mapped to a frequency range of 5 MHz. All of these channels are stacked at the same 5 MHz. It seems as if all data is mixed in a single broadband frequency range. This is just the way CDMA generally works. The key feature is the set of orthogonal spreading codes, which enable the receiver to restore (or decode) the messages that are assigned to them. By the way, spreading codes and broadband technologies are also used in space communication to keep in contact with scientific probes. But why? Even if the power level is extremely low or even below the noise, a very weak signal from outside the solar system can still be identified and demodulated.

OFDM and OFDMA

Complex multiplexing technologies like OFDMA (orthogonal frequency division multiple access) use a huge number of carriers. The carriers (100-1000) are placed side by side with dense spacing in the frequency domain. The carriers represent data channels, and they are realised in such a way that orthogonality to their neighbouring carriers is ensured. This property keeps interference levels very low. How can this be realised? In the frequency domain, each carrier is placed in the minimum of their neighbour's spectrum, and so on. That's the method that keeps interference low.

OFDM (orthogonal frequency division multiplexing) uses all these carriers to take a broad and fast data stream and divide it up on several OFDM carriers before transmission. On the receiver side, the data from all single carriers is summarised again into one main data stream, and the originating signal can be restored. Compared to OFDM, data streams in an OFDMA system start with different originating data streams from several users, which are assigned to a lot of different carriers, as described above. This reveals TDMA characteristics combined with some kind of FDMA characteristics. On the receiver side, the process of packing and organising has to be run just backward in the other direction to resolve all data streams from the initial users. Those packets of data carriers, as in OFDM(A) can handle large data amounts in a short time, and this means a technology for high data rates, just as we need them for modern applications. DAB (digital audio broadcast) and meanwhile, DAB+, the digital broadcast system and follower of the good old FM radio with a countrywide analogue coverage, uses OFDM. All home radio sets and car radios will be substituted by a digital variant sooner or later. We are now in an extremely long transition phase with both systems, analogue and digital. Shortwave has its own digital system, too. For several years, stations have been testing DRM (digital radio mondiale), which could be a follower for the analogue radio broadcasts. The audio quality is amazing when we consider the long ways of propagation when a radio signal comes with fluctuating radio levels to our home. And it really works for intercontinental links over a sky wave.

Prominent examples of OFDM as a multiplexing technology can be found in telecommunications such as ADSL, WLAN, LTE, and DVB-T. The advantages and bandwidths compared to other operating modes are overwhelming. DVB-T and the current DVB-T2 also use COFDM. In the signals of digital radio (DAB+) as an example, a really huge number of carriers, each with roughly 1 kHz make a total signal spectrum of 1.536 MHz. The equivalent streams for digital radio are 1.3 kbit/s on a single carrier, building a total data stream of 1.136 Mbit/s. DVB-T signals have an overall signal bandwidth of

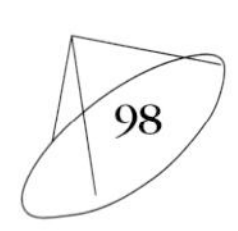

6/7/8 MHz, also with a huge number of single orthogonal carriers. As we know, our expectations have become really high, and we want fast internet everywhere we move. And we want one single telephone sockets for 200 TV channels and the voice and data traffic of the whole family.

Propagation Time

We might think that propagation of waves and signals takes no time or that they can be neglected at least. But when we look closer at the daily TV news, we will just realise the opposite. Listening to an interview that runs between people with a long distance between them, i.e., one in Washington and the other in Berlin, we often notice a long pause after a question until the answer follows. It seems that the interviewed person doesn't really react to the question for a while. The pause, however, has its cause in technology and nature. It is the long propagation time which is needed for the long propagation path. Especially satellite links use a path of 72,000 km from the source over the satellite to the receiver. That means a delay of approximately 0.25 sec. And almost processing time in the systems between has to be added. This can mean 1-2 sec. of delay all in all, and this is what we see on the screen when we wait for the answer to a question. When signals are sent from A to B, it will always take its time. In many cases, we will think that anything within the transmission and receiving processes is running at the same time. But when we have a closer look, it is more complex. In the past, radio and TV networks used analogue technology and almost short transmission paths.

This makes us feel that anything happens at the same time, without delay. But today we often use digital systems and satellite links, and this will add propagation time to the communication process. This time is almost so long that it becomes a relevant delay. It is a long way from a ground station transmitter to a satellite in its orbit. Electromagnetic waves need some time to travel this way. With a propagation speed of $3 \cdot 10^8$ m/s (the speed of light in free space) and an orbit of approximately 36,000 km over ground, it will take 0.125 sec. of time to get up to the satellite. For a complete link to a receiver on earth, it will take 0.25 sec. plus the processing time on the satellite and on earth. Processing times depend on the systems, but in analogue systems, this can mean a few microseconds, and in digital systems, even more. For a correct calculation, we have to consider, that ground stations are mostly not located just under the orbit near the equator, but almost at a distance of some hundred or some thousand kilometres away from the equator.

This causes a certain elevation angle for the satellite dish and makes the link a little bit longer. But this doesn't make a great difference in our considerations. In coaxial cables and optical fibres, electromagnetic waves also propagate at high-speed. To give a rough value, it is about 70 % of the speed of light due to the material used in the cables and in the optical fibres. When a signal is transmitted over a range of 100 km through the air, it takes about 0.3 ms. Latencies almost don't matter in a world of analogue systems because radio waves propagate at the speed of light. When a complex signal transformation by digital signal processing is not needed, delay times will be low. In digital systems, we have to care for latencies much more.

Network areas, that cannot transmit in real time may cause issues if transmitted voice data cannot be sent, for example, in a case of emergency. Digital networks always need time for processing, even if it is very little time in most cases. But we have already learned that even the fast 4G technology comes to its limits in certain applications, especially in real-time applications like medical operations, both remotely and in real-time. Even for autonomous driving, it is better to trust a 5G system than a 4G one. 5G ensures timing that is amazingly close to real-time, similar to some embedded real-time systems for automotive use or for aircraft systems. 5G integrates data transmission and operation, even within radio networks, and makes applications more safe.

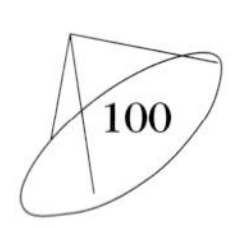

Technology and the People Behind

Let's have a closer look at the people who are behind technologies and what they are doing, and let's also have a closer look at the users. One fact is evident: Each of us is a user of technologies that are based on electromagnetic waves. We find wireless transmission systems everywhere. We use them in all situations of daily life, private or in our job. Most of us are not aware of it. Looking at the different groups that are using wireless systems is quite interesting. It will be clear, how deeply everybody is involved in all those processes in a digital and wireless world, even if we just don't think about it.

Users

They are coming from all areas of our population. Attorneys, doctors, and construction workers are using wireless systems. It's not only the personal mobile phone, which is popular. NFC technology on credit cards, which is needed for paying, or NFC for unlocking cars, there are many applications in different areas. Bluetooth technology is needed at home and in the car for connecting devices such as printers and audio equipment like headphones. This can avoid many cables at home, and it makes life more comfortable. Consumers benefit most from Bluetooth and other short-range wireless systems.

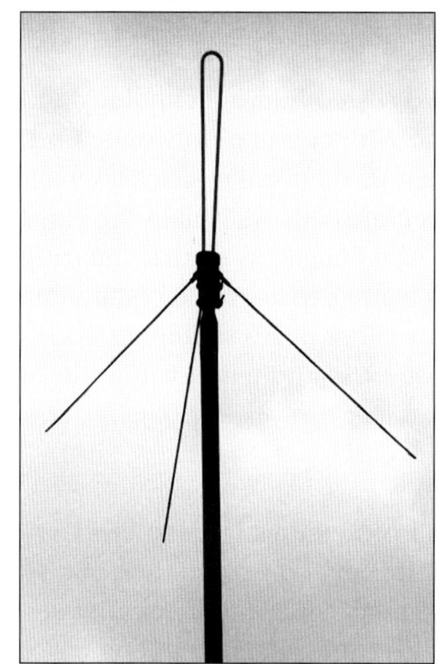

Figure 27: Antenna for professional radio. Thousands of these ground plane antennas for 2 m and 70 cm Band once installed for voice traffic operation in analogue mode for police, fire departments, ambulances, industrial use, taxi, ...).

Now, who is behind all this? How are these technologies created? Who has implemented these technologies? The answer is simple: Many engineers and computer scientists from the development departments and test departments of tech companies implement all those complex systems. They take care of correct planning, implementation, and testing to enable a safe operation. Together with their colleagues from construction departments and prototyping departments, the whole staff of technicians and sales people will get the products to the market in time.

But there is still a special group that deserves to be mentioned. They are very passionate about using their systems, and almost in a special way. They analyse and implement systems that are based on electromagnetic waves and their propagation. They do not only operate the systems; they also combine components and subsystems for new applications. And they do this almost in addition to their daily job.

Radio Amateurs

They were once pioneers in the gold rush of radio technology and wave propagation phenomena in the 20th century. They once used self-made devices and operated them in their radio shacks, which sometimes reminded of a space craft control desk. Even in the Third Empire after 1933, some of them were engaged in that amazing hobby. But without a licence, it was a big risk to use radio equipment, and it was punished severely, even by jail or death sentences. Today, things have changed, and it is almost a trivial offence when illegal radio operators are identified or when illegal eavesdropping is done. Except in nations with dictators. Take care!

Today, radio amateurs come from all professions in daily life. They are company owners, kings, truckers, and construction workers. They come from technical fields and sales departments, employees, pensioners, and young people. What do they all have in common?

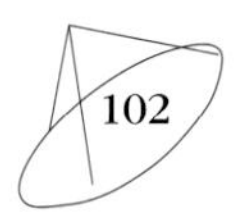

They got a licence for radio amateur operation, and they have an immense interest in radio and wave propagation. All these technologies are meanwhile split into different types of systems. All the enthusiasts use radio communication and wave propagation as a hobby with good technological knowledge of circuits, antennas, networks, digital modes, and much more.

Still today, radio amateurs with a licence are pioneers in a time that is driven more and more by digital technologies. And who else should build up repeater networks with new digital voice modes? And who else would maintain them? Repeaters in a whole network mean a new category of links that can connect radio amateurs on different continents without using shortwave, transmission. Instead of shortwave it is the internet, which is used for some of the connecting links within the network. Sometimes in the evening, we can listen to conversations on VHF with 2-3 OMs on different continents who talk about their shacks or about what‘s new in their region. Their devices are almost handheld radios with a telescopic antenna or with short rubber sausages on the radio, as we say in German. Used radio links can be the link to the next DSTAR-repeater while the OM is sitting on his terrace listening to his contact in New Zealand, for example. In most cases, digital voice traffic is not done in direct mode from mobile to mobile but over repeaters. When no repeater can be used, OMs connect to their internet access node at home to contact other OMs in the digital network D-STAR, DMR (Brandmeister or DMR plus). That means not a pure radio link between the users but a contact with the help of additional network links as the world wide web. Alternative voice modes APCO P25, TETRA, and C4FM.

Today, it is a coexistence of mature technologies for 100 years and completely new concepts for devices and networks. A real OM can tell stories about amateur radio in the last century, and he even uses telegraphy in CW morse for that. Today, we find more and more free spirits who focus on network technology and digital solutions like digital operation modes. We can hear many seniors on the amateur radio bands. The average age of users has increased; young people are using other hobbies, and only a few come to amateur radio. This can be seen as a warning sign, which affects areas like choirs and other traditional clubs. Among radio enthusiasts and radio amateurs life remains exciting. You can listen many hours to the QSOs in the evening or at night, sometimes also at daytime. No matter how many OMs are talking to each other, they can tell a lot of exciting stories about their lives or about amateur radio. Sometimes they discuss details of building up a repeater network, and sometimes they discuss how they can arrange with their neighbour because many people don't understand what amateur radio is, and they often don't accept antennas near their home or garden. Other QSOs are about green repeaters which use solar energy. All in all, it is an amazing world that can be interesting for everyone. Regular bulletins, sent out in voice modes, reveal all the personal engagement of the radio amateurs, who come together regularly in their local clubs. And they appreciate every newcomer who is interested in their fantastic hobby. What are the propagation conditions? When can greyline dx be done? Many questions can be discussed with the experts, who will give answers and good advice for every challenging situation. They all hope that the enthusiasm of the older generations will spill over to the younger people. When older technologies will have their place even in the future and new technologies will give new perspectives and new challenges to the younger generations, this will be a perfect coexistence of technologies and generations, which is really needed.

Radio amateurs have great knowledge in antenna technology, matching, and balancing, but also in electronic circuits, repeater networks, digital configurations to connect devices, and a number of different operation modes, analogue and digital. CW morse is one of the most notable modes. Amateur radio without CW morse is unthinkable. For some people, CW is old stuff, and maybe they don't have the ability to read fast and write dots and dashes. But even enthusiasts

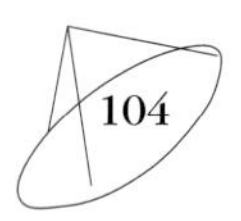

without CW ambitions can find access to this mode by using a computer. Why shouldn't a machine do the job automatically, what do I have to learn in many hours with high concentration? It is no great deal to teach a computer how to read out an input channel that is fed by a relay with on and off patterns and to show the result on a screen. The only thing you need is a little knowledge of electronics and an interest in finding solutions. Realising your own concepts and not giving up just after having started is a good way to get familiar with new technologies.

Beyond voice. There are some exciting data operation modes for digital traffic that require extremely little bandwidth and rf power for transmission. They can be used on critical sites with restrictions on antennas. The modes are Lora, WSPR and FT8 or FT4 and the list will never be complete. They have some extraordinary properties, and this enables operation below the noise floor, for example, 5-10 dB, some people say even 20 dB below the level of noise. For radio amateurs with bad conditions for using proper antennas and with obstructions around, this could be a set of options for making worldwide contacts even with low power and small antennas. Distances of 100-1000 km are possible with high probability and a wire 1-2 m out of the window should be a good practical start.

Digital vs. Analogue. When radio amateurs talk about digital voice modes, we can sometimes observe controversial discussions. Some people doubt, wether the use of internet links as a part of the whole end-to-end link really belongs to amateur radio. They say that amateur radio will always consist of pure radio links. But when the new digital options for voice are considered an enhancement and an evolution of technologies, nobody needs to fear any robotic future systems that can only be handled with huge computer knowledge. When new technologies are developed, it has always been natural that new systems would be beneficial anyway, but it has also been polarising in many cases. Evolution and new technologies are not contradicting the use of traditional systems and methods at the same time.

In our domestic culture, such profoundly changing processes need more time for adaptation compared to the Asian Pacific area, where it is usual to pay by card wherever you are consuming products and services. This is only one example. Let us look at the introduction of mobile phones in the nineties, which had a lot of opponents. When the first people used their mobile phones, you could often see some others who shook their heads in disbelief or made dirty gestures while passing. Was it a sign of rejection of posh products or even rejection of progress? We don't know it. don't we have a hard time, too, with cashless payments? When innovation and progress get their time, the introduction should be no problem, even when we introduce new technologies.

Digital Modes for Data Transmission

For a long time, radio amateurs have been working with digital data modes, rather than with digital voice modes. Data transmission modes play a special role just like other special modes such as SSTV and ATV, too, even if they truly belong to the analogue modes. There is always a lot of movement in the field of digital transmission. It's not only RTTY and Packet Radio that can be used for writing or for chats. There are some interesting new candidates that are available for some kind of automatic mode for chatting from machine to machine. When CQ-calls with minimum rf power are sent out automatically and the answer is stored on a server to rate the contact and the propagation, this is really a new type of communication from station to station.

FT8 is one of those operation modes. FT8 and the derivative FT4 can be used on shortwave. They have a timing grid of 15 sec for transmitting a message from each station. FT8 is synchronised over NTP (network timing protocol) and it works with 8 or 4 symbols with an extremely narrow spacing between frequencies (spacing of 6,25 Hz), which is needed for transmission. Forward Error Correction ensures that it is a robust mode, even when the atmosphere

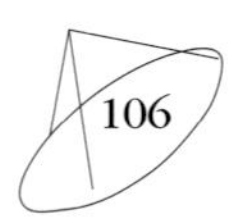

and the propagation is fluctuating. When the received signal level is 10 dB below the noise, operation is still possible. OMs can achieve enormous distances even with small and short antennas. The sensitivity of the systems is so high that they will compensate for the loss by antennas with poor performance and an obstructive environment. FT8 started in 2017 and has become very popular.

WSPR is used on shortwave, too. It has also become very popular, and signal levels can also be identified when they are below the noise level. WSPR (weak signal reporter) sets a kind of beacon for the operator. While the beacon is sent out, you can wait until your signal is received anywhere in the world, and another OM will send you an answer, which goes to a server and is provided by the world wide web. This is a comfortable method to wait for feedback on your beacon signal.

It runs with 30 mW RF power, even if some people prefer 5 W. Generally, it has been developed for low power applications, where 5 W means a lot. WSPR modulation uses symbols with 4FSK and a very low bandwidth of 5.9 Hz. The transmitter and receiver have to be synchronised. WSPR time intervals are 2 minutes, which is longer than in FT8. On VHF and UHF WSPR is a good instrument to use the station as a beacon and to get an idea of where it can be picked up, in the neighbouring city, in another European country, or anywhere else in the world.

LORA. This is the right point to mention LORA, another digital mode that uses a star-shaped architecture. That makes LORA a good candidate for wide-area network (WAN) applications. The idea might have its roots in IoT (Internet of Things). It can be implemented immediately with plug and play components. They use ISM bands of 433.05-434.79 MHz to connect individual nodes and enable data exchange between nodes. Their modulation type is a member of the spread spectrum (code spreading) family, with data rates ranging from 300 bit/s to 50 kbit/s.

APRS. This is a typical digital mode for tracing and tracking in mobile amateur radio, and it is not the newest one, but an exciting one. While others talk about their location and explain it to their counterparts in words, APRS users exchange site data automatically over an appropriate system. APRS uses a web-based server that collects geo-coordinates from all APRS-connected mobile stations. Location data is sent over the air from mobile amateur radio stations to special nodes, which are connected to the APRS server. From here, a map is filled up with small symbols for each station, and this map can be viewed over the World Wide Web.

Professional and Commercial Users

Professional radio applications still use conventional and robust systems, especially if they are used in safety areas. NAVTEX on 518 kHz (medium wave in the mode RTTY 100 Baud) still supports safety matters in international sea areas for shipping. In other areas, such as police and fire, the old analogue systems have been substituted by digital systems, which are working similar to public mobile networks but with different network levels and higher safety and security requirements.

Eavesdropping by unauthorised listeners is not possible any more. Technologies such as APCO P25 (USA) and TETRA (Europe) or similar systems like TETRAPOL are needed because they can also integrate and support different services such as ambulance, fire, and police operations within the same system. They meet today's security requirements much better than analogue systems, and they use switching subsystems similar to those of public networks but with higher safety and security standards. But we should not forget that all these beneficial features can only work when the local sites of the system cannot be destroyed by floods or thunderstorms. Construction has to follow sharp and strict requirements to avoid disruptions at the network level. How vulnerable new technologies as TETRA can be, we observed during the Eifel flood in 2021 in Ger-

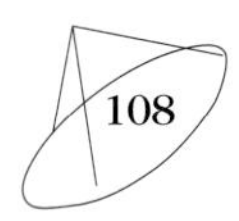

many, when many buildings in different cities were torn away by the water of the little river Ahr. We could learn that there is still a lot of design work to do, simply to increase the robustness and reliability of digital technologies. They are obviously more vulnerable than the good old police radios, which did not need any network access or network links with complex configurations.

They could simply be switched on and taken into operation. To avoid misunderstandings: TETRA radio technology, is surely as robust as analogue technology and it has more and better features for coordinated operation, but there are new aspects to take into account as redundancy of systems, availability of 7/24, power supply of base stations even in emergency mode, and network access in extremely challenging situations such as thunderstorms and floods. The transmission sites need to have a robust design and implementation, too. TETRA itself has proven to be robust for many years in many countries and in many scenarios. After this big disaster in Germany 2021 some additional topics should find their way onto the agendas of safety authorities and network operators.

Figure 28: O'Hare Airport in Chicago: Pilots, air traffic controllers, and many others use wireless devices: bus drivers, baggage vehicles, tankers, security staff, cell phone users, runway radar, primary radar, secondary radar.

Thinking about professional radio and services, we enter an interesting world of different and exciting systems and users who cover special areas in the frequency spectrum. To keep order in that huge frequency domain with all their services and to ensure that standards are valid for all users, regulation becomes a governmental task. They fix allocations and frequency use, and it is really exciting to see who is legitimatelly allowed to work in certain frequency bands as a primary user. State responsibility is most important for public broadcasting and frequency coordination with neighbouring countries. Strict regulatory guidelines are also important for domestic services, such as police, and for military use. Other areas are the regulation of public mobile phone networks and air traffic control. 2G/3G/4G/5G are mostly coexisting services that need a huge frequency range, as well as air traffic services, with a huge demand for frequencies for radar, radio, satellite services, and others.

Example: Aviation

Let us look a little bit closer at the work of some professional users. They all use systems based on electromagnetic wave propagation. When the Lufthansa Jet in Figure 28 has reached a flight level at 12,000 m height above the Atlantic Ocean at 2 a.m. in the night, all systems have to run without errors, even if the outside temperature is -60°. The aircraft has a lot of sensors for several parameters their connected systems have to work without disruption or restrictions. Sensors that use electromagnetic waves are numerous, and their proper and trouble-free operation is urgently required for a safe flight. Air traffic controllers and pilots have to work very concentratedly and they must be extremely familiar with all functions in their systems, especially in the event of any unexpected phenomena. To be honest, when we look at their job, it often seems like an interesting job, that can be done playfully and with ease. But it is the safe and professional handling of all systems, radar, radio, trained voice and data communication with ground stations, communication with passengers, everything has an impact on a good and safe flight and

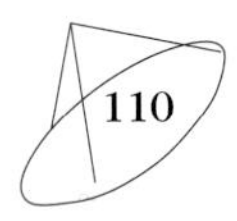

a good feeling as a passenger. And every time, all passengers‘ lives depend on it.

There are still some more professional groups we must rely on when we use an aeroplane. Who makes sure that radar systems and data exchange per CPDLC (controller pilot data link communication as an example) will really work as they should? There is always a big staff in the background who ensures the correct operation and functionality of all systems. Engineers, data experts, construction teams, all those people make systems run without disruptions and errors, and they care for a hardware that will also run in critical situations and that aircrafts are in a good condition. Any malfunction in aviation can mean danger for hundreds of lives.

Example: Authorities with Security Tasks

Some users are dependent on trouble-free radio systems day and night, especially authorities and organisations with security tasks. Disturbances and outages can threaten human lives in the case of warnings that don‘t reach the addressees and in the case of emergency calls. It is important for 100 % that such warnings and calls come to the receiver immediately, without delay. Police officers, paramedics, and firefighters must always be sure that their emergency communication systems work without fail 24/7.

To make this sure, there is a technical staff with engineers and computer scientists working in the background. They care for trouble-free operations day and night. In civil protection, it is extremely important that all disciplines can collaborate with a central coordination on the same radio systems to ensure misunderstandings and delays, which may cost lives in the worst case during a flood or a thunderstorm. TETRA as a radio system meets all the requirements for a clearly coordinated operation of all forces. Beyond mature technology, operators must ensure that all subsystems are configured correctly and that the whole staff has got the right training for their role in the communication system (operational staff, coordinator, administrator, ...). TETRA even has the feature of using patrol cars as a relay station to enhance the range of radio communication. On many sites, TETRA uses linear antennas for radio coverage, but in special areas, it is possible to use a site with three sectors that are equipped with panel antennas. In the 70 cm band panel antennas can be much bigger than antennas on public mobile networks at higher frequencies.

The new digital systems are really more complex, and users need additional training for proper operation. That's a big challenge. While almost everybody was able to build up radio communication sites for the recent analogue police communication systems, digital communication sites need much more care. Every site with digital base stations needs for example, emergency batteries and a method to avoid radio coverage outages for a complete district. Engineers and experts in different areas of expertise are on duty, to maintain the systems and provide full functionality for all radio systems, especially in the case of catastrophes.

Figure 29: Telecommunication tower with antennas for radio broadcast and mobile radio. Microwave antennas are mounted on the lower platform. Close proximity of components from different technologies causes no issues due to strict standards and installation guidelines.

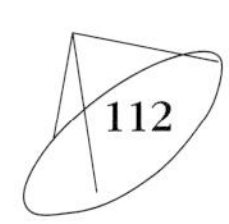

Example: Public Transportation

Beyond the authorities and governmental organizations, many other companies in the public transportation sector are affected by regulations and guidelines. Taxi companies and private bus services also need to follow the rules for the use of frequencies, maximum output power and more. In the transport and logistics sector, wireless devices play a big role in meeting all the expectations of their customers. Some of them use central coordination over their professional radio systems. Dispatchers can talk to every included car, bus or taxi in the whole fleet.

Many vehicles in a company's fleet are still equipped with an analogue radio set. Taxi has to provide 24/7 availability to offer their service to passengers at railway stations and airports. Older people still know the sound of the 5 tone selective call, which addresses exactly one taxi for an order. Patients have to be transported to medical institutions and doctors. Tramways are coordinated throughout the city to avoid delays. This was all done over the good old analogue radio system. Some of these reliable and mature systems are still in use. One reason might be that they are easy to operate. And they are really robust and almost safe against failure and outages.

Today we observe that many users in the field of professional radio change their equipment and use more and more of the infrastructure of public mobile networks with all their features. Radio coverage in 3G and 4G networks is so good that weak signals and disrupted links are very rare. The coverage is even better when we consider that it is still working in neighbouring cities. In the past, professional radio users often lost contact when they left their home area. Software solutions for vehicle fleet applications have become popular, and now orders for transportation of passengers can be transmitted by data telegrams to each taxi separately. This is efficient and works without delay, but on the other hand it is the missing voice contact with other taxi drivers that makes every user more isolated today.

Small talk and the exchange of private matters are missing, and the dispatcher‘s voice in the speaker disappeared too. That‘s the price we have to pay for modern communication services. In this example, it is the price that taxi drivers have to pay. In other areas, the price is much higher. We come back to the consequences of isolation at the end of this book.

The process of substituting an analogue radio system with a new digital radio requires sophisticated planning and good expertise in radio communication systems. Substituting a system with only quick and dirty process steps and unqualified staff will result in a rude awakening. Responsible managers and customers will learn about outages, time-consuming system changes, and plenty of lost time and money. The reason can be in the software system, which probably does not know all reasonable options to make the right decision in a certain unexpected situation.

But in the end, it is not the machine or the automatic system that fails, but the designer, who did not think about everything that was really relevant for an error-free operation. However, all this cannot be completely avoided when a new technology is invented, implemented and tested. Human designers will always learn from the past, and they will implement better releases within optimisation cycles. Machines and automatic systems are rather stupid without input from their human designers and programmers. Nobody can expect that machines will take reasonable decisions in every situation, except when their programmers have looked far into the future. Then machines can do just what they shall do.

Developers and Engineers

It is not an exception when engineers and technicians have been active in the field of amateur radio, CB radio or electronic circuits in their early years. The list of exciting jobs is long, and when we write down all the jobs that are somehow affiliated with electromagnetic waves and propagation. Some of them we have already presented, but there are still some more. We can ask ourselves: Who had the right idea for a product that was later designed, produced, and put on the market?

Construction experts have to define the mechanical basics and the right materials to make sure that a machine really works. There are many scientific disciplines that come together when a product is designed. Physics, chemistry, material science, vibration theory, and within the development process and the quality assurance phase, everything makes sense, according to what we have learned. At the beginning, young students have to learn about the importance of all the different disciplines, which will help them later develop products. Today, there are many interesting jobs, such as those of mechatronics engineers, who combine knowledge of mechanics and computer science. During the first years of education at school, some important fields seem to be missing in many schools. Students are generally curious, but when they don't get any insight into physics, electronics, and computers, there will be a lot of potential left unused. Where or who are the triggers for starting a career in technology? That's hard to understand in a society where everybody uses technology for many hours per day.

To understand how solutions can be created, students need a wide set of scientific basics. When computers are supported, their operators need to formulate mathematical expressions and equations to enable a machine to solve the equations. Numerical methods are a good approach to working with linear systems of equations. With the appropriate method, those equations can be solved quickly by using parameters, which can be varied. Then the solution can be calculated within a few minutes by a machine, even with parameter changes made some hundred times. This process seems to be complex and time-consuming. But a scientific approach like this is an efficient method to use algorithms for computer simulations instead of building expensive prototypes, which have to be constructed, tested and optimized. All this can be done by computer simulation using appropriate methods. That will save time and money and provide much more flexibility.

Beyond engineers and scientific experts, there are several professions and jobs that are needed to support the whole product life cycle. Many systems based on electromagnetic waves can only be created and brought to market, if qualified employees from many professions will collaborate. Complex technologies need, for example sales people with good communication skills to talk to the customer, because complex products often need additional explanations and even training. Trainers who provide workshops and installation staff and service engineers with a huge knowledge about complex products have to play their role just as the development engineer does in his profession.

There are important departments whose experts are sometimes forgotten in the product development process. Service people are indeed the real experts for products that have been in use for a long time. This practical business field needs motivated experts with good skills in talking to customers installing machines, and to configure them. It is a pity that they are sometimes forgotten when knowledge, for optimization is needed. It is a real waste of internal resources, when service people are not integrated into the product development process. Service people have a lot of application-based knowledge, and they can operate their machines really well, sometimes better than engineers from the development departments. Some decades ago, when TV sets used tubes and transistors were still in their developing phase, it was always very exciting when the TV-technician came, opened the back cover of the TV-set and we could see what was inside a TV-set. Most exciting were the glowing tubes. Where did all the voices and pictures come from? To be honest, the TV-technician as a job has almost disappeared completely because today only a few modules and boards are exchanged. Mostly, a new TV-set is bought. Nevertheless, the importance of technical staff in the service area remains high. Today, managers have recognised that practical knowledge has to be integrated into all processes that optimise a product. This is the real benefit for each company in the competition with others. The early use of the feedback knowledge of all available experts, even by the customer, is the key to better products.

SWL

A lot of different user groups are familiar with radio communication devices. But there is one special group that should not be underestimated. This group is called SWL (Short Wave Listeners). They are almost radio enthusiasts without any licence for transmission, but their enthusiasm is the same as that of radio amateurs. They are all passionate radio users, even in different disciplines.

For many, it is fascinating to listen to news from Radio Australia, or Radio New Zealand, or other overseas stations without using the internet, only by using their own antenna and receiver. Radio signals can propagate thousands of kilometres to the other end of the world, where they can easily be received and demodulated. During the Cold War and political rumours anywhere in the world it was always a very important source of information to receive unfiltered transmissions in a direct way, even if some of them were propaganda. People in any country in the world have used radio for a long time as an information source.

The foreign services of China, USA, Russia, England, the former Eastern Bloc States, and the Balkan States, and even some African stations used shortwave broadcast transmissions in different languages to make their view clear to everybody in the world. Still today, many of them do so. In the Cold War of the last century, almost the Eastern Bloc and the Western Alliance used many different languages in their foreign services to convince people all over the world from their point of view with respect to different political and economic contexts.

When we listen to the stations on shortwave today, we observe an increasing activity of Radio China, the governmental voice of the communist party. China uses the languages of the target regions in the world for their transmissions, which can even be English and German. And the use of high power transmitters with 250 kW or even 500 kW rf power.

Meanwhile, some western countries have stopped their shortwave transmissions and dismantled infrastructure on former transmitter sites. But not all. Despite internet broadcasts of news and music programmes and satellite transmissions, shortwave remains exciting. It is a medium where each listener can decide for himself what he wants to tune into and which source is the right one to form their own opinion. With only a little technical ability, everybody can install their own little station, and it makes them even a little bit proud to find their own way to listen to the world. You only need a simple shortwave pocket radio to start. Later, with a good portion of enthusiasm, the receiver can be a portable radio or even a communication receiver with some metres of wire when the telescopic antenna cannot provide the power for an upper-class receiver.

But it's not only news and political comments that make shortwave listening exciting. In the evening, some bands open for additional radio reception from the Caribbean and from Africa. Domestic music and rhythms from these regions are amazing. And they sometimes provide talks with people from the region. During the sunspot maxima of the last 11 year cycles, you could listen to stations from Kenya, Ghana, Botswana, South Africa, Cuba and some more countries. The signals have been strong enough even for the use of low-cost equipment. The best time was in the phase of increasing numbers of sunspots, just as in 2022 in the current sunspot cycle. When dusk begins, radio signals from far away will be available, and a lot of stations can be heard, especially in the range of 4700 kHz to 5100 kHz. Just now, in 2022, during the ascending branch when the sunspot number increases weekly, it is the best time for radio listeners to identify some new stations. It is folklore and music from distant and unknown countries like the hot rhythms of the Central African countries, that makes this fascination so great. In the early hours and during the day, on certain frequencies, we can hear some foreign singing and calls from Arabic countries. They come from muezzins and Arabic prayer leaders. It seems that their religious rituals will never end when they send out their prayers to all Muslims in the world for hours. Between 13 MHz and 22 MHz Saudi-Arabia is broadcasting for followers of Islam on different frequencies. Foreign transmissions come from all over the world, such as Thailand, Indonesia, India and many other countries. Most of them can be identified by their jingles coming to the news every hour. For many people, radio will always mean a piece of liberty. In many countries minorities are oppressed and persecuted. Engaged radio listeners can experience a certain closeness to the real world as it is in foreign countries. In autocratic states with dictators, where the internet is controlled by the government, this independent information flow is not, of course.

In a time of sunspot maxima, attempts to receive overseas stations are worthwhile. Listening to clandestine stations is always an extraordinary event.They are modern pirates who operate their small stations and studios by their own means and without financial support from others.They broadcast colourful and interesting programs. Some of them transmit above the 49 m band for international radio stations at 6200…6300 kHz. Successful reception is probable on weekends.The mother of all pirates, Radio Caroline, made this type of radio transmission popular, and still today, people love it. In the 60s a small group of broadcasting enthusiasts started with their ship Ross Revenge in the North Sea [6] and transmitted their extraordinary programmes with progressive music, while the BBC and others were still broadcasting their conservative programmes.Transmitting radio programmes this way was not conformal to regulations and authorities‘ opinions in every point, but for listeners, it was a phantastic new kind of radio listening. After some time, the pirates were hunted because they used illegal frequencies and because they ignored regulations and tried to escape governmental control. But in the end, they survived, and the brand Radio Caroline still exists and can be heard on shortwave, operated by people with the same old spirit. There are many exciting stories about clandestine radio [6] in different countries. They operated in a semi-legal area, and they were quite successful in providing progressive and liberal stuff in the rather grey landscape of radio programmes. We could learn that a little bit more liberty and good music would be an improvement for the regulated and governmental-controlled world.

Radio as a Hobby

Listening to radio amateurs all over the world. Tuning into radio stations from the other side of the globe. Isn't it exciting? Today, we can use digital radio technology and cheap equipment in our own shacks, and we can listen to OMs in many different foreign countries around the world. D-STAR, P25 and further operation modes provide the right instruments for us to listen to day and night, wherever we are.

But what type of people are they? Who is it in the radio clubs like ARRL, DARC and ADDX, in the radio shacks, and what is driving them to spend a lot of time listening to radio stations? They come from different ages and from all social classes. Some of them are workers and employees, others are bosses and company owners. What they have in common is a great fascination within the fields of radio technology and wave propagation, circuits and transmission devices. Some of them are already experts in rf-technology, antennas and radar. Unfortunately, their number has decreased in the last few years, and it makes sense to have a closer look at how people can find access to that amazing hobby. Some of the disciplines can be started at once without a licence and with only a cheap radio for the Freenet range in the 2 m band. Offering exciting projects to young people, this could really be a start into a career that refers to electronics and RF technology, simply by soldering a circuit or a radio.

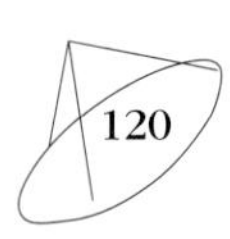

Those who find their way into their job on another path, i.e., as a craftsman, butcher or a teacher, can also get their licence as a radio amateur when they are a little ambitious and share that ham spirit. The learning effort for the different licence degrees is not so high, and during the last years difficulty levels have been reduced to make the licence more attractive and popular. This angers some older experts who have had their licence for a long time, when requirements were probably higher. But why?

The radio hobby has several disciplines, and it is often underestimated as a useful leisure activity. This hobby can even be much more. Last century in the 70s when people had no internet, it was CB radio communication, which connected people in a way and with an intensity that was never seen before. CB radio was new, and suddenly a huge pool of talents and enthusiasts entered the stage, which was completely underestimated at the beginning. Still today, some of the older radio amateurs do not understand that a high potential is waiting there for more action, a rather large group of young people who only need the right instructions for dealing with radio communication and electronics. Sometimes the youngsters from the CB are ignored or even defamed, although many of the licenced amateurs themselves once were "only" CB amateurs.

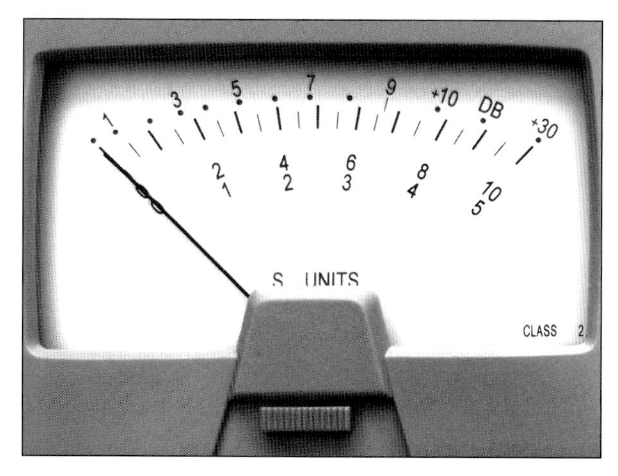

Figure 30: Analogue S-metre with surprising precision in displaying relative signal strengths.

Today, it is hard to inspire young technicians to persue the radio hobby. Too many competing hobbies like internet applications, take the talents away who once were interested in electronics. The number is seriously decreasing.

In the early years of CB radio, it was like a wild west spirit. The first radio operators were allowed to use only 12 channels of AM and 500 mW rf output power. Looking many hours at the S-metre, almost a rather inaccurate tool in the radio devices, many contacts could be confirmed and sometimes even confirmed, by a QSL card. Those who preferred more accurate radio instruments used an external S-metre with their typical 9-level scale. See Table 8 for more.

External S-metres (Figure 30) provided a higher relative accuracy, whereas the smaller internal S-metres have only a 4 LED display for the relative field strength.

Once the German authorities, "deutsche Bundespost" released severe restrictions for CB radio operation, this prevented efficient operation. Compared with today's guidelines, it was forbidden to use amplifier microphones, it was forbidden to use an appropriate rf power, the number of channels was extremely low, no SSB, no FM, only AM, it was a disaster.

Therefore, some radio operators found a set of improvements for their own stations. Some amazing tools and devices, such as the legendary "Turner+3 microphone" allowed strong modulation and reasonable operation of the radios. Especially in addition to the later "President Jackson" this was the ultimate combination when the operator had a good antenna like a ground plane or a tall antenna like the IMAX200.

And there were more legendary devices from other manufacturers, such as Grundig, Zodiac, DNT, Stabo, Albrecht, Sommerkamp and so on. Each one who once lived with this CB spirit in the last century can still feel the warm shiver running down their spine when those devices are reactivated.

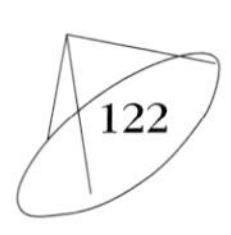

But if it really was a kind of wild west spirit for radio enthusiasts, who the hell was the Sheriff? Woe to those who were caught! The "German Federal Post Administration" had enhanced their staff for monitoring tasks. They had to keep things in order by hunting illegal radio operators.

Figure 31: "President Jackson" was a legendary CB radio for the 11m-band with a strong modulation, also for SSB.

Many were caught during police controls of vehicles or even by inspections at home without announcement. Operators who transmitted for a longer time could easily be found with direction-finding antennas. Some of them were even caught in the act. And it could easily be proved in some cases. When the power amplifier, which was almost 20/50/100 Watt was found and it was still warm, a certain punishment could be expected. At least, however, the device was taken away.

Hundreds of radio sets of all types filled the evidence rooms of the authorities, which is today called "BNetzA" or Bundesnetzagentur. Meanwhile, there are only a few illegal operators. The reason is simple. Today, most of the interesting and required features are legalised. We have a high number of channels, even for SSB and FM and even the allowed output power has been increased to several watts. And those who want more than just talking in the 11m band can even use the 2 m band on the Freenet channels on 149 MHz. There is also an option to operate in the 70 cm band in the PMR range (public/private/professional mobile radio).

In the last few years, vehicles of the authorities with measurement equipment and direction-finding antennas have gotten another focus and other primary tasks.

They will come in the case of interference from receivers or consumer electronics, for example, and when the source of the interference cannot be identified by the user. The technicians coming from the "Bundesnetzagentur" have high-quality equipment like

Figure 32: The professional and official measurement service of the BNetzA on a mission to scan the spectrum for interfering signals with a spectrum analyser.

spectrum analysers and they can find nearly all causes of interference, such as a defective heating control, a poor shielding or malfunction of a SAT antenna splitter or amplifier, or even an illegal radio operator in the neighbourhood.

Do-It-Yourself Projects

Radio operation with or without a licence is a very valuable thing, especially when you discover your inspiration for soldering circuits or even complete electronic devices. Young people can find their profession here, and in some cases, only one little spark is needed to ignite a huge enthusiasm for a new hobby or to discover a new path into an exciting job.

But even if people are only curious about what is behind these electronic circuits, for example, it is very useful, and they can learn a lot by building their own circuits and getting them into operation. A good place for that is in the local club of radio amateurs, but it will also be useful to do it at home. What really matters is the inspiration and motivation to get something done and to be successful. In the local radio amateur clubs, there are always people who can help.

Do-it-yourself projects can be various electronic circuits, such as a microphone amplifier or a complete radio. Any type of circuit for an application is possible.

Figure 33: RF stage of a 70 cm transceiver with a quartz filter and two serial resonant circuits for 430 MHz. The inductances were made of a thick bent wire, and the capacitors were trim capacitors (self-made project from the 1980s).

What is almost missing is the ultimate trigger. But when inspiration comes, you can find all types of circuits, digital, analogue, with LED, with amplifiers, it is a large field, and you will be rewarded by your own success in every new attempt.

Antennas are an extraordinary and exciting field of knowledge. Starting experiments with antennas will always be exciting and surprising. Try to build a J-antenna as a promising project. This antenna is also called Slim Jim and can be made of a parallel two-wire cable. For the 2 m band, it is 1 m long with an additional matching section of approximately 0.5 m. This antenna provides good results. It is effective due to the good matching options.

Electronic circuits can help discover one's hidden talent. This is valid, especially for the younger generations. Some companies offer kits with different circuits to assemble. Building and soldering such a kit is a real adventure with a high probability of success.

They can be flashing circuits or even complex projects such as transceivers, a complete radio with a receiver and a transmitter on a single PCB (printed circuit board), which can also be self-made. PCBs can be realised by a photo process or by a process with an etching fluid ($FeCl_3$). The raw PCB with copper on one or two sides can be painted with a waterproof marker. These paintings will be the conducting lines after etching. With some $FeCl_3$ in a little bath with cold water, the visible copper will disappear during the etching process. After that, the conducting lines have to be cleaned from the marker, and the PCB is ready. The next step is drilling holes and the electronic components such as capacitors, transistors, inductances, and relays can be soldered. When the components are connected and everything is done, the test phase can start. The circuit can be put into operation.

Figures 33 and 34 show some discrete components in a circuit. This type of implementation of components on a PCB was used from 1950 to 1990. Later circuits became smaller, and the density of com-

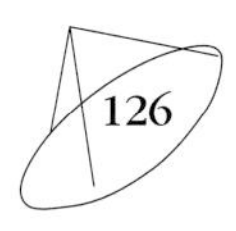

ponents increased more and more. Integrated circuits replace groups of components and let the whole circuit shrink to a minimum.

Most of us still know that mobile phones in the last century were large and heavy, and they could only be used for phone calls. Continuous progress in technology and manufacturing has made the dimensions shrink. The packing density in integrated circuits and in microprocessors increased and single resistances, capacitors, and inductances became as small as breadcrumbs. This step was possible

Figure 34: Modulator stage with a varactor- or capacity diode BB105 and a transistor with filters. Here, the audio of the amplified microphone voltage modulates the signal. The voltage of the modulated signal drives the first transmitter stage.

by SMD technology. SMD components (surface mounted devices) are smaller than the varactor or capacitance diode in Figure 34.

Today, we often observe that there is little interest in technology and electronics. We are all consumers, but only a few are ready to go deeper into the stuff, invest their energy, and fight for an adequate certificate and a job. Some say that girls would not choose this path at all. They are wrong; this is not true.

But where does the lack of experts and engineers come from? For many years, economists have been complaining that we don't have enough engineers and computer scientists. I have personally known this everlasting topic for 30 years. Why do industrial and governmental authorities not start to counteract? All those who are complaining again and again should simply look at where and how new experts can be motivated, recruited, and educated. As long as nobody cares for opportunities and for qualifying and coaching young people to an exciting career, the problem will not disappear. We need some experienced experts who are ready to share their knowledge with young people within a given frame.

This frame can be at school or in private initiatives, also triggered by radio amateurs. Without motivation and coaching and without the proper political frame, all the complaints will go on another year. We need to understand that this is a pro active process and that anybody must get active here.

While technology and science continue to be a small, unloved field, high-tech products will be developed in overseas countries, and we only look at this from afar and are amazed. Sure, we also develop high-tech products, but let us be honest. Which mobile phones come from Germany? Apple? Galaxy? How many antenna manufacturers came from Germany, while millions of GSM antennas were installed in the 1990s on our roofs and poles? I think one: Kathrein!

Example from Research and Development

Readers who give up in this chapter due to a lot of mathematical terms may skip and continue with the next chapter. It is a better option than being bothered by mathematics. This chapter is for all those who try to have a deeper insight into the work of engineers and computer scientists by one example. How can mathematical and numerical methods be applied for calculating the behaviour of high-tech products that are used in space technology, like in this example?

The first question is: How do scientific engineers work at all? What do they develop? And why should this be an extraordinary field at all? In this chapter, we get an impression of engineering by using computers. Exceptionally, the amount of mathematics is larger, but the basic considerations can be followed easily. When we look at approaches we can use, we draw from the full. Mathematical and scientific foundations have existed for a long time. It is an advantage to use what is already available as an instrument, like Ampère's law.

We know Ampère as an inventor. Today, his name is used as a unit and occurs on batteries and fuses in the domestic power supply. His name is very popular all over the world because Ampère is used in the international unit system "SI" for the amount of current.

Our practical example has a close relation to Maxwell's equations. Maxwell found out that Ampère's law had to be enhanced, and so he formulated the following equation:

$$\text{rot } \vec{H} = \vec{J} + j\omega\vec{D}$$

The first part means that an electric current through a conductor is the cause for a magnetic field, which has magnetic field lines like circles around the conductor. Maxwell recognised that any charges on a capacitor would not be considered in Ampère's law. Conse-

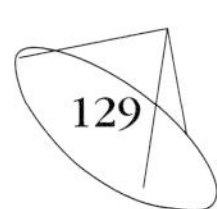

quently he added the term on the right side. Now the electric field or, in other words, the electric flux density D can also be considered, and one of the famous Maxwell equations is written. Maxwell defined these relations in 1861 to 1864.

But now we look at a special engineering task, including some details, and we also consider the environment for this work. In the next subchapters, we will see a planar reflector structure, i.e., for satellite antennas. The planar structure has frequency selective properties and is called FSS = frequency selective surface.

Such a surface can be transparent for certain frequencies, while in other frequency ranges, the FSS will reflect electromagnetic waves. This technology can be applied to reflector antenna systems with multi-focal feed systems, i.e., on communication satellites. Figure 35 shows three geometric options for use in an antenna system. One parabolic main reflector with different feed systems in combination with subreflectors with an FSS is also an economical approach. The different feed systems can be placed in the focal points of the system

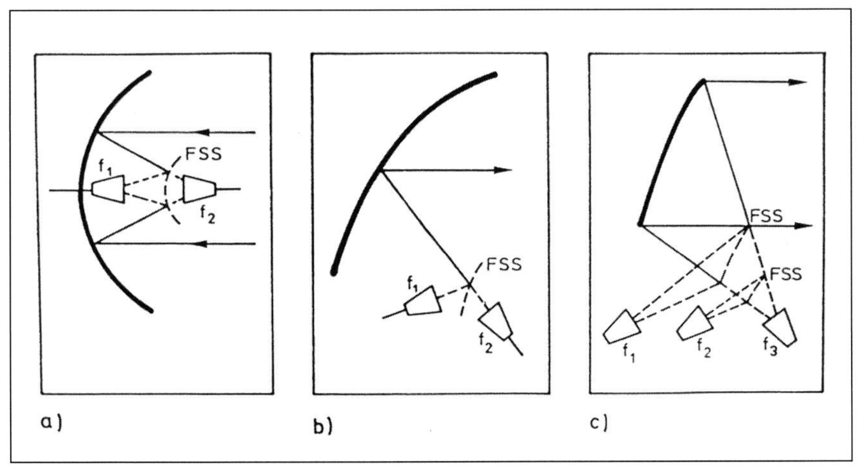

Figure 35: Multifocal antenna systems use frequency-selective surfaces. Two or three feedhorns transmit the frequencies f1...f3. The frequency selective screen (FSS) is transparent for some signals (transmission, f2 in a and b and f3 in c) or they are reflected (f1 in a, b, c und f2 in c).

and the FSS can work as described above, transparent for one frequency band and reflecting another band.

It is right that this technology is also applied when certain frequencies pass the FSS. In this example, the metallic structures are simple square loops and double square loops. But the model can be enhanced for any other structure to be analysed.

The substrate also plays an important role in the analysis, and the incident angle will also have an impact. The carrier material (substrate) and its thickness will influence the electric behaviour of the structure as a whole.

It can be responsible for the loss and absorption of the wave's energy. Absorption can be achieved by the use of a material that has a complex permittivity with a rather high ε_r'' (imaginary part of the complex permittivity). This is the part that causes the loss.

Maxwell's Equations

In most electromagnetic scenarios, Maxwell's equations are the basis for the mathematical description, even if they are not evident at first glance. For the described method of calculation, Maxwell's equations are mandatory because we have relations between electric current, magnetic fields, and electric charges.

For the calculation, however, we need appropriate algorithms that consider the basic equations for the dynamic processes between charges, current, and magnetic field and the geometry of the analysed structures.

If this is formulated properly, the computer can start the job. One of Maxwell's equations says that the magnetic field is free of sources and that there are no isolated magnetic monopoles. Mathematically expressed, we say that the divergence of B is zero (divergence is a mathematical operator, which is applied to a vector field in this case).The operator is used in the following equation:

$$Div\ B = 0$$

Maxwell's equations allow, in this case, to derive a set of equations for a scenario with electromagnetic fields and metallic structures. So the currents on the metallic structures can be calculated. With the help of the exact currents on the structures, we will receive the behaviour in the frequency domain. Between the current in the structure and the magnetic and electric fields, there is a clear relation:

$$\text{rot}\ \vec{H} = \vec{J} + j\omega\vec{D}$$

The rotation of the magnetic field depends directly on the current density. Metallic structures mean, for example, several scattering elements as passive dipoles or square loops that will interact with an electromagnetic wave.

Structure of a Frequency-Selective Surface

In our practical example, we use metallic square loops as a periodic structure. The method used here supposes a structure with infinite expansion in two directions.

For real behaviour, this approach does not include significant deviations from analytical results, and therefore it can be used without restrictions.

The geometry of the square loops is shown in Figure 37. Their centres have a spacing of DX, the width is W and the length is L. All parameters are used as an input for the calculation cycle of every single frequency point in the frequency domain. The complete geometric description of double square loops is shown in Figure 38.

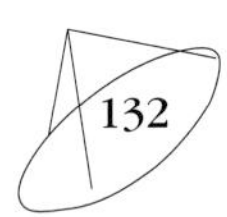

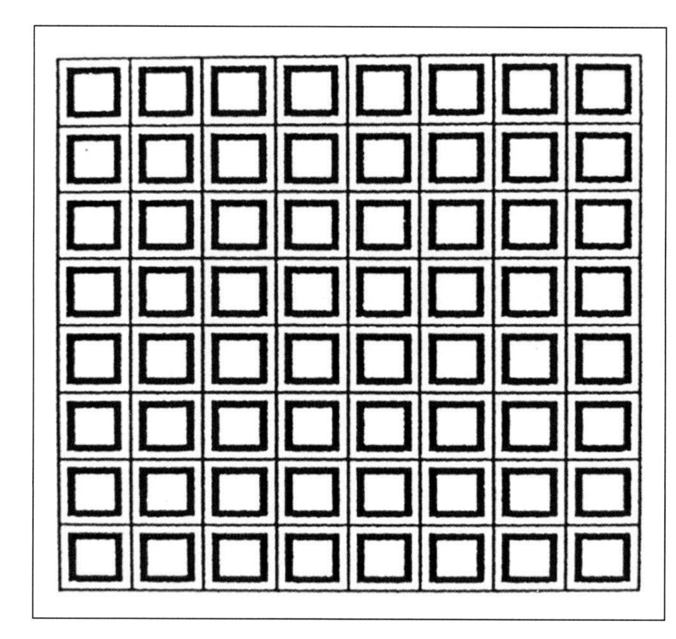

Figure 36: Frequency-selective surface with square loops. Depending on the design, they can reflect or pass signals (transmission). Square loops are independent of polarisation. When dipoles are used, reflection and transmission depend on polarisation of incoming signals.

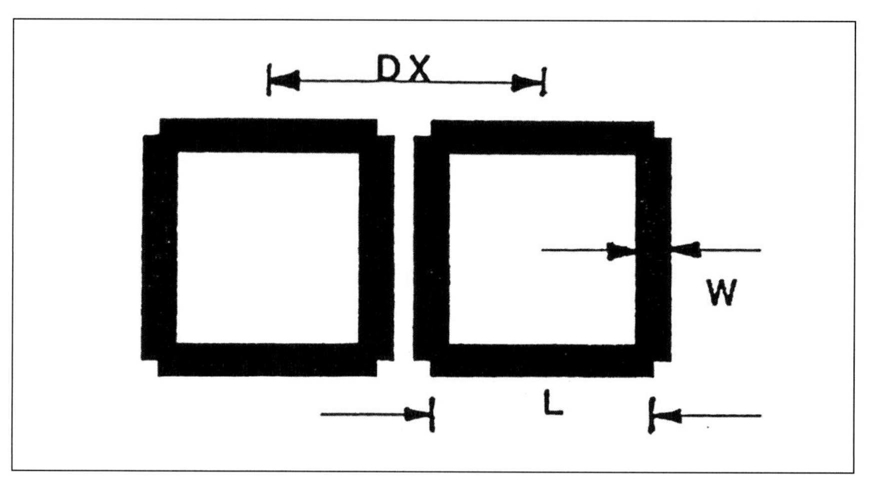

Figure 37: Geometric parameters of square loops as an input for the simulation.

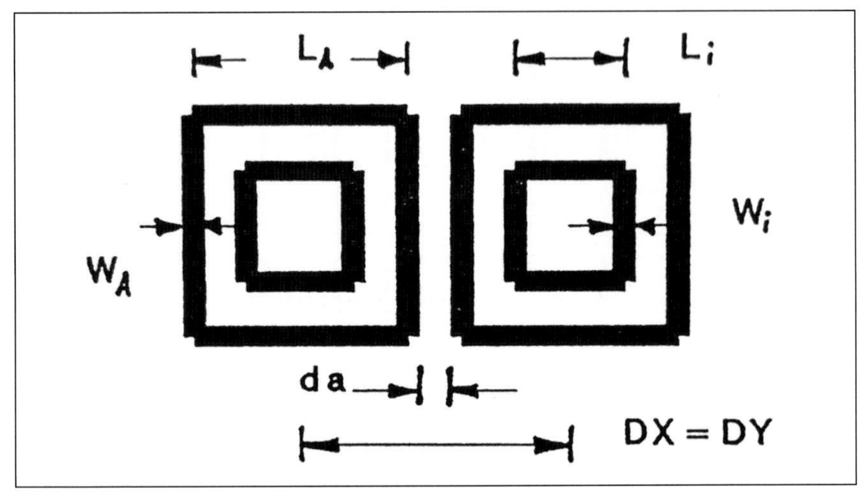

Figure 38: Geometric parameters of double square loops. These parameters are needed to express the structure of the model for further numerical calculation of currents and fields of frequency-selective surfaces.

Incident Transverse Wave

Now the scenario is as follows: A transverse electromagnetic wave (TEM wave) meets the surface of a frequency-selective screen with infinite expansion under a certain angle, i.e., perpendicular with an incident angle of zero.

It will induce a current in every metallic element, and it will reflect a certain percentage of the wave's energy. The other part will be transmitted through the screen. The behaviour in the frequency domain can now be analysed, and we will find out that several parameters will have an impact, such as the choice of the incident angle theta or the thickness of carrier material.

The incident wave will be reflected at certain frequencies. For others, it won't be reflected at all. This depends on the incident angle, material, and also the type of TE wave or TM wave. That means that only the electric field or only the magnetic field will be transverse to the plane of the FSS.

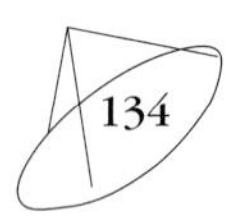

Reflected and Transmitted Wave

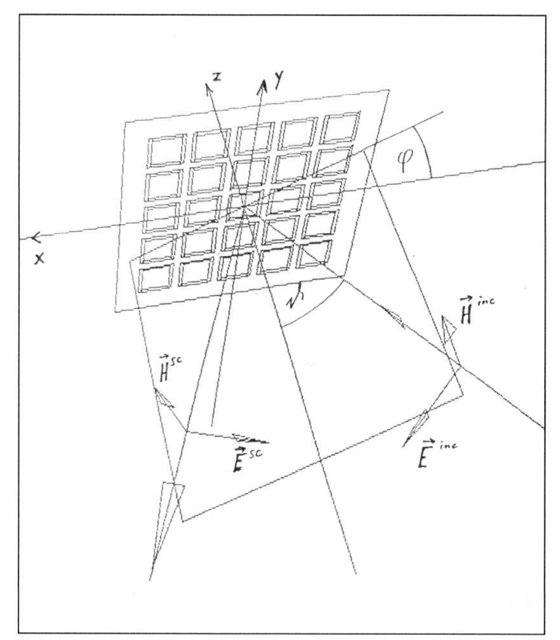

To make the scenario clear, we can imagine a TEM wave (transverse electromagnetic) that runs perpendicular to the surface with a frequency-selective structure. Later, we will briefly discuss the dependency of the incident angle theta (ϑ).

Figure 39: The incoming TEM wave meets the screen and is reflected. We can see the components of the E- and H-fields and the angles theta and phi.

Integral Equation Method in the Space Domain

We have seen that Maxwell's equations are almost the basis for processes in electrodynamics and, therefore, the basis for their simulations by computers. Numerical calculations in our example are done with the integral equation method, which is also based on Maxwell.

$$-\vec{n} \times \vec{E}_{sc}(\vec{r}) = -\frac{1}{4\pi j\omega\varepsilon}\, \vec{n} \times \int_F \left(k^2 \Phi \vec{J}_F + (\vec{J}_F \cdot \nabla)\nabla\Phi\right) df'$$

Integralgleichung der 1. Art (elektrisches Feld);

$$\vec{n} \times \vec{H}_{sc}(\vec{r}) = -\frac{1}{4\pi}\, \vec{n} \times \int_F (\vec{J}_F \times \nabla\Phi)\, df'$$

Integralgleichung der 2. Art (magnetisches Feld);

mit Φ: skalare Greensche Funktion des freien Raums mit $\Phi = \frac{e^{-jkR}}{R}$

R: Abstand zwischen Quellpunkt und Aufpunkt

$\vec{E}_{sc}, \vec{H}_{sc}$: rückgestreute Komponenten des elektromagnetischen Feldes

$\vec{J}_F$: Strombelegung $[\frac{A}{m}]$

∇: Nabla-Operator mit $\nabla = \vec{\nabla} = \frac{\partial}{\partial x}\vec{e}_x + \frac{\partial}{\partial y}\vec{e}_y + \frac{\partial}{\partial z}\vec{e}_z$.

On both sides of each integral equation, we find a cross product or a vector product. These equations provide a relation between the scattered electric and the scattered magnetic field, and the induced current on the metallic backscattering elements. Here $\omega=2\pi f$ is the cycle frequency and f is the frequency of the electromagnetic wave. Additionally, we use rot as a mathematical rotation operator, and G is Green's function. $\vec{n}$ is the normal vector (perpendicular to a plane). In the space domain, the backscattered field can be expressed as follows:

$$\vec{E}(\vec{r})^{sc} = -\vec{E}(\vec{r})^{in} = \frac{1}{j\omega\varepsilon} \text{ rot } \vec{H} = \frac{1}{j\omega\varepsilon\mu} \text{ rot rot } \vec{A}_m$$
$$= \frac{1}{j\omega\varepsilon} \iiint_V \vec{G}(\vec{r}, \vec{r}\,') \cdot \vec{J}_e \cdot dV' .$$

Representation in the Frequency Domain

The following steps are only briefly described because we do not want to give too many detailed scientific explanations. The next step is to derive an adequate representation in the frequency domain.

Fourier series expansion is the right instrument to give a visualisation of the current on the square loops. This is the appropriate instrument, because we can work with trigonometric functions that are orthogonal to each other, such as sine and cosine. With the help of those functions, we can formulate independent linear equations,

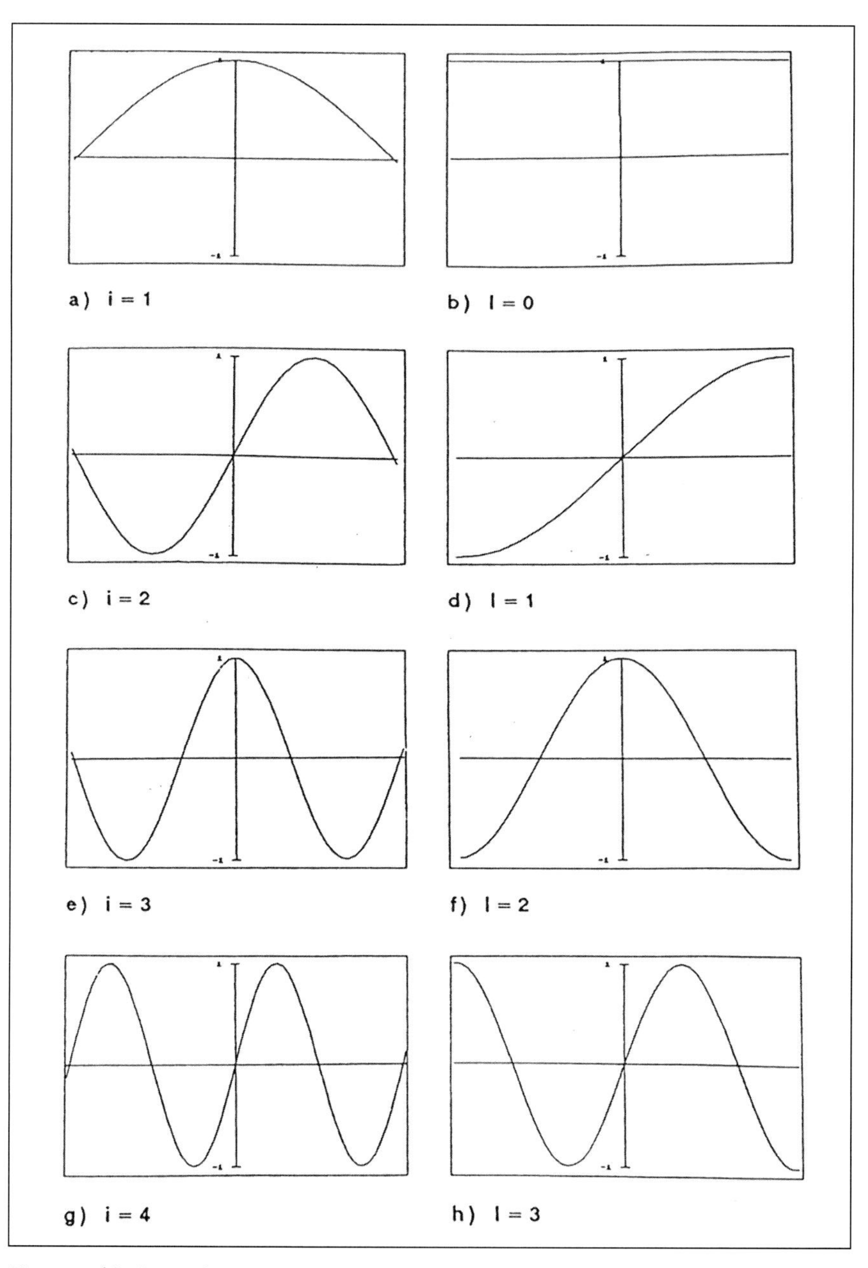

Figure 40: Base functions to express the current, according to a Fourier series expansion with unknown coefficients. Indices number the harmonics (odd and even).

which can be solved easily. Mathematically expressed, the current on the metallic elements is a superposition of sine and cosine functions and their harmonics. Each harmonic has its own coefficient, which is unknown before the calculation. But as a result, the coefficients and the harmonics are assembled into the complete current on the elements. Later, we will apply delta functions and use the inverse Fourier transform (also known as the method of moments). When the current has been solved, we will know the backscattered field, and we can derive the reflexion and transmission coefficients for the whole structure.

First we have a look at the current functions. We need them for the current distribution of each element.

One coefficient for each function is the first step. By doing so, each function and each harmonic are weighted by their own coefficients. The sum of all weighted sine and cosine functions is the total current distribution on one of the four metallic elements of one square. That's the approach and the model for the current in one square loop. And we assume that all square loops have the same current distribution because we analyse an infinitely large screen.

In the frequency domain, the equation for the scattered field looks like:

$$\vec{E}(\vec{r})^{sc} = -\vec{E}(\vec{r})^{in} = \frac{1}{j\omega\varepsilon} \cdot \mathcal{F}^{-1}\{\overleftrightarrow{\Gamma}(k_x, k_y, z) \cdot \vec{J}(k_x, k_y, 0)\}$$

This is the inverse Fourier transform coming from the space domain representation with the "rot rot operation" some lines above. E^{sc} and E^{in} mean the scattered and incident electric field. ε_r is the relative dielectric constant of the material, and ω is the cycle frequency. The product of Green's tensor and the current distribution (current density) J is the term to which the inverse Fourier transformation shall be applied. Green's tensor is derived from Green's function:

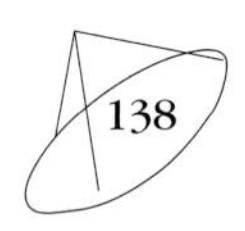

$$G(\vec{r},\vec{r}\,') = \frac{1}{4\pi} \cdot \frac{e^{-jk|\vec{r}-\vec{r}\,'|}}{|\vec{r}-\vec{r}\,'|}$$

Green's tensor in the frequency domain then will be:

$$\overleftrightarrow{\Gamma}(k_x, k_y, z) = \frac{e^{-jk_z z}}{j2k_z} \begin{pmatrix} k_0^2 - k_x^2 & -k_x k_y & -k_x k_z \\ -k_x k_y & k_0^2 - k_y^2 & -k_y k_z \\ -k_x k_z & -k_y k_z & k_x^2 + k_y^2 \end{pmatrix}$$

Tensor means that the included quantities are almost dependent on other additional quantities or parameters. Therefore, it is no longer a matrix, but then it is called a tensor. For example, when the permittivity is not constant but varies with frequency, then we have an additional dependency. As the wave number, we use $k=2\pi/\lambda$ which is a vector in this case.

The current distribution in its complete form, derived from the vector potential, can be written as follows:

$$\begin{aligned} \vec{J}(k_x, k_y) &= \iint_{-\infty}^{\infty} \vec{J}(x,y) \cdot e^{j(k_x x + k_y y)}\, dx dy \\ &= \frac{4\pi^2}{A} \sum_{p=-\infty}^{\infty} \sum_{q=-\infty}^{\infty} \vec{F}_{pq}^{int} \cdot \delta(k_x + k_{xp}) \cdot \delta(k_y + k_{yq}) \\ &= \sum_{p=-\infty}^{\infty} \sum_{q=-\infty}^{\infty} \vec{F}_{pq} \cdot \delta(k_x + k_{xp}) \cdot \delta(k_y + k_{yq}) \\ &\qquad \text{mit } k_{xp} = k_1 + k_p \quad \text{und} \quad k_{yq} = k_2 + k_q \quad . \end{aligned}$$

Now, a clear description of the current on the backscatter elements is still missing. Therefore, the trigonometric functions we are using will be applied by a Delta operator (Dirac function). By this multiplication, we receive a representation with the so-called Floquet modes F_{pq}. Each mode has two indices, and we can consider them as harmonics, that we use for the description of the current distribution. When we run the calculation with too many harmonics, it might exceed the effort for calculations with too many routines.Therefore, we use a termination criterion to stop the further increase of the index as soon as the result converges. Too many harmonics could mean too much time for calculation and increase rounding errors.

Impact of Substrate

In the simulation, substrate parameters are also considered. There will be a slight change in behaviour when using a carrier material because the coupling between the elements and the current distribution of the elements will change. Depending on material and thickness, even the losses can have an effect and reduce the amplitudes of the backscattered wave, and also of the transmitted wave or even shift the resonant frequencies. To consider this in the calculation, each Floquet mode gets their own backscatter coefficient by index p und q . The superscript "+" and "-" are related to the space and mean in front of or behind the frequency selective screen.

$$R_{\bar{m}pq}^{slab} = \frac{\eta_{\bar{m}pq}^{-} - [(1 - R_{\bar{m}pq}^{+}/(1 + R_{\bar{m}pq}^{+})]\eta_{\bar{m}pq}}{\eta_{\bar{m}pq}^{-} + [(1 - R_{\bar{m}pq}^{+}/(1 + R_{\bar{m}pq}^{+})]\eta_{\bar{m}pq}}$$

That's the way we describe all the backscatter coefficients and their dependencies.

Impact of Polarisation

The polarisation is considered by using a polarisation tensor. Both angles we need are shown in Figure 39. Tensor means an enhanced term like a matrix with additional dependencies. The tensor *K* for polarisation is:

$$\overleftrightarrow{K} = \begin{pmatrix} K_{1x} & K_{1y} \\ K_{2x} & K_{2y} \end{pmatrix} = \begin{pmatrix} \cos\varphi & \sin\varphi \\ -\sin\varphi & \cos\varphi \end{pmatrix}$$

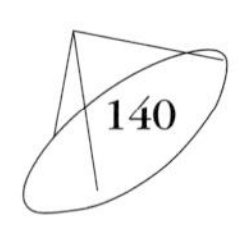

Each Floquet mode has their admittance called mode admittances:

$$\eta_{1pq}^{eq} = \frac{2k_0}{Z_0 k_{zpq}} \qquad \eta_{2pq}^{eq} = \frac{2k_{zpq}}{Z_0 k_0}$$

The impedance of free space is:

$$Z_0 = \sqrt{\frac{\mu_0}{\varepsilon_0}} = 120\pi\Omega$$

The impact of polarisation is:

$$\vec{\kappa}_{1pq} = \begin{pmatrix} \kappa_{1xpq} \\ \kappa_{1ypq} \end{pmatrix} = \frac{1}{\sqrt{k_{xp}^2 + k_{yq}^2}} \begin{pmatrix} k_{xp} \\ k_{yq} \end{pmatrix}$$

$$\vec{\kappa}_{2pq} = \begin{pmatrix} \kappa_{2xpq} \\ \kappa_{2ypq} \end{pmatrix} = \frac{1}{\sqrt{k_{xp}^2 + k_{yq}^2}} \begin{pmatrix} -k_{yq} \\ k_{xp} \end{pmatrix}$$

And the angle of incidence ϑ shall also be considered.

$$E_{tan}^{TM} = \cos\vartheta \cdot E_0^{TM}$$

We can see the angle in figure 39.

$$r_{co} = R_{100}^{slab} - \frac{1}{A}\frac{1}{\cos\vartheta} \sum_i c_i \left(\eta_{100}^{eq}\right)^{-1} \vec{F}_{ipq}^{int} \cdot \vec{\kappa}_{100}$$

$$r_{cr} = R_{200}^{slab} - \frac{1}{A} \sum_i c_i \left(\eta_{200}^{eq}\right)^{-1} \vec{F}_{ipq}^{int} \cdot \vec{\kappa}_{200}$$

These are the final co-polar and cross-polar backscatter coefficients, which already include the propagation direction of the wave and the material parameter, expressed separately for each Floquet mode.

This is a rather detailed description, but it is all needed to set up the linear system of equations with respect to all dependencies.

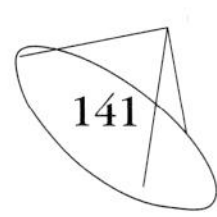

By resolving the system of equations, we will find out the coefficients for the current on one of the four elements and, consequently, the whole current on the passive elements. Knowing the current, it is rather easy to calculate the backscattered and transmitted fields.

Now the behaviour of the FSS is completely described, including all dependencies. Input quantities are almost the geometry of the screen, material parameter of the substrate, incident angle, polarisation, and a termination criterion for the maximum number of Floquet modes that are really useful (convergence).

Linear Equation System

To define the coefficients, we need an equational system. The calculation is done with standard mathematical operations, which are applied to a linear equation system. The use of a matrix representation describes how we find the solution for the vector with all coefficients c_i by inverse Fourier transformation.

$$\begin{pmatrix} [1+R^{slab}_{\bar{m}00}]\cdot E^{i}_{\bar{m}}\tilde{F}_{001} \\ [1+R^{slab}_{\bar{m}00}]\cdot E^{i}_{\bar{m}}\tilde{F}_{002} \\ \vdots \\ [1+R^{slab}_{\bar{m}00}]\cdot E^{i}_{\bar{m}}\tilde{F}_{00j} \end{pmatrix} = \frac{1}{A} \begin{pmatrix} \sum_{\bar{m}pq}\eta^{-1}_{\bar{m}pq}\tilde{F}^{*}_{pq1}\tilde{F}_{pq1} & \sum_{\bar{m}pq}\eta^{-1}_{\bar{m}pq}\tilde{F}^{*}_{pq2}\tilde{F}_{pq1} & \cdots \\ \sum_{\bar{m}pq}\eta^{-1}_{\bar{m}pq}\tilde{F}^{*}_{pq1}\tilde{F}_{pq2} & \sum_{\bar{m}pq}\eta^{-1}_{\bar{m}pq}\tilde{F}^{*}_{pq2}\tilde{F}_{pq2} & \cdots \\ \vdots & \ddots & \vdots \\ \sum_{\bar{m}pq}\eta^{-1}_{\bar{m}pq}\tilde{F}^{*}_{pq1}\tilde{F}_{pqj} & \sum_{\bar{m}pq}\eta^{-1}_{\bar{m}pq}\tilde{F}^{*}_{pq2}\tilde{F}_{pqj} & \cdots \end{pmatrix} \begin{pmatrix} c_1 \\ c_2 \\ \vdots \\ c_i \end{pmatrix}$$

When we have the matrix complete, it can simply be inverted. Finally, we know all the coefficients of c_i, which is a weight factor for each of the harmonics of the trigonometric current functions. And with that solution, we also know the current distribution and the backscattered fields in front of and behind the FSS.

Algorithm and Calculation

We have seen that several steps are necessary to set up a qualified numerical approach that can be used for an electronic calculation with a computer. This approach enables us to analyse the electrical

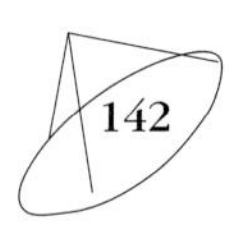

behaviour of a structure with passive elements. It is possible to describe the filter property in the frequency domain, depending on several input quantities.This simulation programme needs the incident angle, the decision of wether to use a TE wave or a TM wave, polarisation, dielectric parameters and thickness of the substrate.These calculations have been done with a HP mini-computer with a reduced instruction set (RISC), much more efficient than a personal computer. The performance in the eighties was really amazing compared to a PC. But even then, one calculation cycle for one frequency took a few minutes. A curve consists of many frequency dots. A personal computer would have taken one or two hours for only one frequency dot.

Some Selected Results

In this subchapter, some simulation results are presented and explained. More results and more details about the equation system can be found in [5].The central point to analyse FSS is the behaviour in the frequency domain while several parameters can be used for understanding their impact.

In Figure 41 we can see the reflection factor and the transmission factor for a defined structure with single square loops. At a frequency of approximately 12 GHz the structure will reflect the whole energy of the incident wave. Pa-

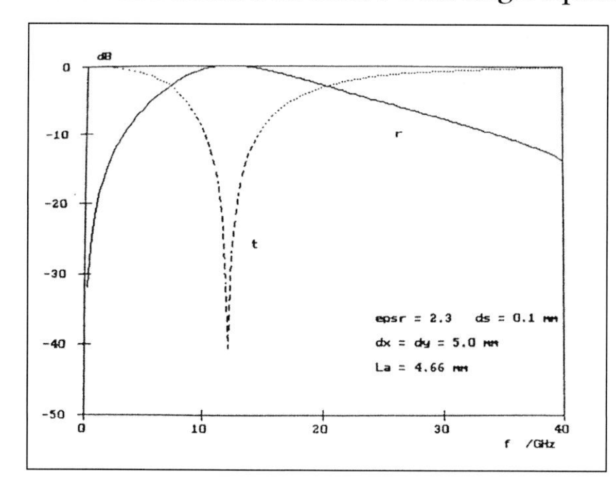

Figure 41: Amount of the reflection factor and of the transmission factor of a frequency-selective surface (FSS). Maximum of reflection: near 12 GHz.

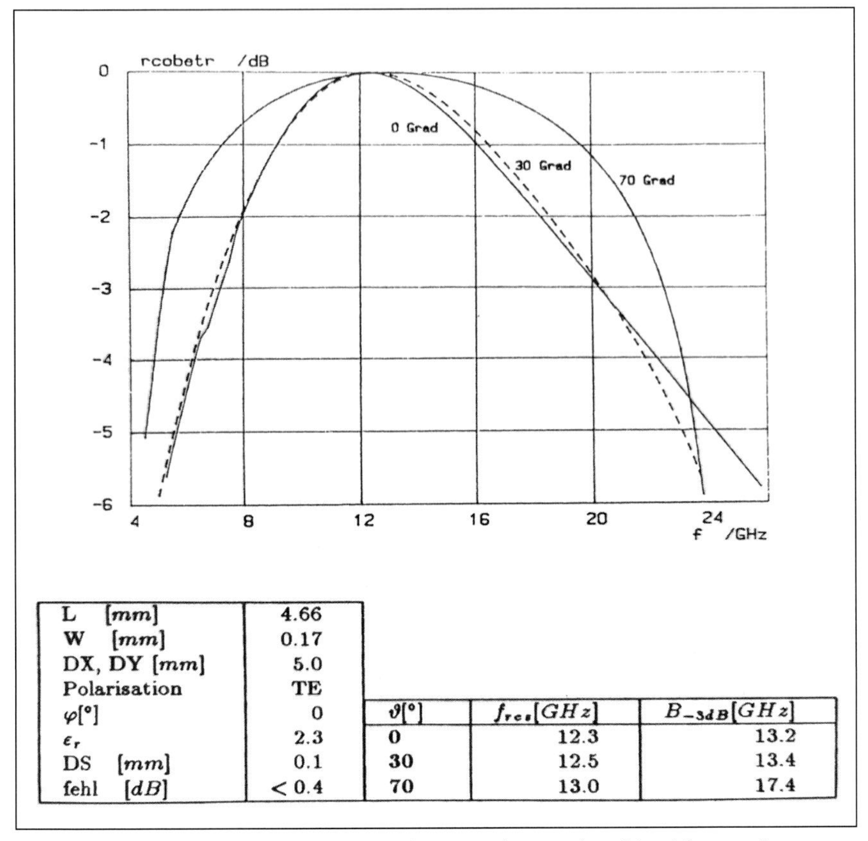

L [mm]	4.66
W [mm]	0.17
DX, DY [mm]	5.0
Polarisation	TE
$\varphi[°]$	0
ϵ_r	2.3
DS [mm]	0.1
fehl [dB]	< 0.4

$\vartheta[°]$	$f_{res}[GHz]$	$B_{-3dB}[GHz]$
0	12.3	13.2
30	12.5	13.4
70	13.0	17.4

Figure 42: Reflection is also dependent on the angle of incidence theta.

rameters can be found in the figure. To understand the performance of the simulation, we shall consider a complete calculation for every frequency point on the axis. Even the fast personal computers of the eighties would have needed 2 days for one reasonable curve. This time-consuming approach was not realistic. But with the help of a fast UNIX machine, the analysis made good progress.

The impact of a H-wave (TE-incidence, hence transversal electric) with two different angles of incidence (0° to 70°) is shown in Figure 42. The resonance has a more broadband character, and the resonant frequency increases by 700 MHz. The convergence para-

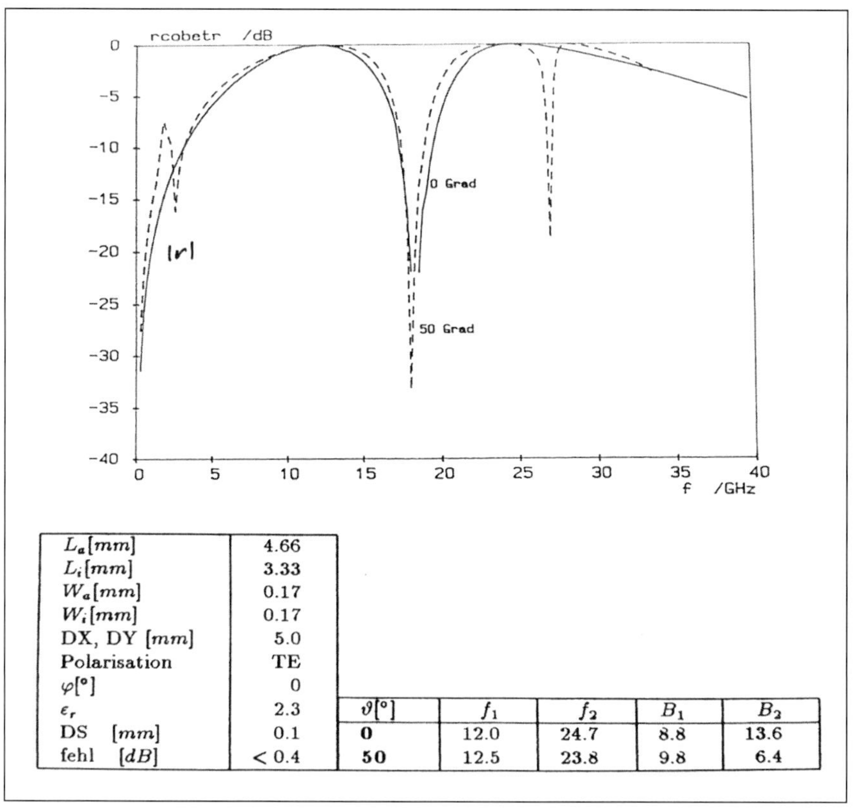

$L_a[mm]$	4.66
$L_i[mm]$	3.33
$W_a[mm]$	0.17
$W_i[mm]$	0.17
DX, DY $[mm]$	5.0
Polarisation	TE
$\varphi[°]$	0
ϵ_r	2.3
DS $[mm]$	0.1
fehl $[dB]$	< 0.4

$\vartheta[°]$	f_1	f_2	B_1	B_2
0	12.0	24.7	8.8	13.6
50	12.5	23.8	9.8	6.4

Figure 43: Double square loops with two maxima for reflection.

meter "fehl" in dB is needed as a termination criterion to stop the calculation, when no significant changes in the result can be expected. Without the termination in time, the calculation effort would exceed too much without a reasonable benefit. Figure 43 shows the behaviour of a planar structure with double square loops as passive elements. Geometric parameters of the structure can be found in Figure 38. Two main resonances can be identified, one with 12 GHz and one with 24 GHz. And when the angle of incidence is increased we can also observe a slight increase of the resonance frequency. While using double square loops a second resonance can be seen and a very narrow transmission range near 27 GHz, too. By this, a third resonance seems to come up above 27 GHz. Simula-

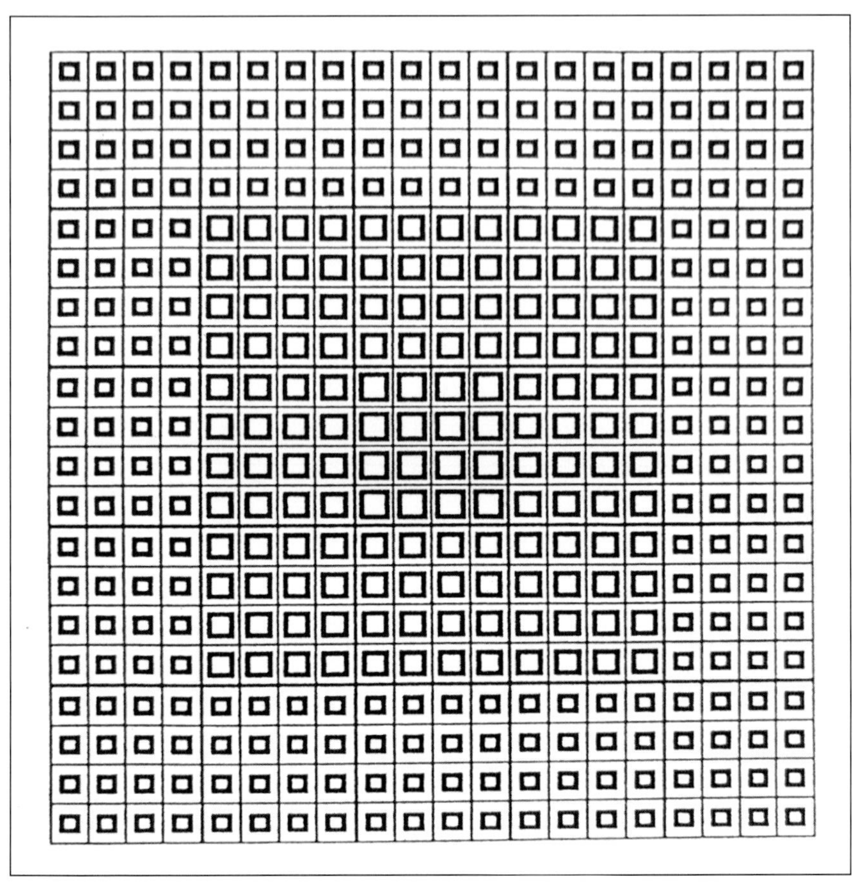

Figure 44: Quasi periodic lattice with square loops, different in size. This structure has an impact on the phase progression of reflected waves.

tions like this can reveal the real behaviour of such a structure very precisely. The effort of construction and the number of prototypes remain low. Next, we can see another variant, how square structures on a planar screen can be designed. In this case, the effect of focusing a wave front will be examined.

Three types of square loops with different geometries are used. With an adequate increase in performance in calculation, the behaviour of an aperiodic screen like this can also be investigated.

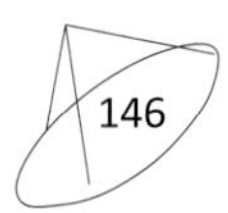

When magnitude and phase of the backscattered field from different elements will interfere, and the whole planar structure is getting new scattering properties.These are the basics for further investigations of aperiodic structures.

Summary of the practical example: In research and development, there is a great need for measurements with different prototypes to optimise their properties. This almost means a huge effort in constructing prototypes, measuring their properties, and investing a huge amount of time and money. Simulations with appropriate software models are a good economical approach to reducing mechanical and measurement effort by using computer calculations. Simulation programmes can even be optimised when results from later measurements are used to improve algorithms and receive more precise results.

We discussed the first steps of engineering in product development. We got a slight impression of how components and systems in industry and space technology are designed and optimised.The people doing this are scientists and engineers, construction experts, computer scientists and RF specialists, measurement technicians and quality experts, and many other experts from different professions. With their knowledge and their work, technology can be continuously optimised, and complex systems remain in a process of evolution and improvement.

Chapter 2: Transmission of Radio Waves

In this chapter, we will learn how electromagnetic waves are used for communication with adequate systems. Good signal and link quality and low bit error rates are essential requirements for the successful operation of communication systems. We need to focus on the different factors that have an impact on the transmission. On one hand, we look at the basic parameters of the systems, but we also have to discuss root causes, which are responsible for interference low-quality links, and even the outage of a system. Some of these disturbing effects are underestimated, and some are hard to find out.

Transmitter and Receiver

When we discuss transmission systems, we can use a simple model that consists of three parts: the transmitter, receiver, and the space between, which is almost filled with any media. The space between them, however, can even be completely empty, which means a vacuum in space. The space between transmitter and receiver means the propagation path includes obstacles like buildings, which also have an impact on wave propagation and signal quality. Example: During the first landing on the moon, the third astronaut circled in a holding pattern with his capsule around the moon, waiting for the landing module Eagle returning from the moon to reconnect the capsule. When the capsule was in the moon orbit, just on the opposite side of the moon, there was no radio contact to earth, because the moon was a huge obstruction like a shadow for radio waves.

For many applications a free line of sight between the transmitter and receiver is essential. Some applications that cannot have a free line of sight use short-wave links (3-30 MHz), where radio waves can be reflected, refracted, and deflected. With a proper radio system, intercontinental radio links can be set up. Radio amateurs use short-wave communication as an own discipline, sometimes even with a very low RF-power for transmission. They call it QRP. The goal is to contact stations far away over thousands of kilometres with very low transmission power (milliwatts or a few watts RF-power). One alternative concept is to set up links over longer distances with relay stations or repeaters. They can enhance a radio link or enlarge the radio coverage of a station over a wide area.

Transmitters with High Power and with Low Power

For many decades, transmitters have been used for communication and to send out electromagnetic waves. Their construction and functionalities are various, depending on the application. Extinguishing spark transmitters were used in a time when ocean giants like the Titanic crossed the seas. Those transmitters meant good life assurance for passengers and crew. When radio broadcasting on long, medium, and short waves became popular and FM on VHF, too, the used transmitters produced an output power of 100 kilowatts up to 500 kilowatts or even more. The energy was almost transmitted over large antennas, reminding me of one or more big curtains made of many different wires and dipoles and supported by gigantic poles. The task was to ensure worldwide broadcast of radio programmes, even if the ionospheric conditions caused additional loss to the signal, varying over time. As the radio signals were almost fluctuating, the broadcaster had to use a certain reserve in transmission power to ensure a good reception in the target countries.

While increasing the transmission power, there was a significant increase in the energy consumption of the transmitters, too. The power supply was generally no problem. Most broadcasters have their own generators or use the public power grid. But there are also applications with very low power consumption. They use almost all batteries, and in some special cases, they generate their own power from an external electromagnetic field used as a power source. RFID-applications do that. Safety-related organisations always need a backup supply system. Diesel generators are almost always used in order to start automatically in case of issues with the public power grid. By doing so, networks can also be run in case of outages in the power grid. So radio links for safety organisations are not affected. For a long time, radio broadcast stations sometimes had to be supplied by their own power plant or by a diesel generator. It was a measure to keep the public power grids safe without additional fluctuations caused by the large transmitters.

Some of the transmitters have a really huge power consumption, but in some applications, this is not the focus at all. Intelligence services need small and unobtrusive transmitters. In the age of transistor circuits, small communication devices used small button batteries, and they still do today. Radar applications consume more energy, because they produce really high RF power levels, almost like pulsed signals. Radar systems for flight control produce RF output levels from 100 kilowatts to 1 gigawatt. Radar waves need those high levels, because their backscattered field some hundred kilometres away will be reduced by a great scattering loss. When it comes back to the receiver, the radar signal must still be detectable. Therefore, primary radar systems need those high power levels to detect even small aircraft. In these systems, special power amplifiers, or klystrons generate those high power levels for millimetre waves.

When we take a look at the other end of the power scale of transmitters, we find applications with very low RF power levels. NFC (near-field communication) shall work for very short distances of approximately 10 mm. It is a technology that is used for authorization and authentication over near field coupling. Data are exchanged over a few millimetres in both directions. This type of operation needs only a few milliwatts. Even if it is more or less a near field coupling system, the developers design it like an RF system with all their parameters for a good matching of the antenna, for example. Some systems are even more economical. RFID tags without a battery use a resonant circuit to take energy from an external electromag-

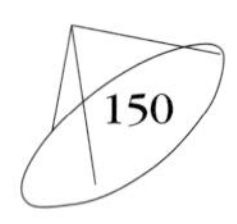

netic fields. With this energy, the tag can operate its own circuit. The external field is the power source and the power supply of the tag. This indeed reminds us of the good old detector receiver, which used a crystal diode to take energy from the radio waves of transmitters with very strong signals. Nearby transmitters could then be heard in an earphone without a battery, simply by a small detector circuit. After World War II it was one of the first options to receive news at all because infrastructure and shops were almost destroyed and you could not buy any batteries or even radios.

Modern radio systems use transistor amplifiers. They provide 1-50 watts for simple consumer applications and professional radio, too. While base stations transmit with 10 W, wireless mobile phones use 100 mW to 1 W. Rf power is generally expressed in milliwatts, watts or kilowatts (mW/W/kW). For calculations, it is an advantage to use logarithmic terms. Attenuations of lines and sensitivities of receivers can be handled more easily in dB and the conversion from absolute quantities is a simple operation. The reference power is almost 1 mW and that corresponds to 0 dBm. Hence, 1 W corresponds to 30 dBm and 10 W are expressed as 40 dBm, 10,000 mW. Calculating with attenuations, the reduction to the tenth part means -10 dB i.e. for the transmitted power, the reduction to 1/100 means -20 dB. In terms of attenuation it is 10 dB or 20 dB. Attenuations are always positive. As a transfer measure, it would be expressed as -10 dB or -20 dB. In transfer function, attenuations appear as negative terms like -10 dB or -20 dB. Those slight differences have always to be considered.

Transmission power is one of the key parameters for the planning and design of radio links and radio systems. One further parameter of transmitters is the phase noise. This parameter is mostly referred to the oscillator. The goal is to achieve maximum purity of the RF signal, with respect to a theoretical signal curve (stable frequency, no deviations in phase relation from a pure sine). Of course a transmitters should never send out harmonics of their frequency because other systems and radio links could be severely interfered. This happened when illegal power amplifiers were used for CB operation with increased power.Trouble with neighbours and authorities were the consequences.

Good Reception Quality and Good Links

Good reception means that a transmitted signal is received with good quality by the receiver, and the listener can enjoy the broadcast. Expressed more detailed. it means:

- The signal shall be stable in amplitude and phase.
- No fading and no distortions (i.e., by multipath propagation)
- No co-channel interference.
- No adjacent channel interference.
- No interference from the electrical environment (man-made noise).
- No interference by their own components with poor manufacturing quality (antennas, cables, receivers).
- Clear and easy to understand modulation and low bit error rates (BER) with 0%.

Interference also comprises all negative impacts from electromagnetic compatibility of the environment. Also, noise can be seen as an interfering effect. When radio links suffer a reduced quality due to this, there are several possible sources of interference that have to be systematically excluded one after the other. In this book, most of the important interferers are presented and briefly discussed.

Noise: Reckon without One‘s Host

Noise is always an unpleasant visitor in many shacks and also in electronic circuits, but especially while receiving radio waves. Noise means a limitation of signal readability and reception. Sometimes a signal gets useless from too much noise. The difficulty is that noise can have different causes and sources. To keep it simple here without extended mathematics, we will look at the practical aspects of radio technology and wave propagation.

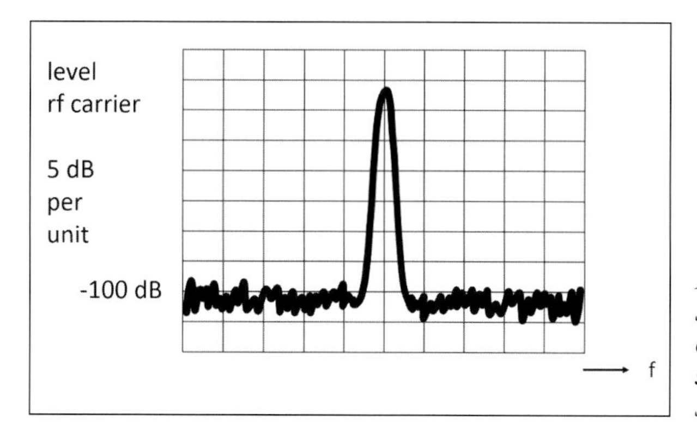

Figure 45: Signal with sufficient or even good signal-to-noise ratio SNR

Noise in a radio system is produced by statistically moving electrons, as in ohmic resistances or other lossy components. That means a kind of noise coming from thermal sources. But there are other types of noise, such as the shot noise in electronics. It is emitted by semi-conductors and electron tubes. This happens when electrons leave their atomic orbit and jump to another energy level. Both types of noise have the property of white noise, which means all frequencies have the same share. Sometimes pink noise occurs, which has another frequency distribution like a 1/f slope or higher potencies. This is a type of noise with almost lower frequencies that is added to the statistical process of electron movement.

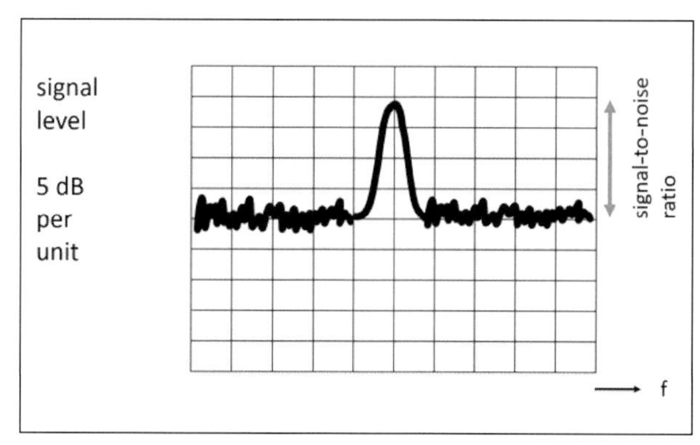

Figure 46: With S/N of 12 dB or less, the readability of AM or FM signals (as for narrow-band FM) are almost always poor or negative. But special digital modes like FT 8 can detect signals beneath the noise floor.

Another completely different source of noise has its origin outside our own systems. Lightning discharges from all over the world can sum up to an atmospheric mixture of noise from thousands of sources. This effect dominates the frequency bands below 10 MHz. This source cannot be eliminated.

That's the reason why receivers do not need an extraordinarily high sensitivity below 10 MHz. Signals are superposed by noise from afar. In the range of 10 to 20 MHz it is the galactical noise that comes up more and more and dominates at some point.

The sources are far away in space. They come from the Milky Way galaxy or even from neutron stars far beyond. This is a type of noise that we can detect on earth. Above 1 GHz further shares come from the radiation noise of hydrogen and water vapor. This is added as well as a noise component originating in our atmosphere from a natural process.

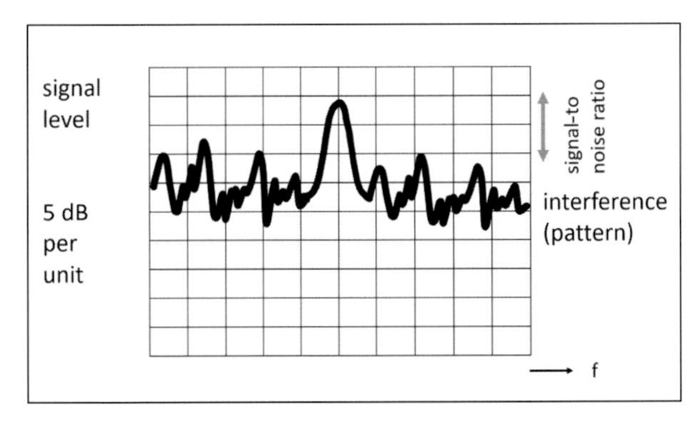

Figure 47: Interfering patterns of noise almost always come from man-made noise and local domestic disturbances.

All those types of noise are components of our noise floor, which we can also measure. When we receive a signal, it is limited by this noise floor and, of course, by its signal level, too. Talking about carrier-to-noise ratio C/N, it means the difference between signal level and noise level, before it enters the demodulator. After demodulation, this difference is called the signal-to-noise ratio S/N or S/(S+N), which can include additional noise from the circuit.

Minimizing Noise

Operators almost suffer restrictions from a noisy environment when they try to receive weak signals. But in some cases, the conditions can be improved. Everybody knows that antennas should be placed far away from interfering sources. That means at a certain distance from buildings, too. Sometimes it is also a good idea to use low-noise amplifiers to improve the signal quality.

In the 90s it was not a cheap matter to manufacture LNBs (low noise block converters) with a noise figure significantly below 1.2 dB at least for frequencies above 10 GHz and by mass production. Today, it takes no bigger effort to produce low-cost amplifier circuits with lower noise figures of 0.8 to 1.0 dB, for good reception of stations far away with weak signals. But these amplifiers are only useful for higher frequencies above 30 MHz, because the atmospheric noise is dominant below 10 MHz.

The noise figure of a two-port is:

$$F = \frac{SNR_{\text{EIN}}}{SNR_{\text{AUS}}}$$

The two-port can be a single amplifier or a cascade of several amplifiers. In a series connection (cascade), it is important to understand,

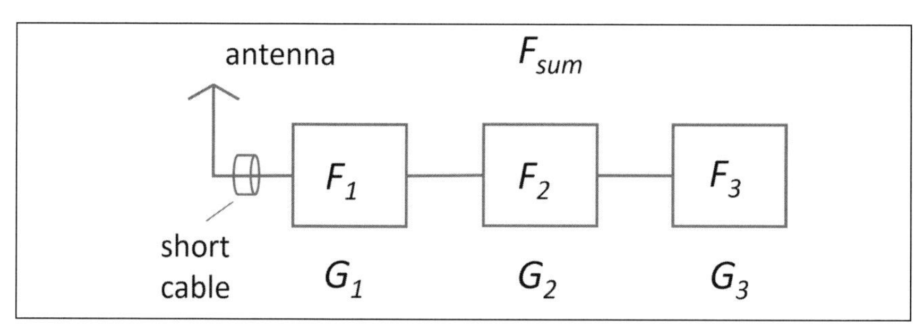

Figure 48: Cascade with 3 stages and different amplifications G and noise figures F as well as an antenna and a short cable with only little loss.

that the first stage of amplification is the most relevant when we want a low noise level (Figure 48 and Figure 49).

When the short cable in figure 48 becomes a long cable with a higher loss (i.e. 3 dB), then the cable acts like the first stage, which means a noise figure of 3 dB. This is not practicable because a noise figure of 3 dB is rather unsensitive for reception.

$$F_{sum} = 1 + (F_1 - 1) + \frac{F_2 - 1}{G_1} + \frac{F_3 - 1}{G_1 \cdot G_2}$$

Figure 49: The overall noise figure of a cascade depends almost entirely on the noise figure of the first stage.

The noise figure is a dimensionless number that has a close relation to the sensitivity of a receiver, the lower, the better. Mostly, the noise figure is used as a figure in dB. The conversion from a factor to "dB" is rather easy.

The use of a low-noise preamplifier can improve reception, but it is of highest priority to place it close to the antenna as the first stage with the highest sensitivity. The low noise amplifier (LNA) will amplify what is in front of this stage, as well as noise shares.

That means that ohmic resistances with high noise production will also be considered, but unfortunately as a noise source. Hence, any ohmic source, like a long cable with some dB loss, shall be avoided in front of the first stage. Consequently, LNA should always be installed close to the antenna.

Any longer cable between the antenna and the LNA will reduce the system's sensitivity significantly. The following equation in Figure 49 explains the noise figure F in the case of a cascade with some amplifier stages and their amplification G.

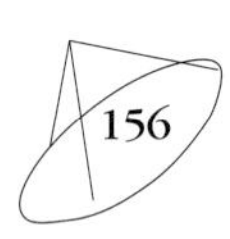

Propagation Loss

Propagation loss in free space is a quite natural effect that can be explained in a two- or three-dimensional model. When energy is emitted from an omnidirectional source, it will be distributed in the space around it. We all know this effect from a two-dimensional example when we throw a stone into calm water and the waves go radially away from the point where the stone fell into the water. We also know the effect, when we observe a point-shaped light source and move away more and more. The amplitude and brightness will decrease, for water waves as well as for light waves. For electromagnetic waves, we can express this in the following equation as:

$$D = 10\,log\,\frac{(4\pi r)^2}{\lambda^2}$$

With D as the attenuation. The wavelength λ and the distance r have a considerable impact. The logarithmic scale is used for easier handling of the great differences in dimensions. This avoids figures with a long chain of zeros.

The equation is a good instrument for estimating received signal levels. The accuracy seems even higher when we have a direct line of transmitted signals on a path that is free of obstructions. For ground, wave estimations, this approach is less useful. Radio transmissions and commercial signals have to pass buildings and other obstructions. This will decrease signal levels and cause additional loss. The path loss depends on the type of obstruction, such as small or huge buildings, the material of the buildings, forest areas, and, of course, the topographic conditions. All this is considered in mature computer models, which use the basic equation for free space attenuation combined with additional terms for rural or metropolitan areas and impact by forest at least. Radio experts will almost have their own approximations for path loss with their immense practical experience over the years.

Frequency Ranges

Electromagnetic waves with different frequencies will show their own behaviour, which is typical for the considered frequency range. It depends on the frequency if a radio signal can really go through the atmosphere into space, for example, to a satellite for telecom-

Figure 50: Meteosat 7. Cloud image of 25th Apr, 1999 12h in the visible range, showing Atlantic and European coast.

munication services or TV broadcasts. Shortwave signals will almost be attenuated or reflected back to earth. They do not have a usable wavelength for space links. Higher frequencies, i.e., in the GHz range, are the better choice for satellite communication. We know the microwave oven as a useful tool to warm up our lunch. It works because water molecules are excited by waves in a certain frequency range. We would need waves with a frequency of 22 GHz for the effective heating of water molecules. As this range is not approved for microwave ovens, manufacturers use a lower frequency range. This might be a surprise, but with enough microwave power, the mechanism will also work at lower frequencies. So we are allowed to use 2.45 GHz due to regulatory guidelines. With the right klystron, the device can provide the needed microwave energy for heating up meals.

Electromagnetic radiation with a wavelength of 10-80 m is not suitable for satellites. But this range is advantageous for intercontinental radio links, as well for broadcast programmes or telecommunication applications. Long distances to other continents can be achieved even with low-power transmitters when the sunspot numbers increase to a maximum within their eleven-year cycle. At this time radio operators can use links from continent to continent for some months. This works even with 10 watts and simple antennas. Reflection, refraction and deflection by the atmosphere support the propagation, process over long distances on shortwave, and sometimes even for medium waves and for VHF. Long waves and very long waves with their very low frequencies (VLF), however, provide a good penetration depth into water. This effect can be used for communication with submarines operating 10 m below the water surface. While using the ELF range (extremely low frequency), the penetration into water increases up to 300 m below the surface. The USA and Russia use 76 Hz and 82 Hz for messages to their submarine fleets. Next, we can have a look at three charts of frequencies up to 15 GHz and also far beyond, where each range has its own characteristics and user groups.

radio LW, MW, SW (150 kHz – 30 MHz)
FM radio (88-108 MHz)
civil aviation radio (110 – 138 MHz)
taxi, bus, tram (145 – 165 MHz)
TETRA police (380 – 420 MHz)

frequency / MHz
0,1 1 10 100 1000
VLF longwave mediumwave shortwave VHF UHF GSM GPS, radar, satellites
Submarines TETRA
Geology

Figure 51:
Frequency ranges and services below 1 GHz.

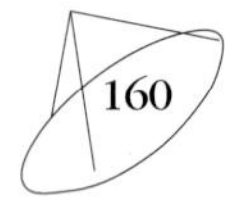

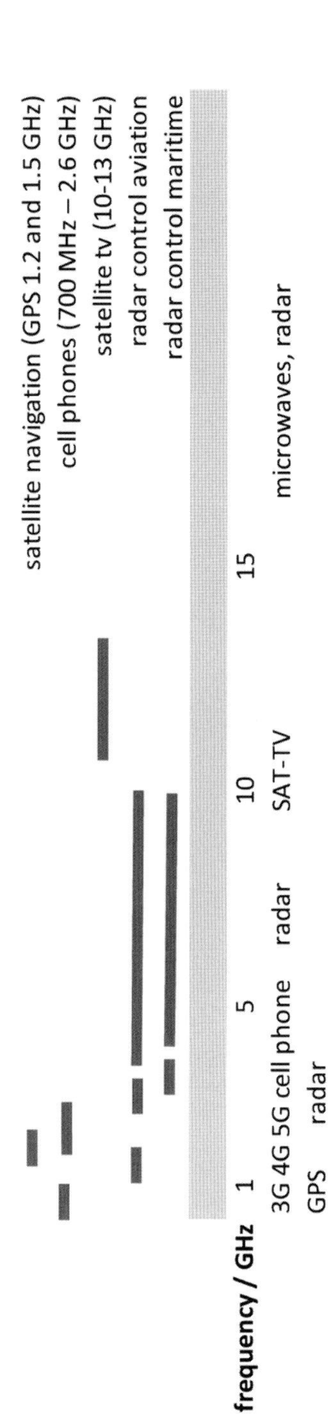

Figure 52:
Above 1 GHz the frequency range is also used intensely.

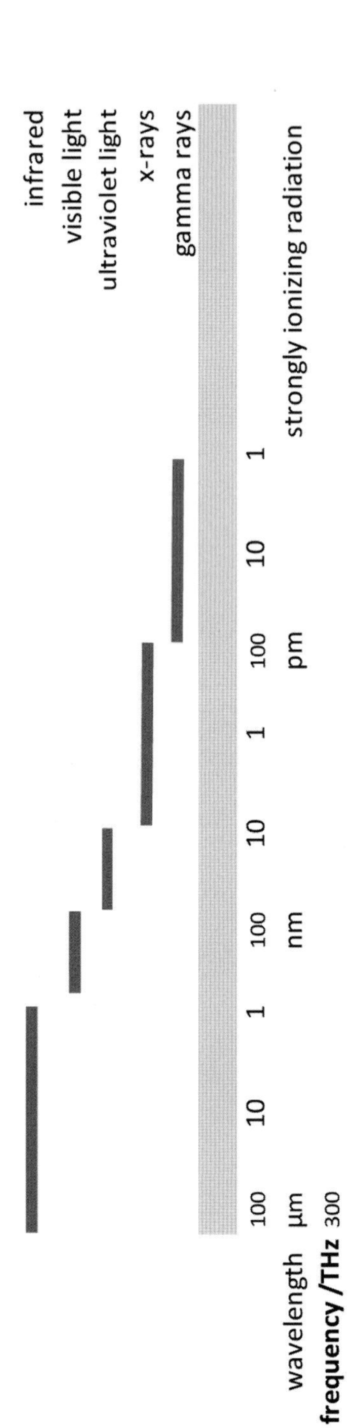

Figure 53:
Above 300 Terahertz, radiation can be dangerous.

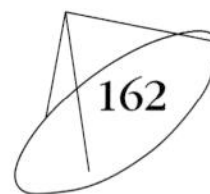

Frequency ranges above 800 MHz with their users are presented in Figure 52.

But the scale doesn't end here. The higher the frequencies, the straighter the path, when we consider frequencies above 1 GHz. Refraction and deflection effects are similar to those of light as we know it. In the Gigahertz range up to 100 GHz satellite communication and radar systems are the preferred users of the frequencies.

Above the microwave range, which is also used by different radar systems, the wavelengths of electromagnetic waves become extremely short. Therefore, it is more practicable to use wavelengths instead of frequencies for an easier handling of figures and to avoid numerous zeros. We assume that the reader is more familiar with wavelengths in these upper ranges of the spectrum.

Frequencies mentioned in the charts may differ in other regions of the world. The exact ranges are part of regulative guidelines by the telecommunications authorities and, changes over the years are certain. The ranges are also incomplete. Any detailed chart would have to consider 100 pages or more of official allocations to services.

The electromagnetic spectrum is really huge, and man has created many applications that are based on the specific properties of each range in the electromagnetic spectrum. Many chapters in this book deal with communication applications and systems. Unfortunately, it is not possible to discuss other topics and aspects with the same depth. Many relations are only briefly mentioned; others are discussed in more detail to give a deeper insight.

When we consider electromagnetic waves for applications in radio systems, this will already be a presentation with many different aspects. This can be really exciting for the reader who is interested in technology. One aspect is for example: Which factors have an impact on wave propagation?

Earth Curvature

Generally, radio waves propagate straight forward. When a radio link is disrupted, the curved earth surface can be responsible. We observe this at higher frequencies (VHF, UHF) over a long distance. This effect is easy to understand when we assume straight-forward propagation and when we exclude refraction and deflection effects near the horizon. There are several equations with approximations or with exact calculations on the World Wide Web that can be used as simple calculation tools. The model behind it is that the earth is curved, but also that there are trees, buildings, or hills, and even mountains, which make an exact calculation very complex. Nevertheless, some interesting observations can be used in approximations.

In a more detailed consideration, means that when a radio link is disrupted, we have several optional causes for that: i.e. distance too large, too many obstructions by trees and buildings, shadowing by topographical impact, and the curved earth surface. When trees and forest areas are not considered in radio planning (i.e., in public phone networks) and test measurements have been done in the winter, then a real surprise in the summer can be expected. Leaves on trees can have a significant effect, when they reduce radio levels by typically 2-3 dB (on 1-2 GHz). The result is that many phone calls in this area can be disrupted or coverage is missing at all. When radio planners use only a little reserve to avoid this effect in areas with forests and trees, any phone call from a car might be impossible due to the weak signal strength. This additional attenuation by leaves can be observed in the 900 MHz range. But in higher frequency bands, as once in the DCS 1800 network in Germany, it will have even more clear attenuating effect for radio links. Some radio planners had their first experience of differences in 900 MHz propagation and 1800 MHz propagation. The surprise: that attenuation was higher and propagation ranges were shorter on 1800 MHz in a rural area. But also in metropolitan areas, reflections were slightly different when they compared propagation at 900 MHz and 1800 MHz.

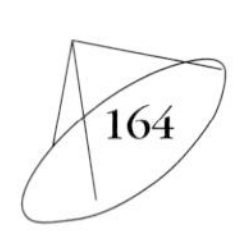

Atmosphere

Even if radio waves propagate straight forward, there are some deviations that are caused by the atmosphere's different properties. Each layer of the atmosphere has its own impact on wave propagation. Some of them can reflect radio waves or act like a waveguide to lead them over great distances before the waves leave their waveguide in the sky and come back to earth (sky wave propagation). Most interesting for propagation are the selected frequency range and the distance of propagation. Reflections, which can happen several times, are the reason why some thousand kilometres of distance can be achieved. By this effect, intercontinental broadcasting and radio communication can be done. Some signals even take two paths in opposite directions around the earth to reach the receiver. This can be proved by the timing difference of two signal shares, which sound like an echo.

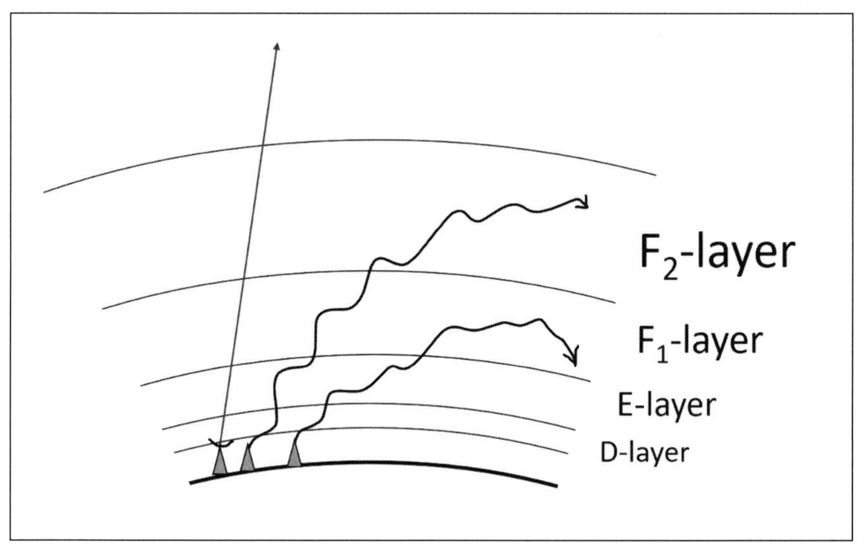

Figure 54: Depending on the used frequency range, the atmosphere of the earth has different impacts on wave propagation. Some frequencies, i.e., 150 MHz and 11 GHz will pass the atmosphere without significant path loss. Shortwave frequencies at 3-30 MHz, however, will be refracted and reflected (height of the F-layer: 150…600 km).

The exact behaviour of propagation in the ionospheric layers, however, is a science of its own and has been sufficiently investigated for decades. Hence, this revealed a certain unreliability for shortwave transmissions. Predictions of conditions on shortwave and stability of links continue to be difficult, or sometimes even impossible. The reasons for this are the dynamic processes on the sun and in the geomagnetic field. All this will have a severe and irregular impact on the atmospheric layers.

Regular Effects. They come from several processes which are totally reliable with respect to radio communication and wave propagation throughout the atmosphere. One of them is the daily cycle of changing radiation from the sun into our atmosphere. Ionisation of gas molecules and recombination alternate cyclically and change wave propagation conditions in summer and winter, as well as at daytime and at night. We observe both cycles. Increased daytime attenuation for example, occurs not only on LW, MW, and SW, but also above 100 MHz. We find a third cycle, the eleven-year cycle, when we look at the sunspots. In 2023, we are very certain to have a year with a significant increase towards the maximum number of sunspots in 2025. DXers all over the world will be happy to enjoy the improved wave propagation conditions and a lot of more interesting contacts from any region in the world. SWLs will also use the improved conditions for listening to new stations, even above 20 MHz.

D-Layer. This is the lower layer of ionospheric layers. It is located in a height of 70-90 km and it provides a high attenuation for longwave and medium-wave signals during the day. Therefore local stations are the only detectable signals. But in the evening and at night this changes completely. This layer dissolves, and additional stations can be heard, sometimes from a distance of a hundred or even thousand kilometres. The best simple antennas for LW and MW reception are ferrite antennas and magnetic loops. By turning them to find the proper angle for good reception, they can be used to reduce interference and unwanted stations with good results.

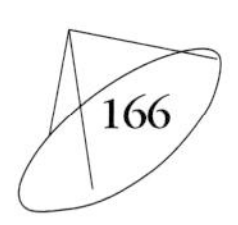

This can be a good method, especially at night when more than one station uses the same frequency and produces co-channel interference. On shortwave, this layer is sometimes responsible for poor reception of broadcast stations and other communication services below 10 MHz.

E-Layer. Located at a height of approximately 120 km, the E-layer sometimes has local regions with a higher ionisation. That enables reflections in frequency ranges of 30-500 MHz and means ad hoc over-range links. This occurs in certain regions and can take a few minutes or a few hours.

F-Layer. This layer has special characteristics with respect to ionisation. It is located at a height of 250 km with a range of 150 to 400 km. Here it has its typical maximum of free electrons. Their maximum is 1012 El./m^3 which is much higher compared to other layers. It is almost the F-layer, which is responsible for long distance wave propagation on shortwave. By doing so, worldwide radio communication is possible. Using large antennas and very high RF power levels for transmission, worldwide communications are stable in most cases. But some dynamic processes in the atmosphere will cause poor signals and a certain unreliability. In some cases, signal levels can fluctuate by 10-40 dB from one day to the next. Extreme signal attenuation can also occur suddenly. This seems so strange that some operators doubt if their systems have malfunctioned. Atmospheric effects from daily changes or over a long time have their relevance too and come almost as an additional attenuation the path loss.

Uncertainties and Imponderables. Unfortunately, we have some issues with dynamic behaviour, that we can‘t rely on. The intensity and duration of a sunspot number maximum belong to those issues. Another uncertainty is how strong the interdependencies of solar activity and radiation with the earth‘s atmosphere and the geomagnetic field will be. Solar flares and solar storms cannot be predicted precisely with respect to their intensity, duration, and consequences.

Meteor showers have their seasons, but they cannot be predicted in detail, even if the Perseids have their regular window in the annual calendar. What we cannot forecast is almost the exact timing and intensity of things that happen in space. That means even the days and the regions with high ionisation in the E-layer are hard to predict. Temperature inversion in our regional weather forecast can also mean overreaching for wave propagation. The estimations about that are considerable, but there are some uncertainties in the forecast. They can be seen in the weather forecast, and they make some wave hunters and DXers happy by monitoring new stations, which cannot be heard under normal conditions. Those overreaches of radio propagation make other people unhappy because, in some cases, co-channel interference by overreaches can be so strong that any reception is impossible.

Solar Effects

The sun has an immense influence on the earth with all that is happening there. On the sun's surface, some processes run continuously, and their consequences for us are not always clear from the beginning. Immense eruptions and solar flares, perturbances and coronal mass ejections, and the sudden appearance of sunspots can affect satellite operations and systems on earth, but even life on earth is affected more than we are aware.

Coronal Mass Ejection. Some of these processes have an influence on the earth's magnetic field. This can be so strong that polar lights come up even in regions far away from the poles. Of course, they are amazing, but sometimes they also announce danger. One of the strongest coronal mass ejections (CME) was in 1859, when a group of sunspots was observed. Such a group is generally not an extraordinary appearance within the eleven-year cycle of the sun. But the subsequent events in 1859 were shocking. The most severe solar eruptions started, and an immense stream of particles directed towards the earth caused a real inferno to systems on earth. When

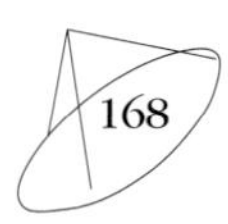

such an event meets us in our modern world, immense destruction can be expected, a real catastrophe for mankind. In 1859 only the telegraph lines and transformers in certain regions of the earth were affected and destroyed. Today, we can assume that it would be hell on earth when telecommunication infrastructure and satellites melt away at high voltages and currents triggered by solar particles striking the earth's magnetic field. It is hard to imagine what this could mean for human life without electronic systems and computers. It's lucky, that the sun is a bowl with many directions into space. This reduces the probability that a severe eruption will be directed just at the earth. What a luck that bad events like this will occur in very large time intervals only.

Coronal Holes, Solar Flares and Eruptions and other processes run slightly below the sun's surface at the beginning. They do not only produce masses of electrically charged particles, that run after explosion-like mass ejections to earth within a few hours. Also, large amounts of ultraviolet radiation and x-rays are produced. When the coronal hole is geo-effective the impact can be measured by scientific instruments, and we can see it when polar lights move through the skies. Electrons from a strong solar wind, which enter the upper layers of our atmosphere, can excite oxygen molecules so strongly that they illuminate the sky with typical green and red veils in the visible spectrum. Oxygen molecules emit green light, and nitrogen emits red and blue light when they are excited by the particles. These effects occur after a time, while particles need some hours or even two days from the sun before we take notice on earth. What we can't see is the ultraviolet radiation and the x-rays as additional effects. During very strong eruptions, this kind of radiation is also produced and will reach the earth within 8 minutes because it comes with the speed of light from the sun without any warning. Solar flares are immense eruptions with the force of some billion nuclear explosions. They occur near sunspot regions and release enormous amounts of energy from x-rays and gamma rays. They may cause complete temporary radio blackouts on earth.

Sun Spot Cycle. Some activities on the sun are prominent ones, and they are considered positive events for radio users, especially radio amateurs and DXers. They occur every 11 years, and it starts with the increased number of sunspots that affect the ionisation of the earth's atmosphere. This improves the conditions for radio wave propagation to another maximum, most probable in 2025. During this time, sunspots become numerous and greater and radio amateurs all over the world are more successful when trying to contact stations on other continents. Radio operation in the shortwave bands gets amazing and exciting whith a highly ionised F-layer. Once in the eighties, one practical test convinced me totally. The conditions were so good that a contact from Germany to England was possible with an antenna of 1 m length, indoor in my room close to the window, with 10 W RF power in the 11m band on 27 MHz AM. The second contact that surprised me was with a Portuguese station in Viseu. It was rather unbelievable that a short DV27 antenna and 10 W output power should enable propagation distances of some hundred kilometres. But all this was real, and the final QSL-confirmation revealed that the contact really happened. Trials during a sunspot minimum with 100 Watt RF-power and hugely efficient antennas are almost not successful on 27 MHz. It is really the ionisation of the F-layer that makes the difference.

Sporadic-E. In our atmosphere, Sporadic-E is another special event. When small meteor particles cross the atmospheric layers, they can excite molecules and produce ionisation. Reflection of radio waves is becoming possible. This enables overreaches of wave propagation, and this can even happen at higher frequencies above 100 MHz. Every year at the same time, meteor showers occur. They have names according to their originating galaxies or constellations such as the Perseids and the Quadrantides and also the Geminides. As the timing of their appearance is regular, radio amateurs are almost well prepared and use the increased ionisation for radio contacts over long distances on higher frequencies. It is almost the E-layer that is affected by these events. Sometimes we are really surprised

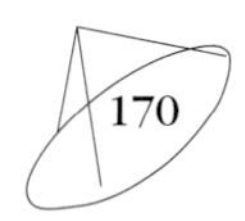

that higher frequencies can also be used. The meteor showers produce isolated areas in the atmosphere where radio waves are reflected, and refracted even in the VHF- and UHF-bands. Those effects, which affect the lower layers of the atmosphere, can also enable contacts over some hundred kilometres. Propagation conditions seem to be improved suddenly. This reminds us a little bit of the improved conditions during temperature inversion in the stratosphere. This causes overreaches of propagation, too, almost in the FM/VHF/UHF-bands and sometimes even co-channel interference may occur. While some radio amateurs are happy with overreaches, radio listeners may be affected by more interference. Especially when one frequency is used by two or more stations. Generally, the regulatory authorities of two countries ensure a safe distance for radio planning. But for overrange effects, the safety distance is sometimes too small. One solution is a frequency change. But all these effects disappear almost within a few hours or a few days, so complex measures and replanning won't be necessary.

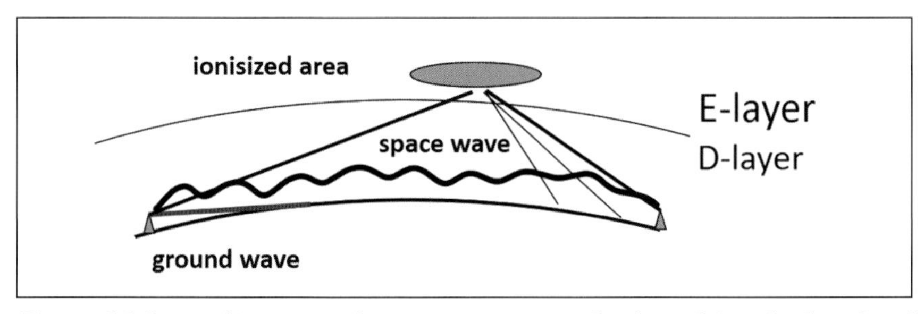

Figure 55: Large distances of propagation can also be achieved when local ionised areas in the atmosphere between sender and receiver reflect signals. Troposcatter is a similarly effect, but at a height of 20 km. Weather and different consistency of air bubbles are possible root cauces.

The processes in the atmosphere are so complex that a lot of fascinating and surprising effects make wave propagation an exciting matter. There are frequency ranges, where the attenuation degrades propagation, i.e., when water molecules are excited. This happens at 22 GHz. That means satellite links would be more lossy, operators should better choose different frequencies in atmospheric windows.

Receiving satellite signals is already possible at 148-150 MHz with a telescopic antenna. In this range, AMSAT Oscars of radio amateurs are operated, and they can be used for worldwide radio contacts, even by FM. There are also professional satellite applications in this frequency range. Russia operates their Cosmos satellites at 150 MHz and the USA transmits over their NOAA weather satellites on 137 MHz. Due to their low polar orbit of approximately 400-500 km reception is rather good with strong signals, but they are only visible for approximately 8 to 12 minutes. In this time, they provide high resolution-pictures with weather information. After a few minutes, they will disappear behind the horizon again. NOAA satellites can be identified by a special sound, like regular pulses or fax transmissions. Even the ISS can be heard occasionally. During ISS overflights, it is possible to catch a signal on 145.800 MHz. Satellite signals can be found at 150 MHz, 250 MHz, 1.5 GHz, 2.-2.3 GHz (S-band), 7.1-8.5 GHz, 10-12 GHz (TV), 12-14 GHz (VSAT), 26 GHz, 31-35 GHz and higher.

By the way, low attenuation for radio propagation is not welcome in every case. Military use on battle fields and the use of special transmissions, which are not intended to be intercepted, need frequency ranges with higher attenuation for propagation. Here they will also operate with lower signal levels to be locally isolated with their radio communication. Between the atmospheric windows for satellite links, there are frequency bands with a high absorption of radio waves (i.e. 60 GHz).

Earth's Magnetic Field

We can imagine the magnetic field as an all-round protection of the whole planet, which ensures our survival. It protects us against charged particles. They will not go straight through the atmosphere, but they will follow the magnetic fieldlines and move to the poles of the earth, and cause polar lights. But who or what threatens us? High amounts of charged particles are ejected by the sun again, and

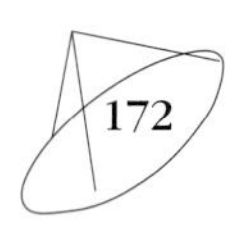

again and they have to pass the earth's magnetic field first, before they might cause damage to our civilization. Mostly, nothing happens at first glance. Sometimes we can see some amazing but strange polar lights. In this case, some regions of the F-layer of the ionosphere are so strongly excited, that we can see wide ranges of illuminated plasma regions. Those colourful waving veils have always been amazing to humans throughout history, and they have always stimulated their phantasy. But behind the veils, there is also a certain danger, what we could find out when modern technology was available. The magnetic field deflects particles to the poles and protects us from strong and striking impacts and damages. Without it, mankind would be helpless because radiation would threaten our civilization severely and continuously. The radiation of particles and the solar wind come with different intensities. When the strength increases and solar storms are directed towards planet earth, the geomagnetic field gets deformed and compressed. During these periods, charged particles will not only move to the poles but also to regions they have never seen before, i.e., near the equator. They have even been seen in Cuba during one of those events.

Field lines of the geomagnetic field start and end in the polar regions near the magnetic north pole and south pole. The magnetic poles are approximately 80 km away from the geographic poles. The field lines also have a defined direction, as we all learned at school when the teacher explained magnetism. Particle streams from the sun with their protons and neutrons also have their own magnetic field, which has an alternating polarisation. When the magnetic vectors of the particle-shockwave meet the geomagnetic field, the particles can be deflected to the poles or they can enter the atmosphere. It depends on the strength of the ejection. Those events are strongly coupled to coronal mass ejections on the sun's surface, and these are coupled to sunspots and flares as a sign of magnetic activities below and above the sun's surface. This all has a huge impact on the geomagnetic field, and it will strongly influence radio propagation on earth.

Our Weather on Earth

Weather on earth can also affect radio propagation and signal quality. Several circumstances play a role: the frequency band, the distance of the radio link, and the quality of the used components. In some cases, weather influence is overestimated, and it is responsible for several effects and observations that have no relation to weather conditions at all. It almost happens while looking for interfering sources, which cannot be explained otherwise. To describe the influence of weather, we have to consider different circumstances.

Heavy rain and thunderstorms can reduce the signal quality of satellite reception in the X-band. One big storm cloud might cause a temporary loss of 2-3 dB. While using a small dish with a diameter of 60 cm the signal level gets too low for clear pictures. Small antennas have mostly no reserve to compensate for losses from heavy rain.

Short outages of reception can be observed for some seconds or some minutes. The better choice are satellite dishes with 80 to 90 cm in diametre. Their signals are more robust against slight fluctuations in weather. Some satellites need even more than 1m diametre of receiving antennas.

Water penetration and humidity in cables and waveguides are the real enemies, because it takes some effort to identify them. Rain, fog, and snow, however, don't mean big problems in propagation in the VHF and UHF ranges, as well as at low frequencies. However, there is one exception: too much snow in a dish. This will always reduce the signal levels and performance of a radio system.

On higher frequencies, the causes of attenuation and their impact change. Fog is a real obstacle for laser transmission in the optical range. Through rain, the attenuation is extremely high, hence the show stopper. Weather radar, however, can detect water and humidity with their special applications, which are often running at 10 GHz.

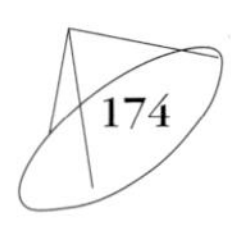

Pilots are warned to avoid areas with severe rain and thunderstorms. Radar applications for cloud detection often use higher frequencies of 96 GHz for example. They are also operated on satellites for earth observation from space.

Radio Weather and Predictions

It is not easy, but very helpful for radio amateurs to know the actual propagation conditions in the radio bands. Predictions cannot be made easily. But in the weekly bulletins and reports of radio amateurs, we can get a lot of useful information about propagation conditions, coming from professional services in the US for example. The governmental NOAA predictions are a very good source of information. They provide detailed and up-to-date charts with overviews for public use. In the World Wide Web when you look for NOAA (National Oceanic and Atmospheric Administration), you will find them. Based on measurements and observations, we can see the major parameters for propagation conditions to use the proper bands for receiving stations on other continents. Useful information on shortwave conditions but also on VHF propagation as for sporadic-E ionisation is available and regularly updated.

We get qualified information about the K-index, which tells us more about the geomagnetic field and wheter it is quiet or disturbed. Additionally, we can observe the sunspots, their numbers, and groups of sunspots. The appearance of coronal holes is also an important indicator of the radio conditions on earth and what we can expect within the next few days.

Solar Flux and Sunspot Relative Number. It was never easy to predict DX conditions on shortwave in detail. But with the help of the solar flux and the K-index we have useful standardised instruments now. Radiation coming from the sun with a wavelength of 10.7 cm is the most common indicator. We measure it on earth on 2.695 GHz and we call it the solar flux. The radiation is expressed in flux units, which can rise to 200 during a sunspot maximum. In times of a minimum, we can expect numbers like 70. In combination with the relative sunspot number we have two valuable parameters to describe or even predict the conditions for wave propagation. The relative sunspot number tells us more about spot groups and their number in general. When no sunspots occur, we can almost expect poor conditions.

K-Index. The geomagnetic field can be very calm, but it can also get more active and even show severe activity. The K-index is the indicator for that. When it is zero, we can be sure that there is no activity, but it can also rise up to 9. Above 5 this parameter indicates a disturbed geomagnetic field. Together with the MUF (Maximum Usable Frequency) and the solar flux, it is the K-index that gives us important information about the propagation conditions on shortwave.

Solar Wind. Nowadays we know some facts about the solar wind. It is a stream of particles originating from the sun, which consists almost of protons and electrons of ionised hydrogen. The remaining share is composed of several ionised particles, such as helium-4 atomic cores, and small traces of other elements such as carbon, nitrogen, oxygen, neon, magnesium, silicon, iron, and sulphur. Non-ionising particles are almost missing completely in the solar wind. This plasma is a good conductor, but in interstellar space, it has an extremely low density of charged particles. NOAA provides data about the solar wind's temperature and speed. Solar wind can even become dangerous, just like a storm. In 2022, such an event surprised scientists and technicians when 40 Starlink satellites of SpaceX were destroyed or got out of operation. They circle around the earth at 210 km height and they were built to provide internet links on earth. Particle streams like this come from the sun, and they propagate slower than light, which takes only minutes down to earth. Particle streams can take 2 hours or even 2 days until they reach the earth.

Time Signals

Without taking notice of them, time signals play a big role in daily life. Living without radio-controlled clocks is unimaginable. The base for our time is provided by the PTB (Physikalisch Technische Bundesanstalt) in Braunschweig and it is transmitted from the long-wave transmitter DCF77 in Mainflingen near Frankfurt. DCF77 uses ASK modulation for that transmission, which can be received all over Europe. This low frequency is advantageous for indoor coverage because it penetrates building materials. Outdoor antennas are not required for reception. Integrated antennas in the radio clocks do a good job, even indoors. And there are further time signals around the world, that can be heard everywhere. One of the originating locations is in Colorado (USA). Fort Collins broadcasts time signals on shortwave at 2.5/5/10/15/20 MHz. While listening to those frequencies, we can hear a time signal that sounds like a clock's ticking and we can also hear additional announcements from a male voice.

The same organisation operates another time signal transmitter on Hawaii. That station transmits according to a similar schedule. But the voice making announcements is a female instead of a male. When we catch a signal on one of the frequencies above, we can tell by the voice where it is coming from. Other countries use similar transmitters for broadcasting time signals.

Antennas, Waveguides and Risks

Now it is time to focus on the different components that are used for the transmission of electromagnetic waves. One of the extraordinary properties is that we cannot see or touch radio waves when they are generated and radiated. While working with electromagnetic waves, we should always take care. Compared to water, radio waves are more complex, and we cannot touch them. We also cannot see them, and everything we cannot see can produce misinterpretations or even risks.

Take care: In some special cases, we can feel electromagnetic waves and this can even be dangerous for our health. Only a few watts at 10 GHz within a waveguide can warm up your fingertip when you put it into the waveguide. Indeed, we can feel that type of RF energy.

It gets really dangerous when this radiation gets into your eyes. This may cause a protein clot in your eyes, and it will make you blind. Take care this is an irreversible process with really bad consequences. Microwaves are used for tracing and tracking flying objects. In the Cold War radar systems were almost used for that. Some soldiers used the radar radiation to warm up when temperatures outside were low. They took a position in front of the radiating antennas, and it felt comfortable while warming up with a pulsed radar power of many kilowatts. The disaster was that this type of radiation was also dangerous due to the immense strength in front of the antennas. Some of them had to suffer bad consequences and severe health restrictions in their lives.

Today we use applications with some milliwatts, and we don't need to worry. Official processes for approval of components and regular monitoring are our protective instruments for safety. Our daily applications have proven for decades to be safe. Their field strengths are so low that impacts are rare. One common impact is individual fear that technology and especially radiation might be dangerous and affect us. And this must also be taken seriously. Generally speaking it is like medication with well-known substances and also natural substances. With the right quantity, they will help. But with a too-high dosage, the medication can be dangerous or threaten your life. Even some plants in nature are extremely toxic and threatening in this sense, like monkshood (Aconitum napellus).

Antennas

Electromagnetic waves, which run in waveguides, for example, are radiated or received by antennas. There are many different types of

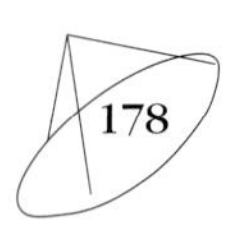

antennas can be used to radiate a guided wave into free space. The basic types are linear antennas, such as dipoles or monopoles, parabolic antennas and panel antennas with arrays inside. Common waveguides are coaxial cables, parallel wires, striplines and microstrips, and rectangular waveguides, as well as those with a circular profile. Waveguides are special structures for guided waves in different frequency ranges. Some of them are both antennas and waveguides at the same time. Coaxial cables with special slots for example, are used to supply tunnels with radiation along the whole length of the tunnel. In tunnels, we also need radio coverage and mobile phone service. For the radiation of waves with circular polarisation, spiral and helical antennas are good choices for some applications.

One extraordinary antenna from 1993 was constructed as a combination of a flat spiral and paired perpendicular slots. A Japanese team led by Professor Goto developed this round-shaped circularly polarised antenna with a 1-2 cm thickness and a diameter of 40 cm. Waves moved between two parallel plates, the upper one with slots as magnetic radiators. The construction was rather simple, but the placement of the radiators was amazing. On a metal plate, there is a virtual spiral as a path. On this spiral-like path, the team placed slot pairs with a perpendicular relation as one circular slot radiator. The slots with a length of 1-2 cm were realised by an etching process into copper. The spiral path considers the phase relation of all those circular radiators on the spiral. By this the slot pairs had a phase shift of 90°, and the paired dipole slots as well as the whole spiral path for all slots produced circular polarisation at 12 GHz for the whole flat and round antenna. This flat antenna received satellite signals for TV.

How antennas are working in detail and how they will be constructed are always exciting questions. Some answers would need a whole book for all details. And they exist indeed. Many answers that focus on antennas are given in [7]. In this book, we will only discuss some basic principles. Additionally some practical experience is presented. Related to some very special topics, even further detailed descriptions are given. People who use scientific approaches and build their own antennas will soon recognise that it is mostly the practical experience that really matters. When you construct an antenna and take it into operation, there might be some stumbling blocks that will cause issues. Some of them are discussed here, but we can say that a successful project is almost based on a huge amount of practical experience.

Always keep in mind that radio amateurs are the real RF experts to solve any challenge related to antenna construction and matching. Radio amateurs work on low frequencies, like 1.8 MHz (160 m band) as well as on high frequencies even in the GHz range. On the short wave bands of 80 m, 40 m, 20 m and 10 m people use popular types of antennas like the long wire antenna, Inverted-V, Beverage, L-Antenna, electric dipole, short dipole, Yagi, HB9CV, W3DZZ, Spiderbeam, magnetic loop, and logarithmic periodic antenna. To be honest, the list is not complete, and it will never be, because radio experts have always been creative.

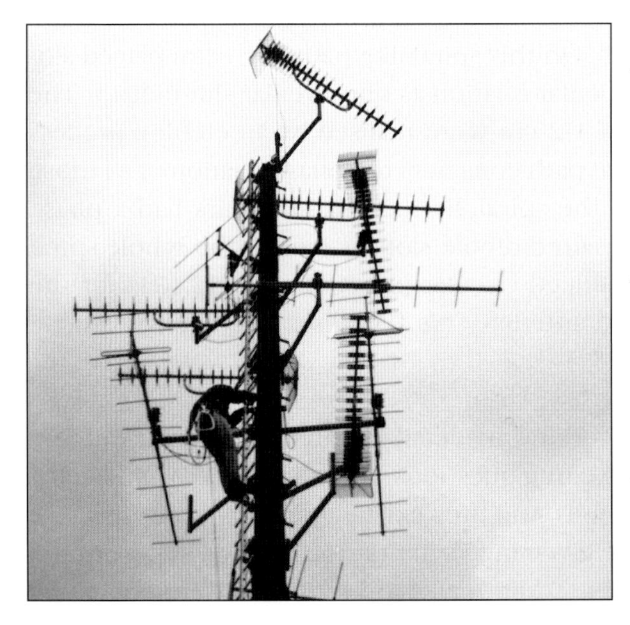

Figure 56: Yagi antennas have an almost simple construction. They were used over many decades by consumers and professionals, for example, in local TV networks as receiving antennas. RF energy can be efficiently focused in one direction. In the 1970s almost every household had two or three Yagi antennas on their roofs.

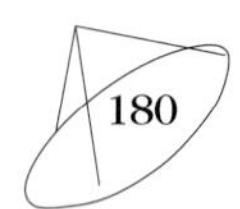

For high frequencies, linear antennas with all their variants are most popular. Directional antennas with different beam widths belong to the preferred components, too. Yagi-antennas meet those requirements in most cases. They have good focus properties, they can be vertically or horizontally polarised, even circular polarisation is possible. Generally, focus and narrow beam widths can be better realised at high frequencies, like in the GHz range. That's because the wavelengths are only a few centimetres. The advantage is that even antenna arrays and parabolic antennas have moderate dimensions and can be handled easily by looking at the mechanical properties. The same antenna types used for shortwaves would have gigantic dimensions as the wavelengths are 10 m - 160 m. Some broadcasters on shortwave know this well. They built up huge curtain antennas with a lot of dipoles, like a great wall. All this mixture of wires and dipoles was constructed between gigantic poles. These large antennas were used for broadcasting radio programmes to other continents and target countries far away. Here are some popular antenna types for higher frequencies: monopole (quarter wave radiator) with ground beneath such as a roof of a car, dipole (half wave), 5/8 radiator, rectangular patch antenna almost as a group antenna or a panel antenna, Yagi-antenna with 3, 5 or more elements, feedhorn, offset- and symmetric parabolic antenna, dielectric rod radiator, helical- or spiral antenna, aperture antennas like magnetic slot radiator.

Basic Types of Antennas

We have seen that there are many different types of antennas on the market [7], which have their own special shape depending on the preferred application. The most important basic types are linear antennas and parabolic antennas. Most of the other types were once derived from them. In this chapter, some of them are presented. In many applications the goal is to achieve high RF gain and low noise for signals from a certain direction. As a common law, we can say that a high antenna gain corresponds to a narrow beam. The sharper the beam the higher the antenna gain. This is even valid for two dimensions. Focussing of energy can be done in the horizontal or in the vertical plane or in both. Table 2 shows some examples of their gains. Starting from an isotropic radiator (point source) focus means that the same amount of isotropic radiated energy is directed by a beam in a preferred direction. The focusing intensity corresponds to the gain.

linear antenna types, feedhorn, panel, dish	gain in dBi
isotropic radiator	0 - reference
hertzian dipole	1.76
dipole	2.15
lambda/4 monopole	5.16
lambda/2 radiator	6.83
lambda 5/8 radiator	8.19
quad antenna	3.14
2-element quad	9
2-element-Yagi	max. 6.5
3-element-Yagi	typ. 4-8 max. 9
n-element-Yagi	max. 20
feedhorn	typ. 10-20
panel antenna	typ. 10-20
SAT antenna TV 85cm	typ. 38

Table 2: Antenna gain is an important parameter, that describes how strongly the antenna can focus energy in one direction. The increased level refers to a reference, almost the isotropic radiator. The table includes the parameters of a panel antenna, a feed-horn, and a dish.

Linear Antennas

Classic linear antennas have great importance. Several types are derivations of the basic linear antenna, and they are also very important.

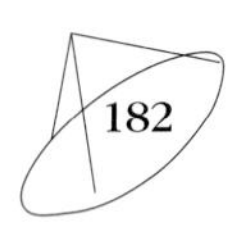

The magnetic slot radiator is the corresponding magnetic counterpart to the electric dipole, even if it is sometimes considered an aperture radiator. Magnetic dipoles can be realised as slots in the narrow side of a rectangular waveguide, for example. Famous linear antennas can be found in portable radios and walkie-talkies as telescope antennas. Monopoles and half-wave radiators, as well as their big brother, the λ 5/8 radiator, and even short derivates such as bazooka, antennas are candidates that belong to the linear antenna group. Ground plane

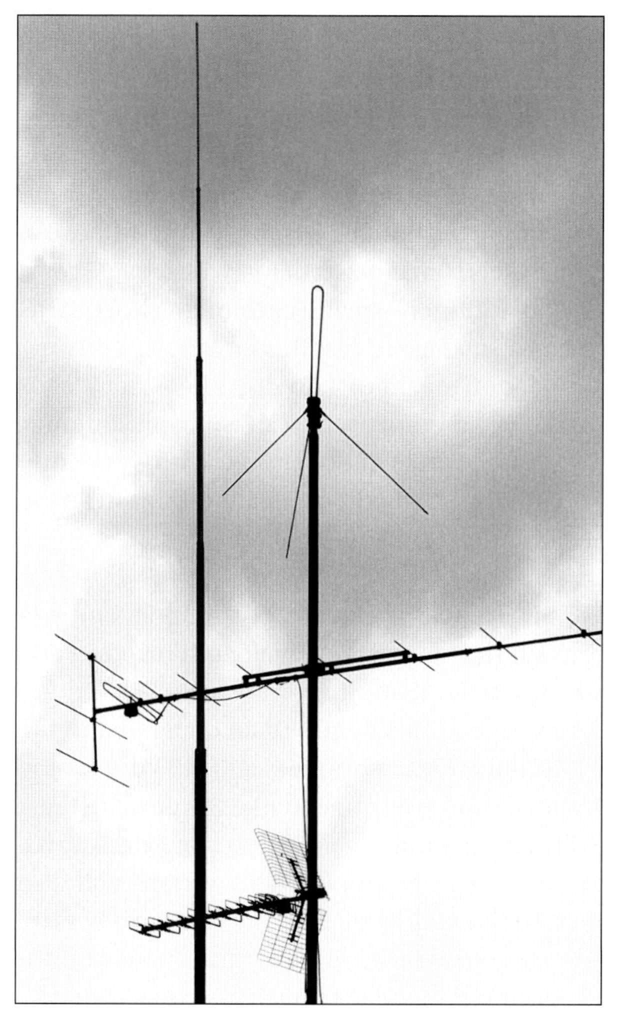

Figure 57: Ground plane with a loop-shaped radiator. They were used in police networks and fire departments on 4 m and 2 m wavelengths. Two different sizes were usual. You could find them on nearly every building with any authority for security. Below are 2 Yagi antennas for analogue TV programmes in the 1970s.

antennas (Figure 27) are very common antennas with a vertical $\lambda/4$ radiator and 3 or 4 radials (sometimes even more). The radials can have an angle of 90° to 135° to the radiator or even more, while the angle can also be used for antenna matching by finding the right angle for best matching. The wave impedance of the ground plane corresponds to the impedance of coaxial cables, which makes the ground plane easy to install without matching effort.

When we consider the gain as a major antenna parameter, we need a reliable fix relation as a reference. Which antenna could be a good reference? It is an antenna that is not real, but just the right model: the isotropic radiator. It is a virtual point source of radiation. Therefore, the gain is 0 dBi. The isotropic radiator is a source that sends out the same amount of energy in all directions in three dimensions. This idealised radiator cannot be bought or constructed because any charges that generate electromagnetic fields will move over a certain length, even if it is infinitely small (extremely short). This length would already mean a small gain. Just this enhanced isotropic type is the Hertzian dipole with a gain of 1.76 dBi. The "i" stands for isotropic and means just the gain compared to an isotropic antenna. The Hertzian dipole is also a virtual radiator, but with very little length and current distribution. Both theoretical antennas can be used as a reference, because their relation is fixed.

Practical antennas have always had a current distribution. The $\lambda/4$ monopole has an expansion that a quarter wavelength matches and the $\lambda/2$ radiator has an expansion that a half wave matches. Due to length, both antenna types have a considerable gain compared to the isotropic radiator. It is 5.16 dBi for the monopole and 6.83 dBi for the $\lambda/2$ radiator. Antennas with a similar expansion like the ground plane antenna (Figure 57) will have a similar gain. Interesting derivatives are the short versions, i.e., of the monopole. This variant will also work with an inductance in the middle or near the bottom. They are popular because they reduce the dimensions of large antennas for shortwave, for example, even if that means a slightly reduced gain.

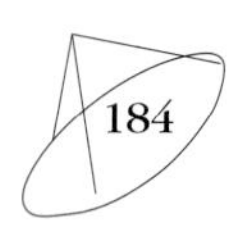

Shortwave linear antennas are often very short with respect to the used wavelength. For short-wave listening, active antennas are a good alternative. They have a length of 1 m for example with an integrated amplifier. They are matched for wavelengths of 20 to 80 m or beyond. The proportion reminds one of a hertzian dipole. Therefore, we can expect only a slight gain of approximately 2 dBi with respect to the C/N, even if the gain of the amplifier is higher. Radio amateurs who have enough space for antennas may use Yagi antennas or even cubical-quads. They perform really well, even for DX. For higher frequencies, Yagi antennas will be a rather smart solution with almost high gain.

Aperture Antennas

The countermeasure against snow in the dish is reflector heating. Warming up the reflector is realised by heating wires, which need a

Figure 58: Symmetric and offset-parabolic antennas with different diameters for single use or for several households. They use 10/11 GHz and have a gain of 36-42 dBi.

power supply on the roof or an additional cable for that. The heating wires will cause a melting of snow and ice, and satellite reception will not suffer additional losses any more. Low-cost variants of parabolic antennas can be manufactured easily. Another advantage is the low loss and high efficiency of focusing electromagnetic waves in one direction. Other planar antennas have almost a certain loss due to stripline waveguides and coupling elements, and they cannot perform with the same efficiency as the paraboloid with a feedhorn in the focus. The horn antenna is part of the feed system, which contains almost a LNB. Horn antennas are used to illuminate a reflector as required. They also belong to the group of aperture antennas. Their gain and beamwidth are selected exactly to achieve a good illumination of the reflector. The illumination is often reduced in the rim area to avoid high sidelobe levels. Alternatives are Yagi constructions, dielectric rods, or even patch antennas to illuminate reflector antennas.

The gain of a parabolic antenna depends mainly on the antenna aperture, the antenna plane perpendicular to the direction of radiation, which defines how large the antenna is. Antenna gain can be estimated as follows:

$$G = A \frac{4\pi}{\lambda^2}$$

Parabolic antennas belong to aperture antennas, and they focus electromagnetic waves effectively, just like in ray optic models. Hence, they have a dominant direction of radiation with a gain depending on their dimensions and the used frequencies.

They can easily achieve a gain of 30 dB or 40 dB, like the antennas for satellite reception on buildings. The typical gain of such an antenna is 38 dBi with a main beam of only a few degrees (3 dB drop). When the equation above is applied to common 85 cm parabolic antennas, we have a result in the logarithmic measure:

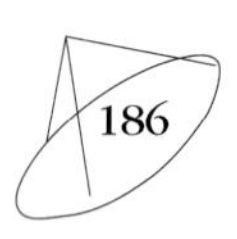

$$G = 10 \log (A \frac{4\pi}{\lambda^2})$$

With A as the antenna plane and λ as a wavelength. That means a theoretical gain of 41.5 dBi. But why does an antenna like this have only 38 dBi gain? Why is there a certain discrepancy?

We might assume that the parabolic reflector is illuminated with the same intensity over the whole surface. But it isn't. From the centre to the rim, there is a reduction of illumination by 6 dB or more. As a result, the gain of the antenna is also reduced slightly. But what is the reason for reducing the illumination intensity near the rim? The reason is an antenna diagram with reduced side lobes. Why is this important? Reflectors with diameters of approximately 90 cm have sidelobes that look into the direction of neighbouring orbit positions. When neighbouring satellites transmit strong signals, they can be received unintendedly by the sidelobes and cause bad interference. To avoid this, it is a popular trick to reduce the level of the first sidelobes. Unfortunately, the effective antenna plane is also slightly reduced when the illumination is not constant all over the reflector.

For large astronomic antenna systems, scientists often use parabolic antennas with subreflectors. Cassegrain antennas (as on page 1) and Gregory antennas are confocal systems, where the main reflector shares its focus with one focus of the subreflector. This is the meaning of confocal. Cassegrain systems use hyperbolic subreflectors. In this case, the feed system that illuminates the subreflector can be installed in the midle of the main reflector surface. Gregory-antennas use an elliptic sub-reflector, which has an opposite curvature with respect to the hyperbolic curvature of the Cassegrain system.

Panel antennas are flat antennas almost with dipole arrays under a protecting radom. They are aperture antennas, and their radiators are placed in rows and columns with an appropriate feed system i.e. by coaxial cables. One advantage of flat construction is that flat antennas can be easily integrated into a spherical or flat surface, as on aircraft, spacecraft, and vehicles. They are also a useful, unobtrusive solution for mobile radio, which is applied to thousands of roofs in people's urban environments. Typical horizontal and vertical radiation charts are shown in Figure 59.

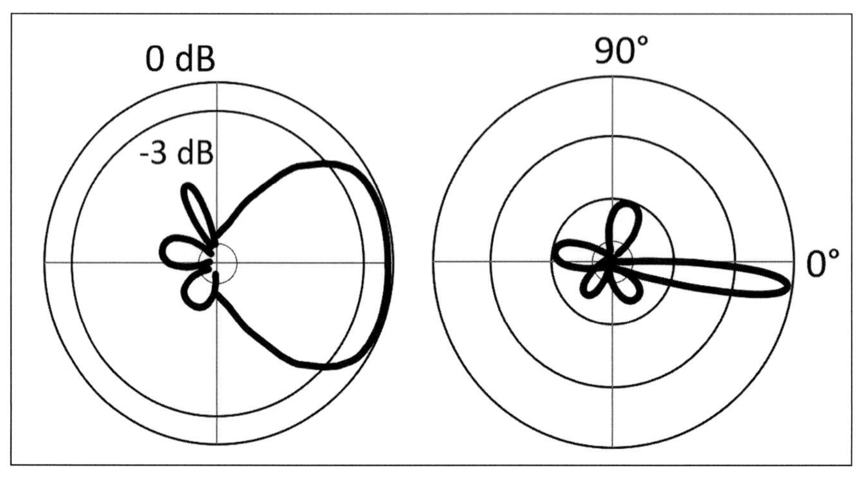

Figure 59: Typical horizontal and vertical diagrams of a mobile radio base station antenna. To avoid interference, they were almost mounted with an electric downtilt. Phase differences between the feeding of single elements cause the tilt. But it is also possible to use a mechanical downtilt.

Directional antennas with a fixed electric down tilt are used when the beam shall look slightly down by using an angle of some degrees. Backward interference shall be reduced by this, and the illumination of buildings can be improved when the electromagnetic energy penetrates and improves indoor coverage instead of blowing energy above the roofs to the horizon. The electrical downtilt is realised by feeding the antenna elements with a certain phase shift from element to element. Mechanical downtilt for panel antennas is also used for the same purpose. Which solution is the better one depends on the radio planner's experience and how he intends to manage challenging coverage scenarios, including interference.

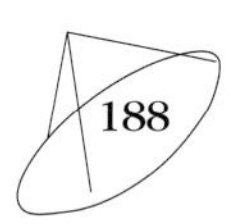

Figure 60: Base station antenna for 1800 MHz with two columns of dipoles. In this case, the beamwidth in the azimuth is an angle of 30°.

Figure 60 shows a panel antenna for public mobile radio. We can see two columns of dipoles for vertical polarisation. Telephone signals from a base station are received and transmitted just over these dipoles. We also see the plastic spacers which shall stabilise the radom, which is removed here. Wires that connect the dipoles as a complete array can be recognised, too. They end up in a coaxial socket for external cabling to base stations or repeater stations. Received signals and transmitted signals run over those dipoles at the same time, just like in a mobile phone. Incoming and outgoing signals are used at the same time, just as in a phone call.

Figure 60 shows a special external grid, which will improve the backward isolation of the antenna. This shielding effect was examined at Ruhr University of Bochum by the team of the working group "Antennas and Wave Propagation" under the direction of Professor Dr. P. Edenhofer. With the additional

Figure 61: Unechoic rooms are used to suppress reflexions and measure the radiation characteristics of antennas.

shielding antennas for repeater operation could be significantly improved due to suppressed coupling. The close proximity of both repeater antennas often caused interference and, this could finally be reduced significantly. The antenna in Figure 61, which is located in an unechoic room, had a horizontal opening angle of 30°.

Magnetic Antennas

They are almost constructed as ferrite antennas or magnetic loops, the category of magnetic antennas. Both are rather small with respect

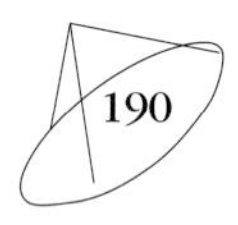

to the used wave lengths (i.e. LW) where they are very effective. Ferrite antennas can even be used as a built-in antenna in portable radios, and they have a good directional effect.

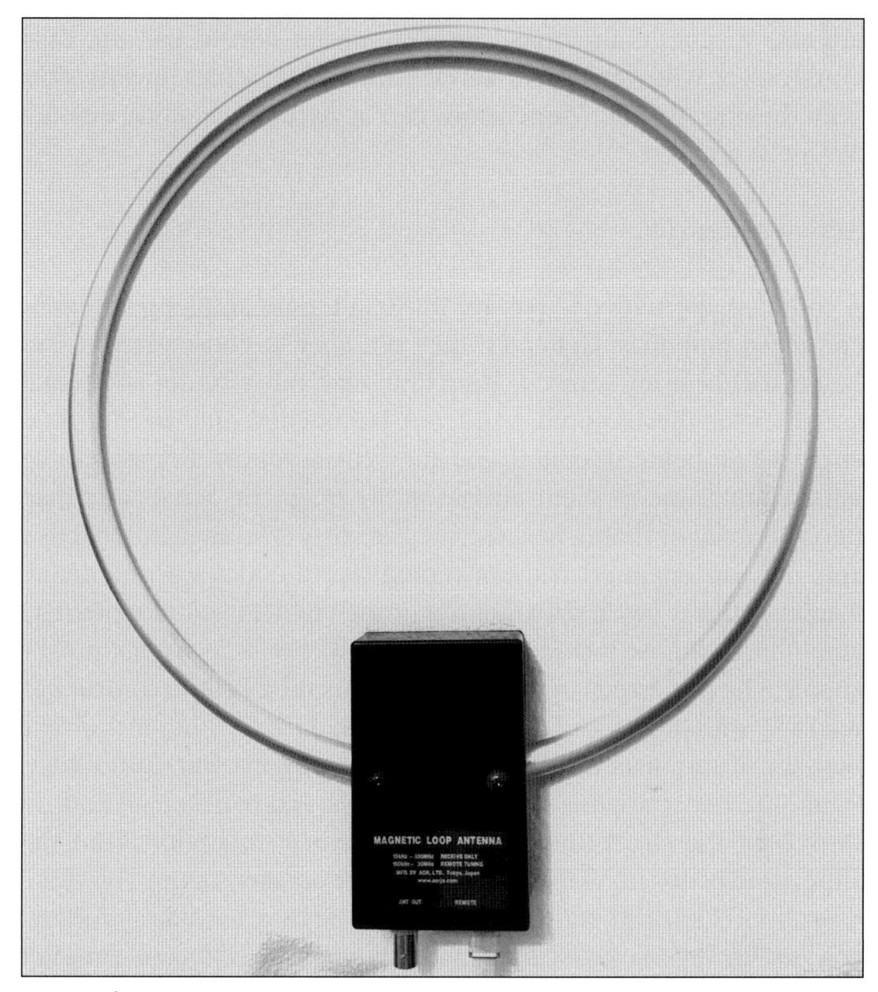

Figure 63: Magnetic loop antennas are sensitive to the magnetic field.

Magnetic loop antennas have a typical diameter of 0.5–2 m. They are excited by the magnetic part of electromagnetic waves, and they are very popular for medium wave and short wave reception. One big advantage is their small dimensions with respect to the wavelength. Additionally, they have good directional characteristics. By turning the interferer into the minimum of the antenna diagram, reception can be improved in many critical receiving scenarios (see also Figure 94). By the way, radiating slots in a metallic waveguide are also magnetic antennas. The slot can be considered the dual counterpart with respect to an electric dipole, which means slots with their appropriate dimensions can be used as magnetic dipoles, almost as arrays.

Active Antennas

A common alternative to passive linear antennas and loop antennas are active antennas. Generally, they are special constructions also based on linear antennas or loop antennas, but with an additional amplifier for the required frequency band. They only need a little space while being operated, and they cover almost a considerable bandwidth. Some of them, almost all the magnetic active antennas, need permanent tuning because they are very selective. Other active antennas are operated without tuning, like the ARA 60, one of the legendary short electrical antennas of the eighties and nineties. The amplifier is almost inside the antenna and needs a proper power supply. The preferred supply is a remote feeding splitter in order to feed the amplifier with DC over the coaxial cable. One of the very best active antennas was the DX One. This antenna was rather huge, with different strange elements and radials, and had a small control device for the shack. But it provided the best reception from the submarine bands on VLF to shortwave without any tuning. The price was about 1000 DM, but everybody who finds a used and working antenna on the market will be happy with that really good product and the results. Some other models, like the magnetic loop of Grahn, which is manufactured in Kirchheim (Teck) in Germany, provided considerable results, especially on the lower frequencies. The concept is based on modules like a loop and some different ferrite elements, with a rather high selection for their band. While the DX One is built for outdoor use, the Grahn antenna can be used in the shack.

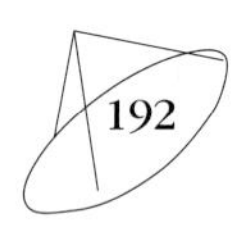

Antenna Groups and Arrays

For special applications, antennas are used as a group, like an array. This can be in flat antennas, as in the panel antennas for public phone networks with rows and columns of dipoles. But it can also be in electronic direction finding antennas for military use or for applications in aircraft navigation, such as VOR, the VHF Omnidirectional Radio range antennas.

There are also consumer products that can be used as directional group antennas. They all use a simple addition from antenna gain of two or more antennas. This works with Yagi antennas in the 2 m range and in the 70 cm range, and even space applications use this on higher frequencies, where antennas are smaller than on VHF. The challenge is to choose the right spacing between the antennas and to feed them with the required phase. This will almost be the same phase relation, except the operator wants to create a phased array with a steering effect in the main beam. With the right distance between two antennas, a small group of two can achieve a 3 dB gain increase. Four antennas can achieve 6 dB without losses. A little loss may occur due to cabling, connectors, and probable insufficiencies.

Antennas in Mobile Networks

In the last 30 years, thousands of antennas have been installed on roofs and rigs to provide coverage for mobile phone operation for consumers. Some sites are located on hills, others are placed at a medium height in urban areas. All the antennas are required for mobile phone operation in cars, on motorways, in cities, at home, in the forest, and even in trains. During the first years of implementation providers used almost omnidirectional radiators on very high sites to cover a huge area with only a few base stations. But when subscribers and traffic increased dramatically, radio planners designed segmented sites with almost three and sometimes two sectors. This means two sector antennas (panel antennas) for each sector and different transceiver units for each sector. That was the standard planning configuration, which provided more capacity for more traffic. Additionally, the operators increased the density of sites, especially in urban areas.

Figure 63: Two sector antennas with a medium height over ground on a building with a flat roof. This type of mounting is typical for cell phone networks.

Transmitter sites are able to illuminate three sectors independently. This also means an increased capacity for calls by factor three. And the RF power was also related to one sector. Compared to one omnidirectional site, the RF coverage was significantly better because each sector had its own transceiver equipment. Antennas had a typical beamwidth of 65°-90° (3 dB width). This provided better penetration into buildings, and in-house coverage made the networks more attractive. Omnidirectional radiators could not provide such a density of RF energy inside buildings. Indoor coverage was additionally improved by using lower sites in urban areas. Radiation from very high sites was almost avoided in the city areas. A typical site is shown in Figure 63. However, in the plain country, it is almost better to use high towers, and radio planners were always happy when they were available at all.

Special Antennas

Over time, some strange antennas could be observed, and while discovering them, there was always the question: Who uses this type of antenna? In which frequency band is it working? Many of them are special broadband antennas, like the discone antenna, which has excellent wideband characteristics. The cone shaping of the lower half enables the use of a wide frequency spectrum. But generally the antenna also reminds us of a vertical dipole or a ground plane.

On the top of a discone, there might be some monopoles for emphasising special frequency bands. But most of them have no monopoles on top. All those special antennas are intended to be used for wide-band applications. That means they are useful for all organisations and operators who use frequencies between 100 MHz and 2 GHz. This can be for transmitting or only for receiving. With these properties, discone antennas are extremely practicable for monitoring on VHF, UHF and beyond. Considering special antennas, they are almost not as strange as we assume. Sometimes they are only derived types from basic antenna types. Radiators for vehicles

and frequency bands as the 10 or 11 m band have a λ/4 length of 2.50 m. This is really not practicable for roof installation or trunk installation on a vehicle. Mechanical shortening of the antenna is a good idea. The length can be reduced by approximately 50 %. An inte-

Figure 64: Broadband discone antenna for frequencies from 30 MHz up to 1.3 GHz.

grated inductance is the secret behind it. The coil can be located in the middle of the antenna or in the lower area. To be honest, the efficiency decreases a little bit, but with 50% of the λ/4 length, the antenna can still be operated with considerable results. When we reduce the length by more than 50 % the efficiency decreases significantly, and the antenna will lose its good radiation properties more and more.

Satellite antennas for ground stations can also have a strange appearance. Parabolic antennas generally have only one focus, which can be used for placing a feedhorn with a low-noise converter, for example. They can be integrated in one unit or consist of separated components. This standard configuration of a satellite dish almost

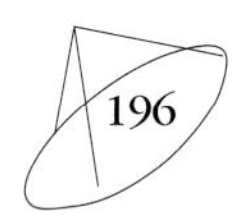

looks at one single geostationary orbit position to receive their TV programs. Those positions are approximately 36,000 km away from the receiver.

If it is possible to construct an antenna with a stretched focus and give the reflector the proper shape, then we will have a new antenna system with improved characteristics. The use of several neighboured feed systems is then possible, and each of them looks at another orbital satellite position.

It is hard to believe, but this type of antenna system has been realised and we can see it in Figure 65. Several dishes can be saved and numerous TV programmes can be received with a single dish yes it is a huge dish.

Figure 65: Reflector antenna with a line-shaped focus area. This special construction uses different LNBs for a number of satellites in several orbit positions.

Reflector antennas like this are not pure paraboloids.Their geometric shape is optimised to focus on the whole orbit. Multifeed systems also exist with smaller dimensions, such as an offset parabolic antenna with two feed systems. One of them is almost placed as usual, and the other is placed just beside the feed in the focus.This combined double feed was a very popular solution for the reception of two satellites in neighbouring positions.This could be described as a squinting effect. But the double use of the reflector costs approximately 1-3 dB loss of received power level (X-Band), that will not be a show stopper with an 85 cm dish. When the second feed is moved from the focus too far away, the received signal level decreases, and reception is no longer possible. Squinting loss is tolerable for the first neighbored orbit position, only a few degrees away. Beyond that, the antenna diagram will be bent more and more, because the focus in the X-band is a small sensitive area which allows only a shift of a few centimetres to achieve the squinting effect without too much loss.

Polarisation

The electric component of an electromagnetic wave can have different directions, but they are almost vertical or horizontal, depending on the antenna and the use case. Generally, the direction is fixed for a certain application, whether vertical or horizontal. Sure, some radio amateurs use special antennas with a control option to choose horizontal, vertical, or even circular polarisation. It is evident that horizontal and vertical polarisation are perpendicular or orthogonal, as mathematicians say. But there is still the other pair of orthogonal polarisations: left and right circular polarisation. They are created by adding both linear polarisations, which have to be orthogonal and additionally have a phase relation of ±90° shift. When the shift is not ±90° the result will be elliptic polarisation.

To release electromagnetic waves from a waveguide system into free space, engineers use almost feed-horns with a circular profile.

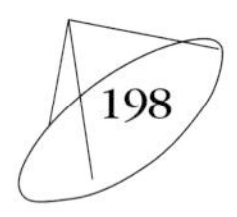

Special applications use Yagi antennas with crossed dipoles or helical antennas. But they need some more mechanical effort for manufacturing and installation. Aperture antennas like magnetic slot radiators can easily use orthogonal elements. That works in rectangular waveguides and also in parallel plate waveguides. A slot radiator is realised as a mechanical recess at the right positions and with the right proportion with respect to the half wavelength.

Bandwidth and Resonance

The transport of electromagnetic energy requires waveguides like coaxial cables, rectangular or circular waveguides, or microstrip lines as well as optical fibres. It is important to take care of the frequency range of the waveguide. Most coaxial cables are tolerant with respect to the frequencies used, but rectangular waveguides are not. They have the lowest usable frequency, which is roughly defined by the length of the longer side. When a half wave of the used frequency does not exceed this length, then the waveguide should work. The choice of an antenna and the appropriate waveguide as a feeding system should be taken with care. One goal is always to keep the loss in the feeding elements low. And it is always important to consider the resonant frequencies of an antenna and their bandwidth. A good $\lambda/2$ radiator for the 11 m band is the Antron 99 antenna or the Imax 2000. Their resonance is not so narrow and sharp, that they can easily be operated in the 10 m band, the 13 m band and the 15 m band without unacceptable loss. This typical $\lambda/2$ radiator will work from 18 MHz to 30 MHz.

Other antennas can have a sharp resonance, indeed. A prominent example is the magnetic loop, with its high selectivity. Some antennas have been constructed as wideband antennas, such as the vertical discone or the logarithmic periodic antenna, almost with horizontal polarisation. Most active antennas are wideband antennas when they are based on a vertical linear antenna. Some vertical antennas, like the $\lambda/4$-monopole have a narrowband characteristic.

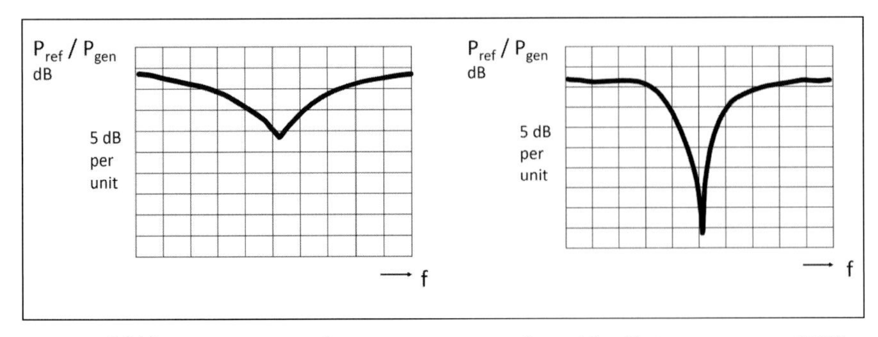

Figure 66: The resonance of an antenna can be wide, flat, or narrow. Different antenna types and differences in the frequency response have a great impact.

Figure 66 shows the behaviour of reflected power related to the generated power with respect to a whole frequency band. The bandwidth depends on the antenna design, and it can be matched to the requirements within certain limits. When the resonance reaches a bandwidth of more than 10 % of the used frequency, the antenna already has good wideband properties. The exact general definition of "wideband" is hardly possible.

The shape of a dipole makes us assume that the properties are closely related to the wavelength. Following this assumption, the dipole will almost be a narrowband element, but with options to make it more wide with adaptations in shaping and matching, the bandwidth can be increased. The operation of antennas will be successful when their input wave impedance corresponds to the wave impedance of the used waveguide. The stretched electric dipole has a theoretical wave impedance of (73+j42) Ω.

A feeder with the conjugate complex impedance (73-j42) Ω is the best choice. Both are complex terms with a real part and an imaginary part. For matching an antenna, this is a relevant term. The feeding cable can be tuned with special methods like the gamma match or with an antenna match box. Folded dipoles have an input impedance of 240 Ω.

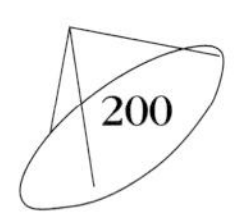

Matching and Return Loss

Antenna matching is indeed relevant, and this is almost underestimated. We can imagine this in a pair of shoes when we wear them. Too large shoes will force us to slow down and move extremely carefully. We cannot run with shoes that are too large. Too small shoes will also force us to go slowly and carefully, and they might be aching. Running is hardly possible; we will stop soon.

Only with comfortable shoes will we have a good match, and we can even run without problems. Maybe the comparison is not fully applicable. Therefore, we will have a more detailed look at some quantities related to antenna matching. Matching can be expressed by the VSWR (= voltage standing wave ratio) and by the return loss. Both are two different scales and terms to express the same. Both ways are practicable and mean the level of the reflected share of an incident wave with respect to the original level of the incoming wave. Please see the conversion in Table 3 with [3] as a source.

VSWR	return loss / dB	P_{REF} / P_{GEN} /dB	amount of the reflection factor $\|R\|$ $=U_{REF}/U_{GEN}$
1.00	50	-50	0.0032
1.02	40	-40	0.0100
1.06	30	-30	0.0316
1.1	25	-25	0.0562
1.2	20	-20	0.1000
1.4	15	-15	0.1778
1.9	10	-10	0.3162
3.6	5	-5	0.5623
5.8	3	-3	0.7079
8.7	2	-2	0.8913

Table 3: The voltage standing wave ratio (VSWR), return loss and amount of reflected energy or power depend closely on each other.

Matching and return loss always play an important role when transitions between different waveguides lead to a sudden change in the wave impedance of the waveguide.

In many circuits and applications, we can observe that a return loss of 20 dB in the used frequency range provides good results. That means a reflected amount of 1% of 100% incident power. There might be some special applications where the return loss in a waveguide system or cable shall be better than 20 dB, but in most scenarios, this value is acceptable. Power is almost reflected unintendedly in transitions and jackets with sudden deviations from the wave impedance. Open gates in a circuit can reflect 100% of the power running from the circuit to the gate. The issue is that electromagnetic waves don't behave like DC. Waves are almost guided by a waveguide like a cable with a defined wave impedance. Transitions have always had to be analysed or focused, because they sometimes mean a change in wave impedance. When this change is huge and sudden, then power will be reflected and may cause interference in a signal, in a circuit, or even in a destroyed stage because reflected power can additionally heat up an amplifier stage of a transmitter, for example. The most evident negative example is to switch on the transmitter while the RF output socket is open, that is, without any antenna or dummy load connected. The dummy load is a discrete broadband resistance of 50 Ω which means a good termination of a 50 Ω line. It changes the RF power at the socket, i.e., for a transmission antenna, into heat. This avoids reflection of energy running back into the transmitter stage and their overeating.

Cables and Waveguides

To ensure safe operation of a system and correct system behaviour, we need to use cable types, strip lines, optical fibres, and waveguides with appropriate specifications. Their impact on system behaviour is often underestimated, especially when false types are used. For cables and connectors the most important electrical properties are

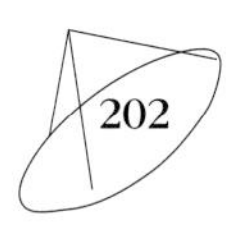

the specified frequency range and the loss over a certain length, as well as the return loss and the isolation. Other mechanical properties such as bending radius, dimensions, and fire resistance have to be considered, too.

To be honest, the list of requirements can be really long, even for a simple cable. Sometimes it is the weight that decides if a cable can be used or not. In the case of rectangular waveguides, it is also the lower cutoff frequency, which depends strictly on the dimensions of the waveguide.

Although cables and waveguides are boring topics for many people, this is a field where often a false decision is taken, sometimes with severe consequences. A lot of trouble can be avoided by looking at the specifications of components. Let us consider two examples: RG58 cable and PL plug.

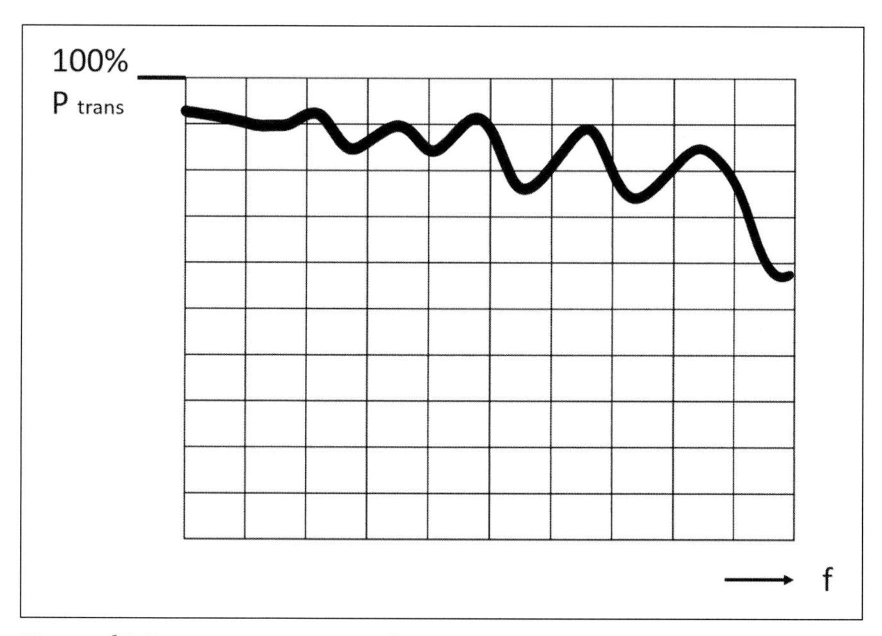

Figure 67: Frequency response of a PL plug on frequencies above 30 MHz. The quality of matching decreases significantly for higher frequencies. Reflections at the transition with PL plugs, or sockets occur.

Those who intend to use a RG58 cable should consider the frequencies used and the attenuation of the planned cable length. RG58 is lossy even in the VHF- and UHF bands, and with some metres it is mostly not a good idea to waste large amounts of power. That means, in detail, a loss of: 5 dB per 100 m on 10 MHz and 55 dB per 100 m on 1 GHz. Aircell 7 is a cable with better quality. On 144 MHz the loss is 7 dB and on 1 GHz the loss is 20 dB, related to a length of 100 m. The isolation against interfering fields outside the cables is 85 dB for a low-cost cable and more than 100 dB for those with better quality.

Optical fibres have a typical loss of 1 dB/km. Due to the permittivity of the material, light will run at a reduced speed through the optical fibre, i.e., at 230,000 km/s instead of the speed of light. For coaxial cables, the relations are similar and the signal speed through a cable will also remain below the speed of light.

Plugs and Adapters

It is almost a bad idea to ignore the importance of a plug type. PL-plugs are very popular, but their performance is rather limited, especially when they are used for higher frequencies. Above 30 MHz this plug becomes more and more a step (a sudden change) in wave impedance, with all its consequences. In VHF applications we have a point of reflection just where the PL-plug is located. To avoid unneeded attenuation and reflections in the system, you should make another choice.

N-plugs can work at higher frequencies, and they are robust even in a challenging mechanical environment. They are available for different cable diametres, and they cover a good share of the GHz range. SMA plugs are smaller and more filigree with small diameters. We use them in GHz frequency bands. TV sets and receiving components require F-plugs. Frequencies up to 2 GHz are no problem for this 75 Ω type. BNC plugs are good for VHF, but for UHF the performance

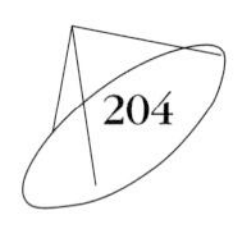

comes to the limit. All in all, we can say that it is always beneficial to use the right plug or adapter because the wrong choice means attenuation, loss, and reflections, and even interference in a system when we don't take care. To be honest, some more types exist, and they all can be combined with the right adapter. Numerous models for all applications can be purchased. It is better to take care while choosing the right components for a project than to waste a lot of time searching for errors at the end.

Practical Example Planar Antenna

The previous chapters were almost theoretical, but they have a close relation to practical work. The following example of an X-band receiving antenna shall prove this. It was designed for the reception of TV signals in the 12 GHz band.

For receiving satellite signals with TV programmes we almost always use parabolic reflector antennas. As an alternative, flat antennas have been designed, and we will have a closer look at one prototype now. An array with 64 radiator elements shall be the alternative to a small parabolic reflector antenna. Each single antenna is connected to a planar feeding system. All this is integrated into a flat, parallel plate case. The result is a flat antenna with a high gain. The feeding system consists of strip lines as well as the radiators. There are various group antennas, once designed for linear polarisation. Some appropriate stripline antennas are designed for excitation by two orthogonal polarised signals. When a second feeding network is integrated, which connects to the same radiators, the waveguide systems and the whole antenna can work with two polarisations. This is appreciated for TV reception of programmes at both polarisation levels.

Radiators for GHz-Frequencies (X-Band)

For orthogonal polarisation, square loops or rings as radiators are a good choice. They have two planes of symmetry, and this makes them proper candidates. Square loops radiate with high intensity due to the four corners.

All this will provide advantages, as we can see later. Figure 68 shows 4 radiation elements, which are connected internally. This is done by trapezoidal line elements with good matching. These short stripline elements connect the radiators internally as a group. The lines have a wave impedance of 50 Ω. The narrow lines have been designed for 100 Ω. This group of four radiators is connected to the feeding network from two sides, also visible in Figure 68.

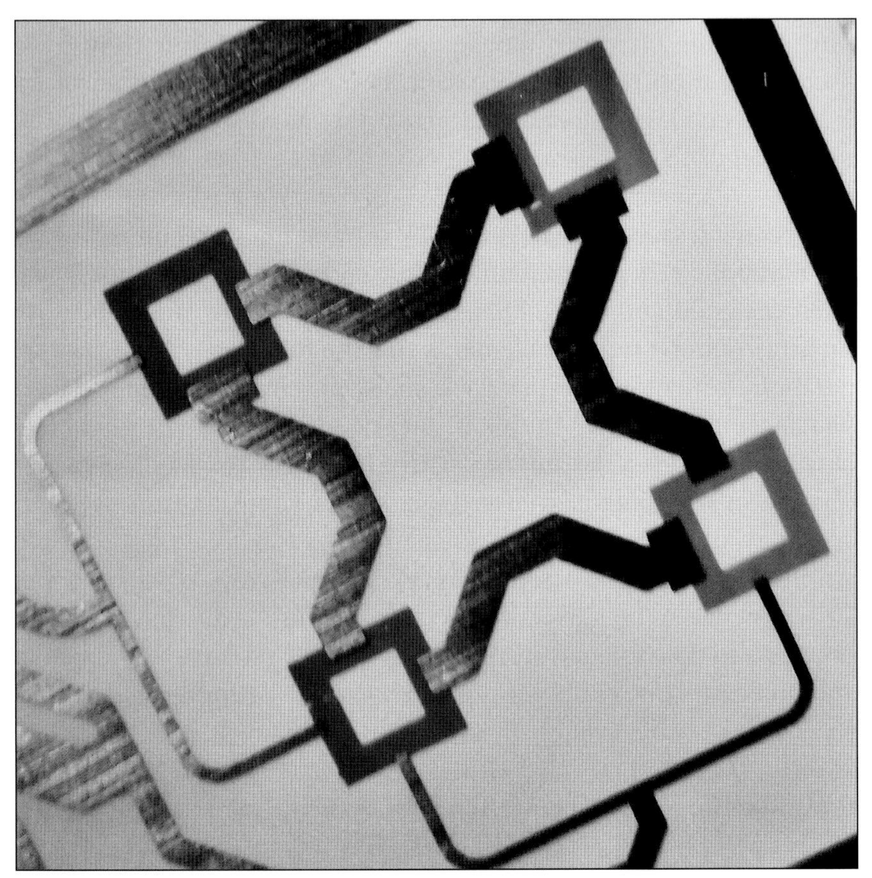

Figure 68: Four radiating elements of a flat antenna for 12 GHz, designed as square loops for double orthogonal polarisation.

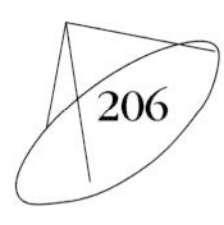

The radiators in this group are excited by a feeding network, which is a kind of double feed system. Two separate networks on the same slide are used for exciting the radiators from two sides independently with an angular relation of 90° (from two sides). Therefore, we need radiators that can be fed from two orthogonal sides.

The wave impedance of the feeding lines coupled to the elements is higher than the lines for internal coupling between the radiators. The feed system has no galvanic coupling to the radiator groups. As the slides have copper layers on both sides, we can use a capacitive coupling between the radiators and feed system, which provides the required coupling impedance for a good match.

Feed System and Coupling

Strip lines are appropriate waveguides for 10 GHz or more. When used on a thin slide with copper lines which is placed between two metal plates, they will have a low loss.

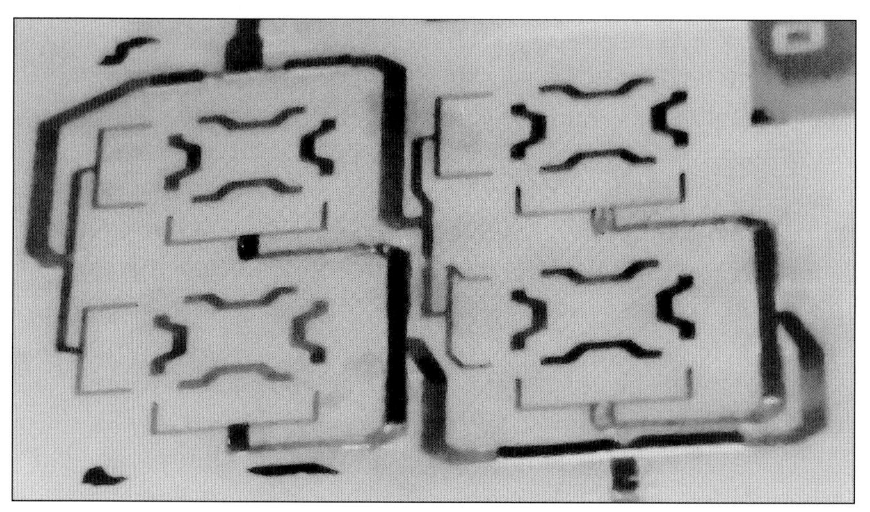

Figure 69: Two feeding networks for orthogonal radiation or excitation of two independent signals, one signal assigned to each polarisation. Square loops can manage both signals simultaneously and independently if the polarisations are orthogonal.

The substrate layer made of foam with a large share of air inside will support the low-loss concept for the stripline even over 10 GHz.

Polyethylene foils with copper layers on both sides are the ideal solution to feed the radiators and to find a good coupling while matching the stripline to the radiator. Capacitive matching is more flexible because it enables us to tune the coupling by optimising the overlapping zones of radiators and striplines. See Figure 69.

Linear Polarisation

Most antenna types that we find are derived from the very few basic types, like the linear dipole. And most of the derivatives use linear polarisation, too. This is also typical for stripline dipoles. In the case of square loop radiators their polarisation depends on the point where the feedline is coupled to the radiator. When the radiator works with one feed, then we have linear polarisation as a result. This is valid in receiving mode as well as in transmitting mode, what we will consider here only as a theoretical option. As we can use the second feeding point, too, which has an angle of 90° with respect to the first feeding point, we can even use two polarisation planes. That means a single radiator can receive TV programmes with horizontal and with vertical polarisation. There is a good isolation between both signals, what enables good reception in both planes without neighbouring channel interference. We will consider now incoming waves with TV signals, which have to find their way to the low noise converter.

As the antenna is reciprocal, it is allowed to consider both directions, receiving and transmitting, with the same validity. As we talk about TV reception, we look at two incoming waves with orthogonal polarisation. Incoming waves can excite the radiator depending on their polarisation plane. The angle of incidence will also be responsible for the coupling into both feeding networks. When the antenna is aligned properly with respect to the satellite signals and their po-

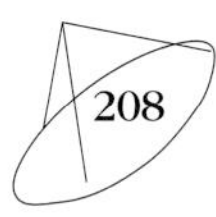

larisations, then h- and v-polarisation will feed energy into both networks, each plane into their assigned network. We will see later that the antenna can also be used for circular polarisation when we combine it with an additional grid.

Circular Polarisation

We have just discussed the use case of two linear orthogonal polarisations with one radiator type and two independent feed networks. But how can we receive circularly polarised waves? Once in the 1990s, the satellite TV-SAT transmitted on 12 GHz with a circular polarisation. One method could be to use both networks and modify them slightly to add their signals with a phase difference of 90°. But there is another smart method to receive circularly polarised signals by directing each rotating E-vector (the left and right circular) of a wave to their assigned network.

This avoids any change at the network level. We simply use a grid as an overlay plane before the waves meet the radiators. The grid shall be placed above the radiator plane on a substrate and consist of parallel stripes of metal on a stable pcb layer, for example. When we choose the right distance of a few millimetres between radiator level and grid layer, then we will observe the following effect with the 45°-grid (Figure 70 and Figure 71):

A wave with a rotating E-field-vector (i.e. left hand rotation) will pass the grid when this vector is perpendicular to the grid stripes at timing point t1. The wave moves on, and the electric field vector rotates exactly 45° until timing point t2. Then the E-vector meets exactly one of both polarisation planes of the radiation elements.

And the E-vector excites a current in one of both polarisations and couples into only one of both networks. We call it network 1. The grid‘s distance from the radiator plane is responsible meeting exactly the rotating angle of 45°.

Right-hand rotating waves do the same. But their E-field vector is turning in the opposite direction.The wave passes the grid at timing point t3, turns the E-field vector by -45° and meets the radiator element at timing point t4. As the direction of the E-field vector is perpendicular to the left turning wave, the energy of the right turning wave excites just the orthogonal plane of the radiation element.

By this the wave couples into network 2. When E-field vectors have a parallel alignment to the grid, they will be reflected completely. Consequently, the left-turning wave cannot couple into network 2 and network 1 remains locked for the right-turning wave.

Antenna, Converter and Receiver

We consider a number of 64 radiators, which couple their energy into the feeding system. By this, the antenna accumulates a huge

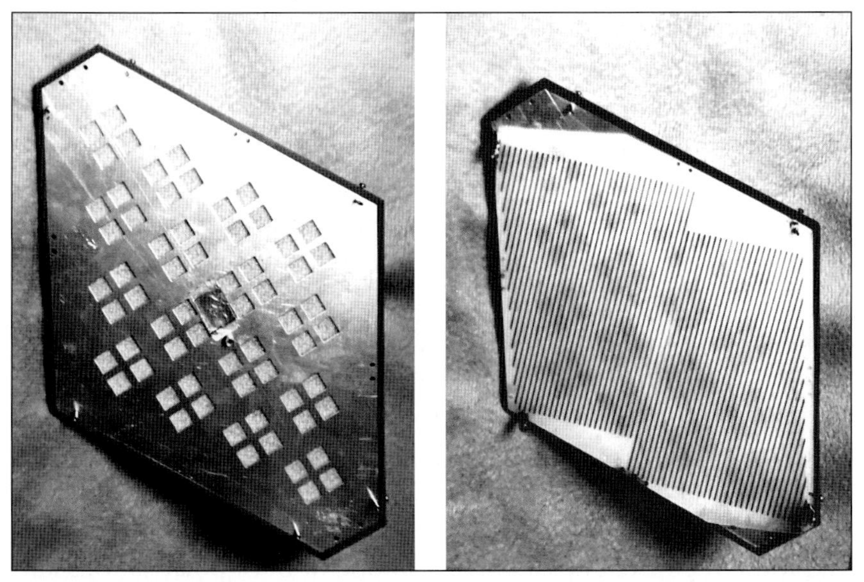

Figure 70: Under each small window one radiator element is placed on a slide with the feeding networks. The supporting foam layer can be seen below the metal plate.The grid on the right side is mounted above and enables circular polarisation.

amount of energy that will result in a high gain. Compared to a parabolic antenna with the same aperture, the gain of the flat antenna will be slightly below the gain of the reflector antenna.

What is the reason? The feeding network of the planar antenna is not completely free of loss. The efficiency is generally a little lower when we compare both antenna types. But with respect to the great number of 64 single antennas couple their energy into a waveguide with a circular profile, the loss is still amazingly low. A rough estimation of the gain is 20 dBi minus a little loss caused by the networks and their transition.

Figure 71: Planar antenna: Right down, we can see the LNB and the short circular waveguide. The feed networks of the antenna excite the H_{10} mode in both planes, and so both signals (h+v) can be coupled to the LNB.

When the received signal's energy is fed completely into the stripline network, it will be conducted to the coupling slots in the waveguide, with a circular profile. From the waveguide it is only a short distance of a few millimetres to the converter's coupling elements, similar to monopoles on a PCB. The converter has a flange for circular waveguides. This is required because we are using both linear polarisation planes in the waveguide. How can the waves enter the waveguide? The feeding stripline network has four partitions (subnetworks), two for each polarisation.

The waveguide connected to the LNB has four little windows for each end of the stripline subnetworks. The ends of the lines protrude for a few millimetres into the waveguide. This will excite two linear orthogonal waves (H_{10}-mode) in the waveguide.

The prototype of this antenna provided clear pictures on a TV screen coming from the geostationary TV-1 satellite which transmitted programmes at 12 GHz using the D2-MAC standard. Additionally needed were a low-noise converter and a D2-MAC receiver, which were connected by a 75 Ω cable with F-connectors.

Simulation Models

In the development process of a new product, it is not useful and almost impossible to build a huge number of prototypes for each use case. In the case of a telecommunications network, nobody would build test sites for a whole network. The alternative is a simulation program which gives a good preview of what has to be built, considering all required parameters.

When a radio path is analysed, the required core algorithm for predictions is an enhanced propagation model that is based on free space propagation. Additionally the model considers topographic parameters as well as parameters about buildings and forests. These terms enhance the basic model of free-space wave propagation.

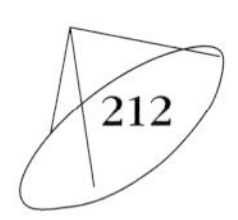

Models are used for predictions of hundreds or thousands of sites, calculated within one cycle. Their results are assumed to be close to the real signal coverage. The software considers different shares of loss which are added, starting with the loss, of free space. It also considers available data such as height over ground, type and height of buildings, and much more. One of the early models was the Walfish-Ikegami-Model, which has been substituted by more sophisticated models, which meanwhile consider a lot more impacts with much more precision.

Other types of mathematical models are used for antenna simulations with their current distribution and the fields around them. Precise calculations have been done for waveguide filter development. Predictive calculations are possible with an accuracy of 1/10 or 1/20 decibel for transmission loss. The methods used for simulations are various. Some simulations use the finite-element method, others are based on the integral equation method. Depending on the scenario, the selection of the right method means more or less effort, more or less efficiency, as well as accuracy. We can see that the use of mathematical models for simulation requires a certain amount of effort starting with selecting the appropriate model. The precise processing of each step ensures the required accuracy and quality of results.

Chapter 3: Users and Systems

This exciting chapter contains a lot about systems with functions that are based on wave propagation. We also talk about people, that means, who is using those systems? A complete overview of all areas and applications would be too challenging due to the number of systems. But as we will see, there are diverse and exciting examples we can take into our focus. Applications are numerous, and they can be found on earth and in space, i.e. in the field of radio astronomy.

Fields of Application

It is always amazing to listen to systems and to watch them work with all their features, which are used all day and night for signal transmission. Somewhere in the city, a taxi can be waiting deep in the night for its last order coming from the dispatcher with a selective four- or five-tone selcall. Most of us still know this melodic sound of the selcall, which is also used for buses and trams. The good old push-to-talk radio systems provided easy handling and robust operation to do all their coordination with the vehicles of the fleet. How else could they quickly ask when a traffic jam will be over and what the alternative route is? How else could the dispatchers quickly inform customers about breakdowns or outages?

Every minute, thousands of aircraft are on their way through the stratosphere, and most of them are connected on VHF to their control stations. Flying over the ocean, they had to change to shortwave or SAT because of the limited range of VHF radio systems. Localising by radar is also an application based on electromagnetic waves as well as the exchange of data telegrams transmitted by transponders to identify airliners and provide further data such as height, heading, and more.

Without radio wave propagation, space travel and the use of satellites and probes would not be possible. Even if there is no medium in

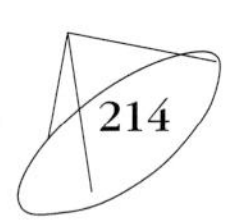

space, it works. In space, wave propagation is possible for electromagnetic waves, but not for acoustic waves. Rain and fog in the case of a laser beam transmission line would restrict the application very soon. Attenuation and refraction would be a showstopper for laser applications after some hundred metres.

Now we come to a quite different example: When you cross the exit of a store with an RFID chip that is not released, you might hear a loud alarm signal. This will identify people who did not pass the cash register where the chip is switched to silence mode when you have paid. In active mode, the chip is detected by an electromagnetic field, which reads out the information on RFID chips, i.e., that this article has not been paid. Then the siren starts.

Many more user groups can be listed that use functionalities based on electromagnetic waves. A very prominent group are radio amateurs, who can be found all over the world. They were once the real pioneers in showing the amazing options that amateur radio can provide to enthusiasts and also to professional radio. They have a great knowledge about radio systems and their operation. Commercial users can be found all over the world, too.

Everybody knows the police departments and fire brigades, as well as ambulances, who save lives, extinguish fire, and enforce law and order every day. They are all dependent on appropriate radio systems, which are the backbone of communication all day and night.

Remote sensing is a discipline in space exploration that is also strongly dependent on radio wave propagation. Localising by radar and medical technology, especially imaging methods in medicine, have become hugely beneficial areas for us. Contactless examinations and analyses of human bodies and tumours are possible now. Electromagnetic waves penetrate into human organs and bones to identify diseases without cutting with a knife to open the bodies. The development of technologies and methods, that use refraction effects with the help of electromagnetic and magnetic fields, have made really good progress. Not only human bodies are examined, but also comets are explored by transmitting radar waves on them and by analysing transmission and response patterns.

New medical technologies are the best examples, with enormous benefits for patients. When a doctor wants to look into our body, he can choose between different methods such as x-rays, Computer Tomography (CT), which is also based on x-rays and MRT (Magnetic Resonance Imaging or Tomography). MRT uses magnetic fields in combination with electromagnetic waves in the range of 64 MHz to 700 MHz. All three methods enable doctors to identify pathological changes in the human body. Many lives could be saved by an early identification of tumours and other severe diseases. Without those modern imaging methods, medical diagnostics would not get the required density of information for the doctor. Without that, they had to take the knife more often and an operation would be more risky. Without imaging methods, many diseases would remain unidentified and the death rate would increase rapidly. This exciting chapter deserves to be studied with more literature. Our short overview can only be an inspiration to go deeper into the stuff.

Radio and TV

In the last century, it was a basic requirement, that radio and TV-coverage be ensured all over the country by terrestrial systems and an effective grid of transmitters. The sites were the result of good radio planning, a method used for radio coverage as well as for public telephone signal coverage.

Even the government's professional radio planning for TETRA requires sophisticated methods and tools to provide good signal coverage. During the first decades, transmitter sites for radio and TV-coverage were limited to a few towers with a good height above normal (Figure 29). Those sites were placed on hills and mountains

or they were at least installed on a big television tower. Some examples in Germany: Zugspitze, Hornisgrinde, Nordhelle, Feldberg, Olympiaturm München, Berliner Fernsehturm (Telespargel), telecommunication towers as in Wesel, Köln, Schwerte, Hamburg, Hannover, and some more as well as television towers as in Stuttgart and Dortmund. To be honest, a lot of smaller transmitters had to be added to optimise reception all over the country, especially in areas with poor signals.

Public Mobile Networks

Operators of public mobile networks need a gigantic number of sites for their transmission equipment to provide their service all over the country. We talk about thousands or even ten thousands of sites operated by one single operator. Some of the sites are used for special tasks, such as microwave link collectors. Others are simple radio sites for exchanging telephone signals with subscribers. Some special sites are operational centres used for monitoring network functions in a whole area and giving alerts in case of malfunctions of any network element (Figure 72). Some radio sites provide space that is shared by two or three operators (Figure 73).

Network infrastructure is rather complex and has different levels. At the beginning, radio planners define hundreds of possible locations, that can be used for radio coverage of phone signals. This is done with special simulation programmes and very fast computers. The next step is to connect the sites by a radio link or by wire (telecommunication line).

By the way, radio planners and radio link planners are very interesting jobs, that require some experience. When a microwave link cannot be realised, i.e., on sites without line of sight, the provider has to book a line from the network of a local telecommunications provider. The radio links can be part of a star network or part of a ring network. A whole area requires additional radio link collector sites and

Figure 72: Radio tower of an operator for mobile services. We can recognise different antennas. The tower is also used as a microwave node with antennas for the transport network (upper area). Between the upper and lower platforms there are some panel antennas for GSM-1800 mounted.

nodes for the network level above. This is a brief description of how the planning of a public phone network is done on different levels. There are a lot of additional elements and steps that play a role in the development of the network, such as the operational and maintenance centres where the status of all elements is monitored. For a closer look at all these topics, the interested reader needs additional literature.

Some central nodes of the network are combined sites for radio links, radio coverage, and monitoring the network with all its elements and their status. The staff in operational centres monitors outages and coordinates maintenance activities. The crew is responsible for network operation without interference all day and night. In the case of an alert, they will send a team to the interfered site to solve the issues.

Professional Radio

Still today, many companies and institutions use analogue professional radio on VHF and UHF like in the last century. For many decades, it has been state-of-the-art wireless communication for their fleets of taxis, lorries, trams and buses. Even the staff of power plants and highway maintenance departments used this simple way of talking to the mobile stations. Communication did not require a network. In the central office, the coordinator talked with drivers of cars via radio sets, which have always been easy to handle. The antennas were almost placed on top of buildings, see Figure 27.

This kind of communication is still used today, even if it is an old but mature way of radio communication. Frequencies for voice communication can be found at 148 to 152 MHz, 154 to 156 MHz, and on UHF at 455 MHz. Some of the above mentioned user groups have changed to public mobile networks and mobile phones. Nevertheless, there are still clear reasons why professional FM-radio is still in use. It is very easy to operate, you have only to push and talk (PTT,

Figure 73: Left hand, broadcasting antennas for public mobile radio (LTE, 5G, ..). On the top, we can see the lightning protection. Right hand: Two omnidirectional antennas of the TETRA network for police radio. In the panel antennas, the receiving path is improved by polarisation diversity. The TETRA installation gets this improvement in the receiving path (uplink) by using two different antennas with a certain spacing and some kind of space diversity (SDMA).

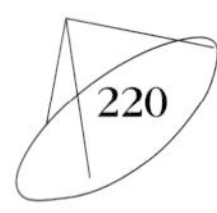

push to talk) without complex configuration steps. The other reason is that the required equipment is cheap, easy to handle, and easy to install.

Later efforts by industry and authorities to install a new radio system were dealing with trunked radio according to MPT 1327 as a standard. This worked for a while, but only for a selected circle of companies. Trunked radio has some advantages, like enhanced coverage by additional sites and automatic coordination of frequency use. But the operation of the system was more complex and required a network provider, who had to care for components and react in case of outages. This implementation was not as easy as simple analogue radio, and it was more expensive because any provider had to be paid for their maintenance service by the network users.

Digital Radio

For decades, analogue radio systems have been a good choice for professional radio communication. In the 1970s wireless phones were in use at home and for mobile operations. They had some disadvantages with respect to modern digital systems. Spectral efficiency was poor, power consumption was high, and there was no protection against unwanted eavesdropping.

Digital radio could help to avoid the important disadvantages of analogue radio and to meet the needed requirements indicated above. Digital technology is sophisticated and mature, and it opens new doors. We can see some clear advantages: Radio signals are less noisy than analogue signals, and the potential interference is generally reduced. The voice quality is much better and more clear. But to be honest, voice quality is also a matter of individual impression. Codecs are used to transform speech into digital signals, and hence, codecs have a great impact on speech quality. But nevertheless, it is still important to select the proper volume setting with a good microphone. Digital radio has the option to communicate in direct mode, which means from one user to the other without any base station. But in general, the applied mode is communication over a network similar to public mobile radio.

Examples: DMR is a professional digital system that is also used by radio amateurs. They operate their own networks. DMR uses 4FSK modulation with a data rate of 9.6 kbit/s (Figure 88). This is a good match for a 12.5 kHz grid. DMR is a standard in many amateur radio bands. D-STAR as an alternative, is also a popular digital mode for voice transmission. D-STAR uses a bandwidth of 6.25 kHz and transmits 4800 kbit/s. A good share of 2400 kbit/s is needed for FEC (forward error correction) and data transmission. The share for voice transmission in D-STAR is 2400 kbit/s. As a modulation type D-STAR uses GMSK (Gauss minimum shift keying). This modulation type works with a special low-pass filter to reduce the slope to a less steep signal curve. That means a flattening of the steep characteristics of rectangles in digital data. By this method, high frequencies are cut off what reduces signal bandwidth significantly. The direct mode can be used in DMR and D-STAR radio as well as in a TETRA system. But the great benefit of using a device in a whole network is that it provides some additional features. However, a whole network also needs additional maintenance and an additional transport network to connect the base stations. All this needs a lot of effort in the configuration of devices and network elements. Latency almost raises new issues that have to be considered.

Digital radio in a network environment has become the new state of the art in professional mobile radio, especially in governmental applications. TETRA for example, is used by police, fire brigades, and ambulance vehicles, as well as in critical and emergency environments in many countries. New features like communicating in talk groups with users in large regions or even in the whole country are possible. PTT-communication between different authorities provides a great benefit, for example, in cases of life-threatening disaster scenarios. Talk groups can easily be created by this PTT operation

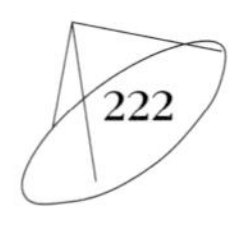

and can be extended over large regions, even all over the country. Local police stations can directly contact other authorities needed in emergency scenarios. In the past, such flexible communication in subnetworks with users like fire fighters, policemen, and ambulance vehicles was much more complicated or impossible in real-world scenarios.TETRA infrastructure and other similar technologies remind me of communication in public mobile networks for cell phones.At least, it has a similar structure of network levels and network elements. The operational characteristics of professional mobile radio networks also have their own features. Eavesdropping security and the use of secure keys to protect voice data streams are two of the new technologies that have not been available in the old analogue world.The general level structure of a professional mobile radio network also has a level with base stations.

On another level, they are connected by transmission lines consisting of microwave lines and a backbone. Operational centres at the third level, manage all these components and use special network nodes with fast computers to manage users, talk groups, and network elements. All this is very similar to the operational characteristics of a network for public mobile radio and cell phones. Also, complex configuration processes for network elements, handheld devices, and devices in vehicles need sophisticated organisation within each level.

When we compare both public mobile radio and professional radio networks, we will also find some serious differences. We already mentioned the feature of using direct mode, which is not applicable on public phone networks. Generally, any user device needs base stations and a network infrastructure with a provider behind it. One of the most evident operational differences is the PTT feature. PTT (push to talk) is generally not needed in public phone networks because users can talk in parallel in both directions. Both transmission lines (uplink and downlink) work in parallel at the same time for each user throughout a call. PTT is another concept, almost used by professionals. Special GSM networks as the German GSM-R (R means railway) integrated this feature too. While using PTT you can contact selected single users or a group of users immediately with one keystroke. The channel is then switched from the sender to the recipient through the different network levels, even if they are located far away over hundreds of kilometres. Even the early walkie-talkies in the 70s had this function. It was funny and exciting to push a key and talk to somebody else who was 200 m away and tell him a story. To listen to the other one, it was surely important to release the key again. This seems like a simple step, but when we look at the functions of a professional radio system, it is a complex switching process while using PTT. Pushing the key means that the radio system starts to connect the users within milliseconds. The switching process includes many steps at different network levels to ensure the correct routing. Base stations, microwave links, user registers, and controllers make sure that the selected users can receive a signal immediately. This works within a city and all over the country, too. It indeed needs a good radio coverage, provided by base stations in each region. In the case of disasters like floods and earthquakes, those networks with their features can help a lot to save lives.

Cell phones, however, do not need a PTT. For a call, both users are connected after dialling a number over all system levels for the whole duration of the call. Uplink, the line from the user to a base station and downlink, the line from the base station to the user, are available only for the duration of the call. The switching process starts at the beginning of the call while dialling. During the call, base station antennas and mobile phone antennas transmit and receive at the same time. The frequency bands for both receiving and transmitting are separated. The switching process within the whole network has a similar complexity as in professional systems like TETRA. Both use separated frequency bands for uplink and downlink. TETRAPOL, APCO25, DMR, and TETRA are typical candidates for mature professional radio network technology. GSM, UMTS, LTE, and 5G are used in public cell phone networks.

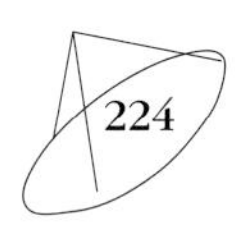

Transportation

Radio systems are the backbone for communication from the office to mobile users in vehicles, vessels, and aircraft. It is extremely important to address a user immediately.The option to contact a person or an instance without delay anywhere in the world, 24/7, is essential for logistic processes. Many different types of communication systems and applications running in the networks provide a mature portfolio of devices to handle worldwide communication. Satellites, network hardware, and software applications were designed for various user scenarios to provide 24/7 availability. And the demand keeps increasing.

Shipping

For more than 100 years, radio systems have provided a great benefit to shipping. At the beginning simple transmitters were used, similar to those constructed by Marconi. When the sinking of the great ocean liner Titanic happened, the meaning of wireless communication became obvious to everybody in the world. Already a few years before that disaster, it was one of Guglielmo Marconi's transmission devices that saved many lives in 1899 when in the English Channel the first emergency call could be transmitted for saving lives from a sinking ship. It was like an initial wake-up call for the shipping companies to make communication systems a standard piece of equipment on ships.

Today, many communication systems and radar equipment are used on ships all around the world. The meaning of electromagnetic waves in an emergency and also for standard navigation has become obvious. Inland shipping uses special VHF-channels on 156 MHz for voice radio. Captains and helmsmen on tankers or container ships need them to contact other ships in case of bad weather, for example, or simply for talking on a long journey and share personal experience. Important warnings can be forwarded immediately to avoid further damage to the route in case of accidents. But there is also official broadcasting available along the rivers. Navigational warnings, water levels, and other news for shipping are transmitted regularly for all ships, still in analogue FM mode. Even phone traffic can be switched by using pairs of analogue duplex channels for uplink and downlink. Captains can make phone calls even from their ships. This helpful system is still in use and works on an analogue base. Official shipping bureaus are operated by governmental authorities.

Several systems for maritime use are helpful or even life-saving. Many of them are based on electromagnetic waves: communication systems (terrestrial and satellite-based), emergency systems, radar systems, and monitoring systems. Some of them observe the vicinity of a vessel, others provide voice communication channels, and a new system category provides data like the current coordinates of a vessel in the context of a global survey to a worldwide network.

Today, AIS (automatic identification system) is a popular tool for visualisation of ship movements all over the world along inland routes. Since the year 2000, digital data like speed, coordinates, heading, and draft, for example, have been provided for global monitoring. The availability of data is ensured by the transmissions of ships when they pass dedicated base stations. They broadcast data telegrams on 161.975 MHz and 162.025 MHz to the base stations and to ships in their vicinity. This is a valuable contribution to safe shipping.

Another system that supports safe transportation on shipping routes are radar systems. They transmit short microwave pulses, which are reflected by objects in their vicinity. We all know those rotating bars on ships and towers. These microwave antennas work similar to slotted waveguides and have a narrow beam in the horizontal plane and a wide beam in the vertical plane resulting from the physical extension of the antenna. The results are pairs of data with angular information and the reflected radar signature. They can be monitored on a screen. In the case of bad weather with fog and rain, the crew on ships can watch what happens around their own ship.

But how do radio communication systems work on the open sea far out. There are no radio towers or installed stations. The answer is simple. For many decades, we have had short-wave communication systems for worldwide voice and data traffic. Sailors once used shortwave radio not only for listening to radio broadcasts but also for telephone calls with their families. Christmas greetings from South America to friends at home or business telex traffic were usable on a ship. For data transmission, CW (continuous wave or Morse code) was a powerful mode to transmit with an extreme narrow bandwidth. Radio operators on ships and at maritime radio stations had to be familiar with the Morse operation mode. What was done manually in the 1950s, could later be solved automatically by electronic devices and, at last, by computers. It was not only CW or Morse, but also FSK or the Baudot telex mode, which ensured robust radio data links. Later in the 1990s the data transmission mode SITOR with duplex channels provided error correction by a special protocol with ARQ, the automatic repeat request. Hundreds of data transmissions rushed over the oceans at the same time in any region of the world.

Meanwhile, technologies have changed. When we travel on a ship, we use almost satellite systems for voice and data transmission and even for the internet. In coastal areas, a ship‘s crew can still use the public phone coverage of the network providers, but far out, there is no LTE signal and no tower with radio communication antennas. Nevertheless, there is a need for safety systems. They shall send out signals in case of emergency. Those signals must be identified in real time, no matter in which part of the world the emergency call is transmitted. GMDSS (global maritime distress and safety system) is that type of safety system for maritime use. It integrates different components under one roof like AIS, DSC, NAVTEX, the COSPAS-SARSAT system, and radio beacons. GMDSS rules and requirements have been committed globally by the countries' authorities using GMDSS. What makes overseas shipping safe, is the option to send out emergency calls immediately, voice calls, and data messages. GMDSS uses standardised DSC data telegrams for that. In coastal areas, ships use NAVTEX transmissions from special sites in different countries. NAVTEX works on medium waves (518 kHz) with data streams using FEC (forward error correction) with 100 Baud FSK (frequency shift keying). Every station transmits in their own time slots of several minutes to avoid interference of signals. Nautical warnings, weather station reports, and weather forecasts are provided for oversea shipping, see Figure 89 and Figure 90. Meanwhile, telegrams and phone traffic are sent over global links, i.e., over the geostationary satellites of INMARSAT which operates several satellites in an orbit 36,000 km above the equator. Ship positions and coordinates are used by AIS on all oceans. Data from ships is continuously scanned by AIS satellites and provided for visualisation. ORBCOMM satellites scan with special methods for detecting data packets, like SDPOB (spectrum decollision processing on-board). The uplink and downlink of the ORBCOMM satellites are located in the VHF band of 137-153 MHz. SARSAT satellites receive worldwide signals from emergency beacons (EPIRB = emergency position-indicating radio beacon) on the international emergency frequency of 406 MHz. Those orbiting satellites use polar orbits to cover any place in the world. Thousands of lives of people in distress, for example, on sinking ships could be saved by that system.

WINLINK is a system for skippers who want to use email on the open sea. WINLINK combines email with modern amateur radio operation modes by using amateur radio transceivers.

VARA can send emails with a TNC interface. This terminal node controller connects the computer with a transceiver. VARA uses OFDM as an operation mode for different carriers with phase modulation. 5 kbit/s of data can be transmitted in a channel bandwidth of 2.3 kHz. VARA FM even works on air channels with a bandwidth of

25 kHz. As an alternative, you can use the VARA SAT which can work over QO-100 as a radio amateur satellite. By using an ARQ protocol, even error correction is ensured.

ARDOP (amateur radio digital open protocol) is another option in addition to VARA. ARDOP can be used without any expensive TNC and needs a 2 kHz bandwidth for transmitting with data rates of 2300 kbit/s. The data streams are modulated by 4PSK, 8PSK, 4FSK, 8FSK, and 16QAM. This cannot achieve data rates of 5.5 kbit/s like PACTOR-IV, but you can save the costs for a TNC.

Aviation

Radio systems for aviation are, of course, based on the use of electromagnetic waves, and they are essential systems for safety. Everywhere we meet those systems, some for voice communication from aircrafts, others for transmitting data during a flight, and also others for identifying objects in the air or on the ground by radar waves. Modern aircrafts are full of systems that a pilot has to operate with his special knowledge. The whole coordination of traffic on all flight levels in the lower and upper airspace is based on monitoring with radar. Official staff in the air control centres as a part of the national flight authorities, is responsible for any movement in the air. Air traffic controllers are a skilled species, professionally educated, and highly concentrated in their work. They use, for example, voice speech in an extremely disciplined and effective way while handling air traffic within their sectors. All this will help ensure that anything in aviation is based on safe processes. This will protect the lives of people on board. During a flight, passengers can enjoy comprehensive and individual service, which meanwhile even includes internet on board.

But electromagnetic waves also carry signals to ensure the proper operation of safety systems in different frequency bands. Navigational data comes from satellites; GPS is an example. But traditional methods are also required. Radio communication is still analogue and uses the AM mode and SSB (single side band) and even direction-finding systems once and now still use radio waves. Some of the systems are really complex, and they are designed to do their job even under adverse conditions. Safety is a major topic within each technical system. They have all been designed and tested by the best engineers and scientists.

Over the oceans, however, VHF radio communication channels will remain silent even if transmissions from aircrafts can be heard over more than 300 km on the ground due to their high flight level. In airspace over oceans, position reports can be broadcast over FANS (future air navigation system). These data telegrams will be detected by satellites, such as INMARSAT and IRIDIUM. As a backup system, pilots can still operate shortwave radio, sometimes crackling and almost a little bit noisy.

Dedicated shortwave channels for USB mode are continuously monitored all over the world from stations like Shannon Radio and Gander Radio, covering the North Atlantic. They use a two-tone selcall to contact aircrafts, i.e., on North Atlantic routes. This has been the backbone of any overseas communication since long before satellites were available. The broadcast of data telegrams is also needed on continental air traffic routes. ADS (automatic dependant surveillance) is a system operated on 1090 MHz to provide flight data such as height over ground, heading, speed, and more. Passengers, meanwhile, use the internet over satellite links provided by the airlines via WLAN even on a flight.

Railway

Railway systems have been optimised for the transportation of people and freight, like containers for example. Just here we find further communication systems and applications, operated by engine drivers and shunt personnel. Many participating instances need to be con-

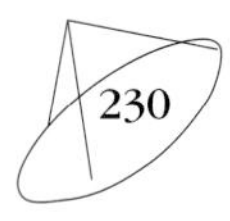

Figure 74:
Airliners use various antennas and sensors, i.e., for voice radio on VHF- and UHF-channels to provide transponder data, weather radar information, and navigational information.

nected day by day on express trains and in freight trains. They all need safe communication lines on their routes and also in train stations. In the big freight yards, logistic processes have to be supported by radio communication. In the last century voice traffic was transmitted with analogue radio using the 70 cm band (UHF), duplex and simplex. Even in modern freight yards, the good old analogue radio is still in use, some on 70 cm others on 2 m frequencies. When express trains are on their routes all over the country, railway needs a country wide network for voice and data communication. In Germany, it is the GSM-R system, a derivative of the ancient GSM with special features for railway use, indicated by "R".

One required feature is the PTT function. When a user in the system is selected, communication devices can be used similarly to analogue radio, simply by "push to talk". GSM-R networks have an implementation of that feature, which belongs generally to professional radio systems. In those modern networks a lot of additional features, like the use of talk groups and immediate transmission without dialling, are available. Network structures are similar to those of public cell phone networks. One level consists of base stations, which are connected by a sub network with microwave links. A core network makes sure that all the traffic coming from the users via base stations, is routed to special controller stations (BSC = base station controller), which act as nodes on their own network level.

MSC (mobile switching centres) work like a big matrix for connecting all instances, which are stored in special registers. The GSM-R technology provides radio contact to all engine drivers and other personnel in trains, in railway stations, and in the control centres. Travellers on modern trains can use their cell phones. But the isolation of train windows with a thin metal cover causes a high attenuation of the signals in public cell phone networks. Therefore, railway providers offer wireless LAN in their express trains and sometimes even in local trains. Today, travelling and staying connected belong together in trains, on ships, and even in aircrafts.

Road Traffic

Since GPS information (global positioning system) is available in cars, everybody has the personal benefit of this huge progress in technology. It was a quantum leap in the navigational support of road traffic, wherever you are driving. The feature, a small display in a car with a map and one's own location on it, reminds me a little bit of the early James Bond movies about 60 years ago. Of course, small screens showing your own position in real time were really designed a long time ago. In the 1960s it was science fiction. Today, we all use this wonderful feature to find the way in regions far away from our own residence. Also in dense urban environments, it is a great benefit when a male or female voice tells us where to go, even if it is only a sampled voice. Meanwhile, we can monitor traffic jams in real time on our displays because many cars transmit their coordinates over public phone networks, while the data is also provided via the phone network for monitoring traffic data. But this is not the end of development. Today, driving assistance systems get their data from various sensors, such as radar, camera or even laser in high resolution and also in the infrared range. Meanwhile, all this is standard equipment of a modern car.

After 20 years in the 21st century, we face a new quality of system architecture installed in cars and for traffic support systems. 5G as an evolution of LTE will accelerate the networking and data exchange of traffic participants. Vehicles run with more sensors, which combine their signals in a complex system architecture. Vehicles and their subsystems interact with other vehicles and with fixed instances along routes. Also, in urban areas, we use a mobile technology based on real-time radio applications with enormous data rates.

But what is really new in 5G? This new generation of mobile radio can run applications with extremely low latency. This feature provides real-time control without delay between two or more devices. Driving vehicles with a remote control, for example, needs a real-time system to avoid delay and damage or even accidents. Without real-time control, any command for action needs several milliseconds or even more to be detected and then executed. This simple example shows that real-time operation is inevitable in many areas. The use of 3G or 4G cannot provide this feature, and that means collisions and other critical issues in applications. Future mobile networks need a design for significantly increased performance. Simple phone and data transmission are no longer state of the art even, if broadband data transmission is still at a very high level, and voice quality of mobile devices has increased a lot.

Signals from Space

Systems using electromagnetic waves do their jobs on earth and even in space. Engineers have built rockets, satellites, and space crafts like the ISS. All those vehicles need safe radio communication for voice, data, and control, like telemetry commands by various links and frequency bands. High-resolution cameras support exploration of distant planets with their orbiting moons and other strange objects like comets. But in outer space, we find sources that emit electromagnetic energy of a quite different spectral quality and in huge amounts that no one can even imagine.

Bright x-ray flashes and pulsed radiation with immense intensity come from strange galaxies. The sources are pulsars and quasars that emit strong electromagnetic radiation of natural origin. When scientists on earth detected phenomena like this, and our curiosity grew more and more over the centuries. But people on earth also receive signals from man-made flying objects in space, which were started to explore space years ago. Space probes and satellites transmit huge amounts of scientific data to their ground stations, which are almost equipped with large parabolic dish antennas. Many data streams carry news channels and entertainment content. They are transmitted over several geostationary TV satellites to cover many countries with a single footprint and to entertain millions of radio listeners and TV viewers.

A well-known family of satellites are the European Meteosat weather satellites. Many generations of this spin-stabilised satellite type have been in use since the 1970s. The first generations used analogue

Figure 75: Symmetric Parabolic antenna for Meteosat 7 signals on 1550 MHz.

signals to broadcast weather data and pictures to their ground stations. Even for private use within scientific experiments, people on earth could easily install small ground stations consisting of a dish and a demodulator feeding an appropriate software, even for animations of cloud movements. Meteosat signals for earth stations came on 1,550 MHz on an analogue downlink, and all transmissions followed a regular timing schedule. Meteosat still provides perfect cloud pictures today. They use three different spectral ranges: one with infrared, one with visual, and one in the spectrum of water vapour.

Today, all pictures come as digital data in high-resolution quality. For this mode, receiving equipment is much more expensive, and it is no longer useful for most radio enthusiasts. And by the way, it is much easier to view current high-resolution weather pictures by apps on a mobile phone.

Galactical Radio Sources

Let us come back to natural electromagnetic radiation coming from distant galaxies to earth after many light years of time. Some of them are extremely far away, having transmitted their light billions of lightyears ago. Who can imagine such dimensions? One of the most exciting sciences for many people is radio astronomy. Huge optical telescopes and large parabolic antennas of enormous proportions are typical for that kind of science.

They detect radiation even beyond the range of optical light. The largest telescope in 2023 is the FAST (Five Hundred Meters Aperture Spherical Telescope in China). It was built until 2016 with dimensions of 500 m in diametre of the reflector. Scientists are looking for sources of radiation that were emitted thousands of lightyears ago. There are still numerous questions to be answered concerning the creation of galaxies and even the whole universe, starting at the Big Bang, which was 13.8 billion lightyears ago.

Scientists analyse frequencies between 10 MHz and 100 GHz, and sometimes the received signals are really weak, some with -260 dBm. But what are we looking for, and what is really examined in the endless expanses of space?

Quasars are a group of fascinating objects. They are extremely bright objects in the centre of strange galaxies far out. A certain amount of energy keeps the gas of galaxies in constant motion, and in those galaxies with extremely bright, shining objects, we almost always find a black hole in the centre.

Other objects in the vicinity are sucked up and swallowed. This process sets free that huge amounts of radiation energy. What we detect on earth is just that strong emission coming over lightyears of time to us. Because of the gigantic brightness and our sensitive instruments, we can see those sources even on earth.

Pulsars are another category of objects in space that are most interesting for scientists in radio astronomy. These rotating neutron stars emit immense pulsed radiation in different spectral ranges, even x-rays. The physical dimensions of neutron stars are typically 10 to 12 km, with a mass of 1…2 solar masses. But this is not all. X-ray double stars transmit extremely strong radiation coming from black holes. This comes from the companion star when it is dissolved over a long time until it is swallowed. They are objects with the highest amount of x-rays. Scientists all over the world analyse them, i.e., with large telescopes like the 64 m antennas of the DSN (deep space network) operated by NASA. Also, flying telescopes such as the James Webb telescope, make their observations drifting in appropriate positions in space. What we know so far: Our universe expands more and more, and the speed of expansion even increases. In recent years, scientists analysed just that process by observing very distant supernovae and their radiation. Supernovae are very bright, shining stars, just exploding and emitting a lot of radiation, also in the visible range of the electromagnetic spectrum.

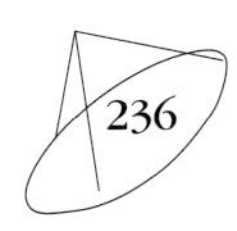

Thinking the expansion through until its end, all matter will expand, and in the end, nothing is left from it.

By the way, the overall amount of matter in our universe is almost overrated. What scientists found out: The biggest part of our universe consists of dark energy, with approximately a two-thirds share. 25 % is dark matter, while only a share of 5 % is the matter we know as stars, planets, and galaxies. What dark matter and dark energy really are, has yet to be explored more deeply, like black holes and anything around them, too. Oddly enough, dark matter and dark energy, as well as black holes, have been proven recently. Einstein's theory has been confirmed several times. Is there probably a need to extend Einstein's theory of relativity, i.e., to understand better what dark matter and dark energy are? Or should we alternatively follow the theory of a holographic universe? Which is the right way?

Still today, we see new phenomena and space objects every year. For scientists they, mean new challenges every time, like recently when a student of Natasha Hurley-Walker's team discovered something that could not be explained. Natasha Hurley-Walker is the leader of a team of scientists in the ICRAR, the international centre for radio astronomy research. They found a phenomenon which cannot be explained with modern scientific methods.

The suspicious celestial object has a distance of 4000 lightyears from Earth and cyclically transmits radio pulses. Scientists observed transmissions three times per hour for approximately one minute. The origins were tremendous energy bursts that could be detected over the large distance. The object appeared in 2018 within several hours and then disappeared again.

During the time of observation, it was the brightest radio source in the sky, while its dimensions were smaller than our sun. Today, scientists assume that they observed a magnetar. The existence of magnetars is supported by theories, and their appearance has been predicted. That slowly rotating neutron star transforms magnetic energy in to radio waves. And it changes magnetic energy much more efficiently than we have ever seen in any other process. What we do not know is: will the object return again, or was it a unique appearance? Another actual event is the discovery of a very strong source of radiation that becomes continuously brighter, according to "Spiegel" in 2022. One open issue was peer review, a kind of confirmation by other scientists. But the scientists of the Australian National University, who observed this radio source, assume it was a sensational event.

The reason for that sensation: The observed radio source is a quasar that did not attract great attention for more than fifty years, although that type of object has always been of great interest and many scientists were looking for them. Many quasars can almost be clearly seen on earth, even if they are extremely far away from us. In the centre of strange and distant galaxies, quasars emit exceptionally bright radiation that we can see on earth even after a billion lightyears. That sounds unimaginable. To put it in other words, quasars are the glowing hearts of galaxies. But in this case like in many others, they are a black hole that swallows anything around. The amount of radiation compares to the brightness of 7,000 galaxies radiating an intensity like our own milky way, while the black hole in the centre devours anything. Each second, the mass of planet earth is swallowed. Meanwhile a quantity of three billion sun masses have been devoured. This actually seems to be a sensation. It is also a serious indication that electromagnetic radiation is not only man-made. Galactic sources and their quantities of radiation go far beyond our imagination, and we should hope to keep them at a safe distance.

Scientific Missions in Space

Looking at space missions, we find a wide area with very exciting things. Here, we can only present a few examples. Our universe has

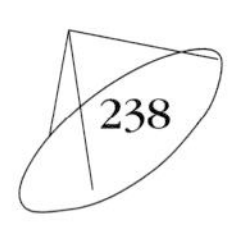

always been fascinating to humans, and we want to find out anything about the mysteries in our solar system and even in outer space. Today, the ISS (International Space Station) moves around the earth in a low orbit and returns cyclically above our heads. We are familiar with spacecraft and the people who work there, doing experiments that will help us in different scientific areas. But how did space travel start?

Once, it started with a surprise a time when nobody thought about spacecrafts or even people in space. In 1957 a small spherical object appeared in a low-earth orbit, and radio experts discovered a beeping signal at different places around the world. It really was a great sensation, when Sputnik and its Russian creators amazed everybody.

Sputnik transmitted radio signals with one Watt from an orbit in 200 to 900 km height. Typical cycles around the earth took 96 minutes. The RF power amplifier used miniature tubes. Signals were broadcast on 20.005 MHz and 40.002 MHz. Sputnik was an early opportunity to monitor the Doppler effect from a satellite.

A typical frequency shift of 250 Hz was observed while it was passing. Sputnik was visible in the sky, and in the former GDR, people even followed this satellite with binoculars. Unfortunately, the operators of Sputnik underestimated the air resistance, even if it is very low, hundreds of kilometres above the ground. After 92 days, the famous ball with a mass of 80 kg sank down into lower orbits and finally burned up in the atmosphere.

After that great sensation of 1957, space travel is still exciting to this day. Electromagnetic waves have always played a big role in all space missions. They were used in many examinations of planets, even on an asteroid and on Mars. In 1969, America's moon landing, the first at all, was a booster for space exploration by scientists all over the world. Since then, astronomy and space exploration have gained new quality and quantity.

Figure 76: Telstar reminds us of Star Wars, but it was only one of the early satellites of nearly 90 cm in diameter and a mass of 77 kg at the beginning of space exploration. Telstar provided the first transatlantic transmission from America to Europe.

Exciting and spectacular missions have been done with two space probes, Voyager 1 and Voyager 2. They revealed what is behind the great, continuously rotating eye on Jupiter. While coming closer to Jupiter, scientists identified that big spot as a gigantic, never-ending storm. Before that discovery, the large spot looking like an eye on the planet was a great mystery; it never seemed to stop moving or to disappear. Both Voyager probes transmitted their signals on links in the S-band and in the X-band. They used RF power levels between 9 W and 30 W. Their RF signals were focused by a parabolic reflector with a diameter of 3.90 m. The dish was made of graphite epoxy resin composite material. To keep in contact with the probes, the large ground stations of the Deep Space Networks (DSN) had to be used. They are located all over the globe, and they can be operated

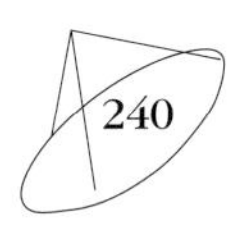

in a combined mode. By this method, the antennas of several ground stations work like a significantly enlarged, gigantic antenna. So even the weakest signals coming from a space probe can be detected, and control commands can be sent to the most distant places in unknown galaxies.

Meanwhile, the Voyager probes left our solar system and followed their path in interstellar space, more than 150 AU away from us. 1 AU (one astronomic unit) means the distance from earth to the sun. Running with a speed of more than 60,000 km/h, their transmissions come only rarely using the telemetry channel with a data rate of 50 bit/s. Even more rarely, we receive data from the scientific channel at rates of 150 kbit/s. Signals need several days through space before we can detect them on earth. And we can expect that their energy inside the accumulators will expire some day in the near future and that Voyagers one and two will send their final transmission to earth.

Transmissions over Communication Satellites

It is hard to imagine how we can live without satellites orbiting around the earth and providing beloved services like TV channels. They also broadcast large amounts of scientific data for agricultural use, for example. In many areas of our daily lives, they fulfill complex operations, send data, and enable us to use knowledge that nobody would ever have expected in the pre-satellite era. Worldwide navigation with high precision and phone calls from any place in the world are only two evident examples.

Sputnik's beeping in 1957 turned out to be a wake-up call for the western world. At the same time, a race for supremacy in space began, and the Soviets and Americans started some fascinating missions up to the present time. Everybody could see what humans were able to achieve. America's first communication satellite for transatlantic transmissions was Telstar. In 1962 the first TV broadcast over the Atlantic Ocean was done. Telstar (Figure 76) moved in an orbit that enabled transatlantic service for approximately two hours because it was moving in an elliptic orbit. A pure geostationary position has not been reached, but in later missions, the performance of satellites was improved simply by using an orbit above the equator. By Telstar, the foundation for generations of communication satellites was laid. Even today's TV satellites, like the famous ASTRA satellites, have their forefathers in those early achievements during space missions. In the 60s Telstar even got its place in the music charts. "The Tornados" created a popular instrumental hit that has not been forgotten even in the 21st century.

Figure 77: Syncom 3 was launched in 1964 and provided modes for links with one duplex call or for 16 simplex transmission lines.

Giotto and Halley's Comet

It was the year 1300 when Giotto di Bondone, an Italian painter, created his famous work, the cycle of "Adoration of the Three Kings". The most famous work shows the bright, shining star of Bethlehem above the stable. We suppose that he never assumed that great celebrity, 700 years later. Who could have known that Giotto's famous work would be one of the brightest symbols in relation to space travel and astronomic research? What came to his mind when he looked up at the sky and saw that bright shining star? Did he watch the same star that once led the Three Kings to the newborn Jesus Christ? What we know today: Halley's comet really appeared short before Giotto di Bondone completed his masterpiece. This can be proven. Halley's comet appeared regularly in previous centuries, and it could be seen with the naked eye. Halley always returns after approximately 75 years moving on a hyperbolic orbit that leads him far outside our solar system into deep space. After entering our solar system again, it is always a fascinating approach we can see on earth.

When Halley passes gravitational forces in the universe, such as from Jupiter's immense gravity, Halley's orbit can be changed. That is the reason for an appearance with a slight deviation. So its time to return can take even 79 years or less than 75 years.

Now we can look closer at the approach: A shining example of remote sensing with electromagnetic waves. Giotto's mission to Halley's comet is one of the very spectacular events in space exploration. But what is it all about when we hear from Giotto, a kind of mysterious probe? At least we know his namesake. The major event during the Giotto mission was the encounter of the probe and the comet. Halley's coma and Halley's appearance were of great interest for scientists all over the world. In 1986, the time for that great event has come: On the night of the comet, Giotto reached Halley's comet and could do all the experiments successfully during the flyby which had been predicted exactly.

Several scientific instruments applying clever methods have been placed on the little spacecraft to observe and analyse the comet at close range for the first time. Cameras with high resolution, mass spectrometers, and other high-tech sensors had their own tasks for providing us with more information about the comet and its coma. A mass of 50 tons of dust and particles is left by the comet every 75 years while approaching the sun. Also, these small particles have been objects of interest during the encounter.

For earth communication, engineers have installed a high-gain parabolic antenna on the probe with a narrow beam directed to the ground stations. The antenna transmitted signals of 70 W RF output from two redundant travelling wave tubes. That high frequency energy then was focussed by the dish antenna with 1.47 m diameter and radiated from the rear side of the space probe, while the spin stabilized probe transmitted data and pictures from the encounter. Microwave signals with a 40 kbit/s data rate were sent over the high-gain antenna. Pictures revealed the core of the comet, which had a shape similar to a potato. While the great reflector antenna transmitted scientific data in the S-band and X-band (2.1 GHz uplink, 2.3 GHz downlink, and 8.4 GHz downlink), two other antennas with low gain were used for telemetry data links. One of them was placed on the top of the tripod mount of the dish antenna, broadcasting with a heart-shaped beam. The other was a microwave patch antenna with low gain. It was integrated into the probe‘s surface near the bumper shield.

Everybody watching the event was very excited and fascinated, knowing that several serious risks had to be considered. Would the power supply remain stable when all experiments and all devices get active during the encounter? The travelling wave tube needed 70 W, the cameras 51 W, and the other experiments on board used 85 W. To avoid that risk, engineers installed additional batteries because the solar cells would have been overloaded. In the end, the whole team of ESA and all hobby astronomers could breathe a sigh

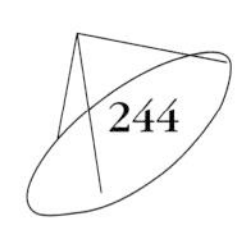

of relief. The mission was successful with all its experiments on board. For the first time, scientists received photos from a comet's core and a lot of new insights about a comet's coma. Even vapourised gas and particles could be analysed. The results from the mass spectrometer, cameras, and other instruments were extremely interesting because they told us a lot about the solar wind and dust from the comet and how they interact. It was lucky that all this happened without critical incidents, thanks to the bumper shield. This protection was a clever and stable construction. But why was the bumper shield so important for the space probe at all? What made the bumper shield a clever construction?

Giotto was equipped with special protection against expected dust and other particles crossing its path. The protecting construction at the frontend was a bumper shield with several properties. It had to be very light in weight and also stable and effective against particele impacts. That was achieved by a composition of several layers made of kevlar (7.5 mm), polyurethane foam (5 mm), and honeycombs (structures similar to those of honeycombs of our bees), and aluminium (1 mm and 40 mm for front- and rear-shield). All this was designed to compensate for the impacts of particles with dimensions of 0.1…1 mm. The bumper shield really made the probe safe against severe impacts. But there was one event that made scientists and technicians tremble. The probe was hit by a particle of approximately one millimeter. The physical impulse caused a precession similar to a gyro and a misalignment of the reflector antenna. The link suffered a signal loss of approximately 20 dB for a short time. It was lucky that the probe could be stabilised again. Those were minutes of great fear.

One of Giotto's experiments was very special. It was the GRE (Giotto Radio Experiment). The question behind it was how to measure a negative acceleration, i.e., by particles while passing the comet tail. The answer was easy. It is the Doppler effect, which provides information about the deceleration of an object transmitting electromagnetic waves. Ground stations should receive a signal with a slightly shifted carrier frequency. And of course, they did.

Remote Sensing

For decades, the astronomers' instruments have been measuring deeper and deeper in space and time. Many missions in space but also on earth, monitor and scan new galaxies, supernovae, and a lot of other cosmic phenomena like neutron stars, x-ray sources, and black holes. When a surface of a planet in our solar system is explored, scientists use cameras, radar, and other multispectral systems for scanning with different wavelengths.

When the Voyager probes started, we could not have guessed that they would leave our solar system and stay on air while continuously broadcasting scientific data to our ground stations. After the shock to America when the Soviets launched their first satellite in 1957 with the dimensions of a basketball, America also put a lot of effort into space technology. In 1958 they launched Explorer 1 into space. Still in the same year, the world heard a lot about the new Pioneer programme with several space probes. In the following years, they were launched. One of them was sent to Venus. It had no impact protection, so it was a hard landing on Venus. It was like a wonder that the probe was able to transmit about one hour of the ground of Venus. Since 1977, the Voyager space probes have made several headlines. Meanwhile, they left our solar system and were still transmitting in 2022 over a distance of 19 billion kilometers. With its dimension of 3 m in diameter, the parabolic antenna achieved a high gain of 42 dBi in the X-band.

We could present many more space missions and focus on their technologies using electromagnetic waves. One spectacular project was the Cassini-Huygens-mission and the landing of its module Philae on a comet. The Huygens-telescope, another famous spacecraft, is doing a great job while discovering distant galaxies and

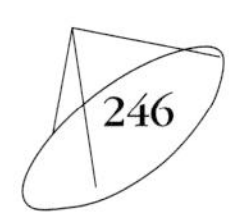

monitoring their evolution. Its follower, the James Webb space telescope, promises even deeper insights while exploring the outer space with a look close to the Big Bang. All the missions are the result of the work of many engineers and scientists all over the world using modern instruments and clever methods for remote sensing and analysing surfaces and radiation. All the effort follows one objective: exploring space back to the world which existed close to the Big Bang. On many missions, Einstein's theory of relativity has been underlined and confirmed. But the process of understanding and probably extending this basic theory is still going on.

Today, remote sensing plays a big role. Many tasks and methods belong to that scientific area, and all challenging new jobs need motivated people facing new tasks. Remote sensing uses two different basic types of methods: active and passive systems. While passive systems only absorb radiation, which has to be examined in different frequency ranges, active systems transmit electromagnetic energy. By analysing the reflected wave and the backscattered spectrum, scientists get deeper information about a surface and its material.

Many applications in space are based on satellite operation with appropriate sensors. Climatological data are more and more important for exploring weather phenomena and the consequences of climate change. The ozone layer needs a special focus due to its protective function for life on earth. Therefore, its changes are consequently observed from space. The impact of oceans is also changing due to temperature changes. Emissions have a significant impact on our atmosphere. Volcanic eruptions and forest fires emit a lot of particles and pollutants into air, as well as unnecessary man-made emissions, like the side effects of a war. Today's modern satellites observe all this by scanning anything that happens below. They help us to receive and evaluate data about cultivated areas and agricultural planes with grain, for example. Earth observation satellites transmit information about the state of the soil, dry or wet. By satellite view, we can monitor different agricultural effects from a space view.

Only a short time ago, satellites recorded a comprehensive topographic scan of the earth's surface. New instruments and new methods enable us to learn more about our own planet. Today, a share of 50 % of all meteorological data for atmospheric and climate models comes from satellites, which are in use on any orbit around the planet.

Various systems, like multispectral cameras and sensors, microwave radiometers, thermal imaging cameras, and radar systems, help us collect data carrying special information to monitor and discover the world around us. Thanks to the ever-increasing performance of data centres with high-speed calculators, huge amounts of data can be used in real-time and give us new insights into nature and about ourselves. The use of all those components shows that engagement and commitment among young generations are very important. Who else can help develop new technologies and systems to understand how climate changes anything, and to predict critical events early enough? What we need is a good deal of constructive and collaborative engagement from young people around the world. To be honest, just to do this, we need a lot of hard work and a good education. In the past, we have seen that humans can achieve much more than most of us ever assumed. Let us hope that there is still motivation for future generations and that they will be ready to work passionately for new solutions, for peace, and for prosperity on our planet.

Worldwide Weather Data

For many years, we receive pictures and data about weather from the Meteosat. Those spin-stabilised satellites have been continuously improved, and they are positioned in a geostationary orbit approximately 36,000 km above the equator. They observe all that is relevant in the context of weather, such as cloud movements and temperatures. Satellites send current weather data in different optical wavelengths with a regular time table. They use infrared scans, others in the visible optical range, and some scans run in the range of water vapour. But how did we receive weather data before any satellite or weather app provided forecasts and station reports? Even once upon a time, long before the satellite era, weather data were essential to survive while crossing the ocean, i.e., while passing a thunderstorm. Data had to be available on ships and also all over the country. But how could this be done, and which technology was able to provide forecasts and weather maps with high-pressure areas and low pressure-areas? In the second half of the 20th century many frequencies were observed almost on shortwave. They were used for broadcasts from governmental authorities to provide weather information for shipping and aviation. Even military stations had their own channels to provide weather data to their troops all over the world. Most transmissions came with a regular schedule and provided forecasts, station reports, maps with anticyclone and depression areas, and also wind speed and wind direction. Some of the transmission formats can still be received today. Before the internet and satellites, all weather information had to be presented and broadcast to any place in the world, just in time, and it had to be up-to-date. A real challenge without cell phones and apps.

Shortwave transmissions were a good solution. In a worldwide network for the exchange of weather information, weather reports and forecasts were prepared and broadcast over stations in many countries. Ships provided station reports from their positions on all oceans of the planet. Transmissions had their own dedicated timeslots, so all receiving stations and ships could receive data at any time around the clock. Weather transmissions had their own worldwide committed formats to transmit efficiently. The metar code is one of the typical weather formats (see Figure 78). Synop is a format using groups with 5 number packets. They strongly remind one of

```
EDDL 160830Z 00000KT 0100 R14/0250N R16/0250V0400U R28/0300D ...
EDDK 151850Z 22010G12KT 190V260 FEW010 SCT025CB BKN060 23/15 Q1010 ...
EDDS 041320Z 29010KT 9999 FEW040TCU 09/M03 Q1012 NOSIG ...
```

Figure 78: Metar code, a worldwide format for weather transmissions

the number codes used by intelligence services for agents, like those of the former GDR's MfS (ministry for "Staatssicherheit", see Figure 84). Synop comes from "synchronous" and "operation". Reports from ship stations and from continental stations had their identifier at the first position in a new line, followed by data about wind, temperature, visibility, and others. All this information was put into a format of number groups with five numbers and a fixed order of the data packets. Transmissions on shortwave almost took some minutes or even some hours and were repeated several times. For transmissions, all data was encoded with the CCITT2-code with 50 baud. This was not fast, but transmissions at 50 bit/s were stable data links and worked with many different implementations of circuits and devices on the receiving side.

```
zczc 449 29301
smro20 yrbk 131200
aaxx 13121
15010 32960 72508 10281 20161 40145 54000 80008=
15020 32965 42304 10324 20141 40127 58017 82031=
15200 32960 62705 10274 20181 40158 54000 80006=
15280 32770 50000 10118 20085 47151 58003 85300=
15410 nil=
```

Figure 79: Weather data encoded by the SYNOP format. Each line represents a station report starting with id which is followed by data about rain, wind, clouds, and more.

Synop reports were used worldwide and turned out to be an effective and robust method to transmit to any place in the world. In the Cold War, increasing broadcasting activities and radio transmissions were monitored. One reason was simply the lack of weather apps and cell phones. For military alliances, the readiness of the armed forces was of the utmost priority. They operated their own networks with high reliability to provide weather data continuously at any time and for any place. Still today, robust formats are in use and sometimes we can observe transmissions intended to be received in all countries and on all oceans.

Military Applications

Military departments have always been interested in new technologies, and consequently, they observed how radio communication could be used within their troops. And as we know, radio communication provides a lot of features. Satellite operation took some time and started commercial use in the 60s. So it was evident that shortwave radio communication as a backbone would still be extremely important for decades. Military and civil aircraft and ships used shortwave transmissions from 10 kHz to 30 MHz as the only communication option for a long time. In World War II, German navy headquarters contacted the submarine fleet at 16.5 kHz in CW Morse. The transmission site "Goliath" was located in Calbe (Milde, Saxony-Anhalt) with an RF output power of 1 megawatt.The groundwave covered a range of 10,000 km, and submarines could receive the signals still 10 m below the surface in the Atlantic.

Military satellites around the globe provide services for communication, observation, and spying, of course, like the SATCOM satellites of the Unites States at 253-263 MHz. Some of them are already out of regular operation, but nevertheless traffic was still monitored. What was behind that? Pirates were observed on SATCOM transponders recently. The signals on 255.550 MHz turned out to originate from Brazilian radio amateurs.

During the Cold War and several years after, many transmissions on shortwave could be heard. Some of them obviously came from military sites and seemed like military aircraft traffic. According to speculation, radio communication was used to keep in contact with a worldwide aircraft fleet that was in constant readiness in the air. There have been rumours about aircrafts with special armour receiving nuclear codes from their headquarters. They were held in the air to react to the first strikes of enemy military alliances. They used a characteristic information exchange that could be observed at any time around the clock. The transmissions were about long

6.604		VOLMET New York
6.617		VOLMET Moskau u.a.
6.685	....	GREEN VALLEY; ..
6.693	GAF	(not conf.)
6.697	RNy	Pitreavie/GB
6.729	....	
6.730	DänAF	
6.738	RAF	ARCHITECT
6.741	USAF	Alconbury/GB
6.742	....	Needshead (?)
6.747	....	ARGONAUT u.a.
6.750	USAF	Croughton/GB
6.753		Berlin IF
6.757	USAF	Croughton/GB
6.761	USAF	SKYKING
6.763	NATO	Geilenkirchen/D
8.864		Shanwick NoAtlContr
8.888		VOLMET UdSSR
8.891		Shanwick NoAtlContr
8.924		Warschau LOT

Figure 80: Military channels on 6 MHz and 8 MHz during the Cold War.

chains of letters and numbers from the international spelling alphabet. As an introduction, some strange commands were monitored, like "Skyking, Skyking, do not answer…".

Whatever was behind those strange messages, they produced some speculation. They seemed to affect a global network, probably for global security. Could this really be information for military use and for aircraft around the world? Frequencies have been in use for 24 hours in USB mode. Were they part of the GHFS, the Global HF-System of the US Air Force? Radio traffic was monitored during the daytime or night. Frequencies around 4 MHz were used at night, while messages during the day were observed between 11 MHz and 18 MHz. Tactical frequencies could be found on the higher VHF-bands on 139 MHz, 250 MHz and 350 MHz in a 25kHz grid (AM). Military aircraft also use civil frequencies when they are flying on civil continental routes, and the voice traffic is then part of the civil air communication on frequencies between 118 MHz and 138 MHz in a 8.33 kHz grid using AM.

Figure 81: Logarithmic periodic antenna of the German Airforce near Münster (Westfalen). From here, military authorities were in contact with aircraft far out during the Cold War on 5691 kHz and other shortwave frequencies using USB.

Military data transmissions from eastern Europe as well as from the Soviets, for example could almost be identified by the Cyrillic alphabet. Some stations were located near the Black Sea, and they were part of the Soviet Black Sea fleet; others were located near Moscow.

Even after the fall of the Iron Curtain regular transmissions could be identified as coming from Odessa and Mariupol. They used CW and RTTY on frequencies between 3 MHz and 18 MHz. Will they still transmit in the future? The current political situation in this region and the armed conflicts since 2022 give no reason to be optimistic about that.

OTHR (over the horizon radar) is a military RF system for reconnaissance. This radar works on shortwave. Hence, it is able to "look" far behind the optical horizon, and it is a real nuisance for DXers and radio amateurs. The system's nickname was the Soviet woodpecker. NATO also uses similar systems with a wide frequency range and short pulses. They still appear on shortwave between 7 MHz and 20 MHz.

Maritime radio systems are used for long-distance links to fleets operating in the far on any ocean in the world. In coastal areas, longwave transmitters can be heard, also transmitting to submarines and other parts of the fleet. One of them broadcasts at 147 kHz below the long-wave radio band and transmitted current meteorological information and nautical warnings. Some other stations also broadcast below 100 kHz.

Frequency bands between the public radio broadcast bands on shortwave are still in use for several other services. Between 3 MHz and 30 MHz, a lot of maritime communication has been monitored using different modes like voice (USB), FSK, and CW. Today, maritime links are almost always switched over to satellites. INMARSAT is one of the providers of services over geostationary satellites. Terrestrial data links used CW and FSK (telex) in the early years of the 20th century. Many transmissions were unclassified for a long time and did not use any codes or cipher devices. Sensitive receivers with proper selectivity did a good job of monitoring. Magnetic antennas like ferrites with coils and loop antennas were efficient on low frequencies, up to 3 MHz. Electric dipoles, with their huge dimensions for low frequencies have not been the first choice because they turned out to be more sensitive to interference in many cases.

Armed forces on land also used wireless communication systems during the Cold War. They were present in frequency bands above the upper public shortwave radio bands and below VHF. Transmis-

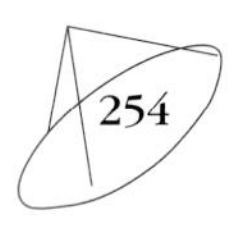

Figure 82: Military microwave relay station

sion modes were AM and FM for voice. Troops also used their own microwave links (Figure 82) and phone networks.

Military applications using electromagnetic waves have always played a big role in many countries, and there are more applications than we might assume. Worldwide data exchange is one of them and is almost based on special military satellites.

Terrestrial data transmission is also available, but there are some new modes in use, like ALE (automatic link establishment), for secure data traffic. That's more effective. Frequency hopping to improve their performance and to keep out people without authorization is a special feature. Keys and ciphers turned out to be the right means to avoid eavesdropping. Plain text is only used for initial link establishment. Message content remains inaccessible.

But there are more systems that are used in the military. When we look at missiles, radar sensors and infrared controls make sure that they find their pre-programmed targets. Another extraordinary application was reported from the USA. Radiation from a gigantic two-dimensional array of dipole antennas just above ground was sent to the ionosphere to analyse the consequences of heating by radio waves. Heating describes the fact that very high power levels of electromagnetic energy were sent to the sky to excite molecules.

Through this project called HAARP (high-frequency active auroral research programme) scientists should find out some more indications about the behaviour of the different layers of our atmosphere. While monitoring this, layers in the atmosphere were excited with electromagnetic energy far beyond the usual power levels of radio broadcast and communication. Sure, we may assume that services from other countries, like China and Russia, executed similar experiments. Roughly said, the huge antenna fields should focus radio waves in special areas of the atmosphere to achieve certain effects that still had to be explored.

Recently, another application with huge field strengths was presented from naval sources. New generations of catapults on aircraft carriers will replace the good old steam catapults. Electromagnetic catapults use a near-field effect with strong magnetic fields to move and launch aircraft more effectively than before.

The near fields couple their energy to the aircraft, and it is accelerated in a very short time to enable starting. Several electromagnetic segments are placed along the runway on both sides. The magnetic fields are switched very quickly, one after the other, during a start.

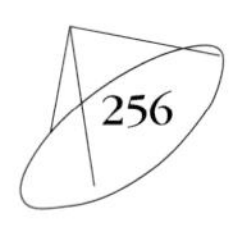

Their field lines are similar to circles that tear on both sides of the planes into the start direction on the runway. Each segment is used first for push and then for pull, changing between both modes within very short intervals.

With the right timing, aircrafts will reach a speed of 300 km/h within one second. The large amount of energy used for the movements of planes are coming from the carrier's on-board generators. One of the first American aircraft carriers with electromagnetic catapults is the USS Ford. Also, Chinese carriers are currently equipped with that technology.

```
. echo : frequences radio demandees -------- a1 :
2789-4199-6298,7-8398,3-12597,7-16797,3
f1 :2847-4150-4154-16 ns th228-6256-831995-
8313-12444,555n2491-12471-16600-16612-
16615-
puissance max d'emission : 400w foxtrott : ravitaillement
en gazole : f 76 ---------- - 60 tonnes pour le l.v. le
henaff. golf : 1/ le capitaine de vaisseau o rr s i n i
command t
a--- nu-b ezecole navale et le groupe ecoles du poulmic
sera
present lors de cette escale.
2/ seront en outre embarques stur l'ensemble s
batiments:
15 officier i e t g officiers mariniers instructeurs.
63 aspirants de l'ecole navale
18 ele es du csen : )3 togolais-2 congolaiss1 came-
rounais-2 malgaches-2 gabonais-2 mauritaniens-1
zairois-2 quat

3 beninois
15 hommes d'equipage en renfort  bt as si nr ta 5670 ewco
ii p 281629z nov 6 fm marine brest to aig 1913
bt non protege mca avurnav nmr/0322 np 2811 - ops/cot txt
t t t avurnav local brest nr 3038 nord gascogne - anse de
benodet explosion sous marine le mardi 2 decembre de 1200
a 1400 locales au point 235/touelle du taro/0,25 nq
)epave sainn  rand  zones dangereuses de 3000 metres de
rayon pour nageus et scaphandriers et 3000 metres pour
navires et embacations centrees sur point d explosion
annuler ce msg le 021300z dec diffusion jusqu au  t
133eenz d   bt as
```

Figure 83: Nautical warnings transmitted by a French naval base with call-sign "FUG" on 4314 kHz (CW) 18.11.1986 at 10:10 pm.

There are still some other applications with electromagnetic waves for military use that are less spectacular. The TACAN-System, which was developed after World War II, was designed to provide lost aircrafts and helicopters with an instrument to find the route back to their home base or to their carriers.

One radar system for military use is the IFF or "friend or foe radar". It is able to identify aircrafts as friendly or as enemy threats. In the air and in the cockpit, fighter pilots are supported by the "military airborne radar" working on approximately 1 GHz. It provides a view of activities in front of the aircraft in a range of at least 150…300 m until 4…40 km maximum range.

Intelligence

Electromagnetic waves have properties that are extremely suitable to run their way across state borders unnoticed and invisible to find their way to the receiver of a secret message. This is just what some special organisations, like intelligence services, have always needed. Electromagnetic waves even penetrate buildings and go through walls, and it is hard to control this type of information exchange. Broadcasting over shortwave to contact persons from the secret services' environment on foreign missions was the right application. For governmental surveillance, it has been nearly impossible to find out who is listening to a dedicated transmission on shortwave anywhere in the world. This was most important in the Cold War especially in Germany, when two military blocs, maximally armed with nuclear weapons on the ground and in the air, considered each other enemies. For radio communication, there was no border that couldn't be overcome by secret messages.

The communication of diplomatic services and their agents was typically operated from a central office to many different receivers. Agents received commands for industrial espionage or military reconnaissance activity. Their intelligence instances used simple unidirectional radio links similar to a radio broadcasting station with a regular time schedule. Receiving equipment was unobtrusive and could be operated anywhere, like ordinary small portable radios. Data communication from foreign ministries to embassies and representations in other countries were also operated over shortwave links. For bidirectional data traffic, they used ARQ-codes with message formats in clear text for headers, addressees, and callsigns. Message content has almost always been transmitted by using codes and ciphers.

Ministries of Interior and state representations used similar systems for classified traffic. What was really noticeable were the huge antennas on the roofs of embassies and governmental residences. Antennas drawing little attention were electric dipoles stretched out on the roofs of high buildings. Even if the dimensions were also huge, they could be operated unobtrusively. One alternative option for high demands were those logarithmic periodic antennas (Figure 91). They were almost rotatable and most efficient, reminding one a little bit of the huge Yagi antennas. There was a big advantage to using a log periodic. They had good directional characteristics, and they were broadband antennas, suitable for any shortwave band. They provided excellent link quality, and the installation of different wire antennas like dipoles for each frequency band could be avoided.

28495	39756	37284	50800	45200	55643
11365	22144	32667	11365	53777	22144
64599	39756	22112	64599	32543	39756
53000	45589	88956	53000	70008	45589
78633	39756	46377	78633	46224	39756

Figure 84: Groups of five numbers were sent in AM, SSB, and CW-mode. This type of transmission was used by special organisations in the Cold War: western countries and Warsaw Pact countries used this type of transmission for secret messages. CW as A1A (silent carrier switched on and off) or as an AM carrier with a tone for dots and dashes was in use, i.e., by the gong station.

How intelligence communication worked in detail has been assumed in several papers. It was the CONET project from the 1980s that revealed many facts and valuable information. While monitoring all suspicious frequencies systematically and documenting schedules, languages, and transmission modes, the CONET community provided a lot. Many facts had been assumed a long time before, and they were then confirmed. Other facts were completely new for the scene, and many radio enthusiasts joined. Direction-finding systems were used to find out from which countries the mysterious shortwave transmitters were broadcasting and who was the operator. The CONET project revealed that several originators in different countries were responsible for the mysterious transmissions. Countries and antenna locations could be identified, and a well-structured data base gave an overview of things that should better remain hidden according to their originators. Still today, any transmission content remains encoded, but now it is obvious who is involved in those broadcasting systems and for what kind of listeners those transmissions were meant. To most people, the result was not really surprising.

When it was getting dark in autumn in the 1970s and you turned the little knob of your receiver for scanning shortwave frequencies in the range of 3 to 5 MHz, you could meet a mysterious signal with a frightening regularity. One station broadcasts signals that sound like a never-ending gong. After some minutes, the gong stopped, and a woman with a stern voice started to tell number groups in German without an end, as it seemed. This strange female voice was obviously that of a military person because she seemed to be very disciplined in talking numbers for hours. For whom were these transmissions? What were they about? From which organisation was this woman, and why did she speak that way, strictly and without any emotion? Today, the transmissions are classified and any station gets its individual id like E05, S02, M11, and more. E stands for English, S for Slavic, and M for Morse code. Some of the stations got additional names that characterised them according to their sound, like the above-mentioned gong station. Those who listened to the gong station or similar channels, could realise that the voice sounded mechanical, just like sampled voice-telling numbers from 0 to 9. Operating frequencies could be found between 3 and 10 MHz, sometimes above. According to several reports, number stations also used frequencies between 5 and 20 MHz during the day, always between the radio broadcast band, sometimes in English. Some station names, which have been assigned later, were:

Gong Station
Swedish Rhapsody
Lincolnshire Poacher
Drums and Trumpets
Skylark
Russian Man
Oblique
English Man
Spanish Lady.

Precise descriptions and documentation can be found the internet today. The examinations have revealed that special organisations of the great political blocs, that is countries affiliated with the USA or Soviet Union, were the originators and responsible for those mysterious transmissions. Intelligence services and other secret organisations were behind the strange voices and morse transmissions. We know today that the gong station came from a transmission site near Berlin in the former GDR (German Democratic Republic) from a department that was part of the MFS. And it was really a sampled voice. Those coded commands and information were meant for agents in western countries, almost active in the western part of Germany. The ministry of "Staatssicherheit" was leading hundreds or even thousands of agents in foreign countries. Still today, some of the mysterious stations can be found on shortwave. But their signals are weaker, and they use unknown foreign languages. We can assume that the origin of those radio broadcasts is in Asia and the Pacific region and less in European countries.

Systems and Components

In this chapter, we will look closer at radio systems and what is typical while operating them. Modules, methods, protocols, and subsystems are designed to play together for safe and effective operation. Furthermore, it is important to meet all specified requirements used for the construction of a communication system. They have all been built for a precisely defined operational environment.

To realise that, it is always most important to select appropriate standards, precise specifications, and procedures for production and quality assurance for the final product or even a whole system. To give an example: It makes a great difference if a communication device is designed and constructed for stand-alone operation or for use in a network environment. According to the scenario, engineers have to decide how communication has to be handled, if it will be secure, and a lot more. When the set of requirements is complete, almost an endless paper with clear specifications, technicians and construction engineers can start the development of sub systems.

When engineers talk about components, they almost always mean single elements, but also subsystems of the whole communication system. This can also be a functional unit to operate a communication system in stand-alone mode. Using terms like this is not always precise and sometimes even ambiguous. For the use of efficient terminology, it is almost helpful to give examples of what is really meant whenever it is possible.

The little transceiver station in your shack can be described as a system with different components like cable, antenna, power supply, and peripherals. Using a deeper level of description, the transceiver can be characterised as an electronic system with software, hardware, functional modules, and external devices. Moving one level up, even a radio network can be described as a big multi-level system. One level is the radio or RF part with base stations and RF repeaters, another level is the transport level, wich collects all traffic and routes it to another level with nodes and controllers. In the end there is even a level that assigns customer data to SIM cards and bills, all executed in large data centers. As we see, it is always helpful to define the context before discussing terms (i.e. system, subsystem, component, or module) which can be understood differently by different people.

Radio system, as a term, almost means a composition of different elements, like subsystems or components. The system as a whole follows a defined operational plan with defined use cases. Radio systems, which can provide radio signals in tunnels, are a good example. Electromagnetic waves do not penetrate deeply into tunnels. Depending on the used frequency, signals get weak after 10 or 100 metres. But how can a tunnel be illuminated with radio signals despite the high loss, and how can it be achieved without an additional expensive base station?

When a radio signal (radio broadcast or public phone signal) is coupled into a tunnel with the help of a slotted cable or a dedicated antenna, radio coverage even inside the tunnel is ensured, and reception with proper signal levels provides a good quality of service. That needs a system that combines two antennas, or one antenna combined with a feeding cable, and, of course, a repeater that provides a huge amplification of uplink and downlink signals. For radio broadcast, the uplink is not needed. Feeding a tunnel without a slotted cable is also possible. In this case, engineers use two antennas. The first antenna (also used in the case of a slotted cable) shall receive the signal from a base station. Amplified by the repeater, the signal will be transmitted by a second antenna, and placed where the tunnel begins. Repeater types in use have selected channels for operation, or they are even band-selective to transmit a whole radio band. GSM almost always used channel-selective repeaters, but for signals in railway waggons, wideband repeaters were the right choice. In any case, it is extremely important to ensure that both antennas are decoupled by a good isolation of at least 20 dB. Below 15 dB, bidirectional repeaters with high amplification factors start to oscillate similar to acoustic feedback when the microphone is too close to the speaker, the end of any useful operation. The isolation between both repeater antennas supports the decoupling of uplink and downlink. We remember that, both work simultaneously on a call, talking and listening at the same time.

Methods and Modes

The use of electromagnetic waves in technical systems always means the use of methods and operational modes applied to different levels. In this chapter, we look closer at some methods used on the message level, on the protocol level, and also on the radio frequency level as well as on the system level. All methods and modes in a system have been selected to support dedicated applications, and all will collaborate. Some of them are really complex, and descriptions in special literature are needed to apply them correctly and to understand any detail.

Communication Protocols

When two machines or computers talk to each other, they need clear agreements, commands, and specifications about how they shall accomplish it and what they shall do in detail at any moment in the communication process. The basic instrument for that is an appropriate communication protocol. Some of them are really prominent now. The internet is the area where they are applied every day by billions of users all over the world, and everybody has heard about them. It is the famous TCP/IP, a pair of protocols in a whole protocol collection for internet use. TCP/IP means Transport Control Protocol, and IP is the Internet Protocol. Other famous protocols are http and its secure derivative, https. Many people, like web developers, are also familiar with UDP, ICMP, and FTP, also from the web environment, and a part of those 500 communication protocols only for internet use. IP makes

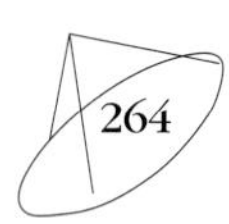

its job on the network layer of the OSI model and takes care that data packets find their way to the designated receiver. TCP, as the transport Control Protocol, makes sure that a link between two instances remains stable, and in case of disruption, the protocol will repeat the link activation to enable complete transmission.

For browser applications, it is the hypertext protocol (http) and the secure https, which define the contacting of web servers and visualization of web sites on PCs. How these components communicate to each other, that's one task for http (hypertext protocol). On the application layer of the OSI representation, machines use the SMTP and IMAP for applying defined processes during email exchange. We meet their names and functions when we install the mail exchange configuration on smartphones.

Information technology executes commands for information exchange anywhere and anytime. Also, radio systems, which are operated for data transmission from ships to shore stations, use similar agreements for communication jobs. We will look at this later. One of the early protocols for wireless communication was established in 1971 by scientists at the University of Hawaii in Honolulu. It is the CSMA-protocol (carrier sense multiple access), which is also called the Aloha protocol. It is the prototype for access and efficient data communication in a local area network (LAN) and even WLAN, the wireless variant used by most of us at home.

Once, data packets were sent from Honolulu to islands in the neighbourhood, and the data were received without issue. While data traffic increased, transmissions became less successful in reaching their addresses. Collisions of data packets turned out to be the enemy. The protocol gained new functionality by repeating the transmission in case of collisions. While traffic still increased, the protocol was extended to a slotted Aloha protocol to avoid too many collisions within data traffic. By the slotted Aloha variant, data packets were assigned to selected time slots, and they were only sent precisely in their slots. By means like that, CSMA has been improved more and more in order to operate with efficient traffic control on the bus, that is in the cable or in the air in the case of a wireless application.

Regulation and Harmonization

Transmissions from radio systems also mean a lot of agreements and rules to be followed. The reasons are: to operate without interference or incompatibilities. Therefore, determinations of channel grids, bandwidths, and transmission parameters are of great importance. Narrow-band FM, a variant of voice traffic as an example, is used in typical RF-channel bandwidths of 10/12.5/20/25 kHz. Worldwide civil aircraft communication was first used within a grid of 25 kHz in AM mode.

Recently, the availability of new frequencies has been enhanced by introducing a grid with 8.33 kHz channel spacing. That means the density of channels has increased by a factor of three. That was appreciated by regulating authorities because radio channels have always been rare for any application in the spectrum. The popular GSM in the 90s used a bandwidth of 200 kHz per channel.

All these agreements with high importance for international applications have to be negotiated and committed by the government authorities of every country. They are the base for a countrywide operation by all affected user groups. An interesting example is the coordination of frequencies in border regions. In cases of frequency use without coordination there might be a channel that is used in both countries near the border. Without a certain spacing there will be strong interference from this channel. Those cases are individually coordinated by specialists for frequency reuse, i.e., by blocking the frequency for one of both countries.

Another level for harmonisation of standards are user devices and radio systems. Manufacturers are very interested in establishing their

own standards in the market. Operation modes for digital radio, like D-STAR, DMR, and NXDN (Digital Smart Technologies for Radio Amateurs, Digital Mobile Radio, Next Generation Digital Narrowband) are examples of amateur radio. Devices for certain modes are developed and manufactured by different providers. Each company tries to push and sell its own products. As regulation affects only the mode as a standard, different system enter the market and cause a huge variety of products. As an advantage, each radio amateur can select and operate their preferred radio. But a side effect is that groups that are using one preferred mode are getting smaller. Amateur radio has never been affected by such a huge diversity of devices and standards. The split into DMR, D-STAR, and other different systems is not a showstopper at all. As long as any group has many followers, radio traffic remains exciting for all users.

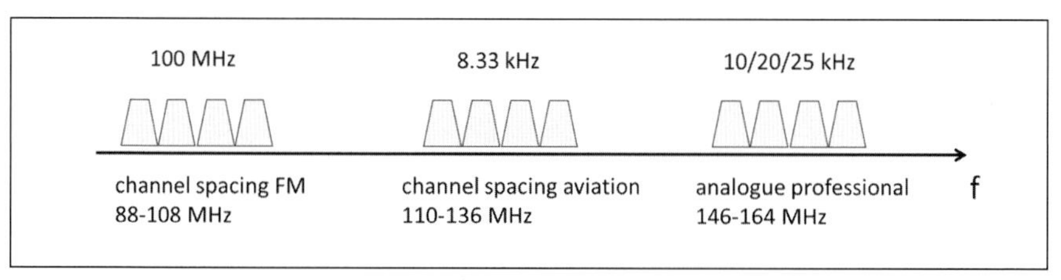

Figure 85: Different frequency and channel grids for FM radio, VHF air radio, and professional radio applications

While designing radio systems, system engineers have to consider which mode shall be usable: single-channel use, duplex, or half-duplex modes. In case the of dual-band modes, the duplex offset has to be defined.

There are a lot of aspects to consider to ensure safe operation without interference. Some of them are about technology and devices; others are about harmonisation and standardisation in an operational context, including coordination with neighbouring countries. International regulation processes must also be considered because worldwide standardised radio technology and operation in aviation, for example, affects all countries.

Amateur Radio Repeater as an Example

Operating a stand-alone repeater in amateur radio needs two frequencies. It works like a relay station with one frequency in the lower band, which receives the initial calls of an amateur operator. The relay station amplifies the signal and converts it to another frequency in the upper band with a frequency spacing, also called offset.

From the frequency in the upper band, the repeater transmits the signal to any listening station on this channel. When a repeater is initially activated for calling other users, it will send out its callsign in CW-Morse, which can be heard by any listener on the relay downlink. Repeater sites are almost always located at a certain height above ground for broadcasting with wide coverage for distances over 30 km or more. From some high locations, even over 150 km and more are possible.

Repeaters receive and transmit simultaneously, but radio amateurs with their own devices will either listen or talk. Therefore, any radio that is used for repeater operation has to switch between both channels: uplink to call the repeater and downlink to listen to repeater traffic. This duplex offset is needed to decouple signals in the repeater while receiving and transmitting at the same time.

Typical duplex frequency spacings in amateur radio are 7.6 MHz and 600 kHz for operation in the 70 cm amateur band and in the 2 m amateur band. The use of locations on hills and from skyscrapers or towers has two great advantages. The achieved distance of transmissions from the repeater site has significantly increased. The second advantage ist that: radio amateurs inside areas surrounded by higher buildings can also reach the repeater, and hence, they increase their own coverage distance by far more than 5 km. That's the reason why repeater operation in analogue voice mode and even in digital modes is still very popular.

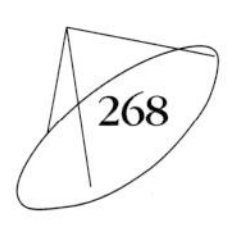

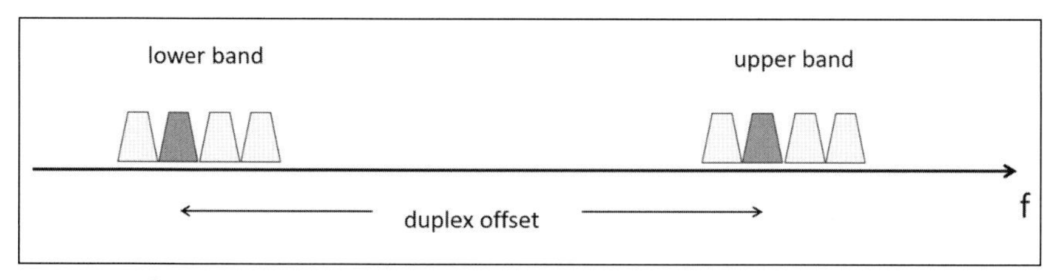

Figure 86: Upper and lower bands for repeater operation. For the 70 cm band in Germany, the duplex offset or frequency spacing is 7.6 MHz.

Also in public cell phone networks and in professional radio, a certain duplex spacing between uplink and downlink is used. GSM base stations need the spacing because both links, FROM users and TO users are active at the same time. And this is absolutely necessary in other networks like 3G, LTE, and 5G, as well as in professional radio like TETRA, which is operated by amateurs, and also by security and government authorities. Broadcasting networks can be operated more easily because they don't need the duplex offset for an uplink. Frequency coordination has to be done for only one band: the downlink or, better yet, the broadcasting band. But even with one radio band, foreign coordination is needed because interference may cause disruptions or at least poor quality of service, i.e., in case of overreach effects caused by the atmosphere. In the US, we recently heard about interference in aviation, which was obviously coming from the new 5G frequencies. Altitude radar in aircraft seemed to be affected, a real safety issue.

ARQ Semi Duplex in Maritime Radio

When we look closer at maritime radio systems of the 20th century, we recognise their huge importance for communication. For a long time, they have been the only wireless system to send telegrams and nautical warnings from ships in the open sea to one of the shore stations. Additionally, they made phone calls possible, especially when personnel on ships contacted their families during Christmas time. Telegrams and business communication had to be managed every day and night on ships. Shore stations like Saint Lys Radio were responsible for routing the traffic into continental phone networks while they used SITOR as a data exchange mode with ships. Phone calls were switched manually at this time. Some famous shore stations were Norddeich Radio, Rügen Radio, Portishead Radio, St. Lys Radio, Bern Radio, Tuckerton Radio, Shanghai Radio, and many others. They have all been ready for operation 24h a day to get in contact with any ship far out in the oceans. For voice transmission and also for data transmission, ships had to use two frequency bands with a defined duplex offset. Shore stations always listened on the lower band where ships were transmitting, like in 8.2 and 8.7 MHz shortwave bands for maritime traffic.

As shortwave has always been a moody medium, shore stations always broadcast their callsign on different radio bands simultaneously. This was a good opportunity to check the link quality and to decide in which shortwave band the call to a shore station would be successful. For broadcasting their callsign, shore stations almost always used the CW mode, which had a minimum of bandwidth and could be identified not only by machines but also by radio operators without any decoding device. Shore stations used repeat loops for their callsign, so they could be observed like radio beacons (Figure 87). According to this example, Tuckerton Radio in New Jersey was on stand by on 8/12/16/22 MHz on defined channels and available for any call. Just this is what the callsign loop tells us. Pairs of frequencies for voice calls and data traffic are separated within the marine bands. In the 8 MHz band, ships from any ocean in the world call on frequencies in the lower band near 8,200 kHz as an example and ships listen to the assigned frequencies in the upper band around 8,700 kHz to the transmission of the shore station.

SITOR ARQ. During the operation of that type of bidirectional transmissions, only one conversation partner was the master. Only the master was able to transmit content in packets with a defined length. The slave sent only a short control or status feedback after any master packet. The answer could have been "ok" or like "please trans-

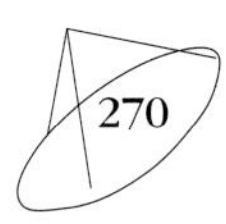

mit again" in case of error.As we can see, the semi-duplex mode was used, was an error-correcting code. Errors were identified when a received data packet did not have a 4:3 relation in the data words with 7 bits (0 or 1). Extra status information or commands were available for start, for the end of transmission, and for toggling between master and slave state. So both conversation partners could alternately transmit data packets with content to each other. While monitoring both conversation partners during a transmission, the master could be identified because he transmitted the longer packets, three data words in one packet. The slave answered every received packet with only a single data word as feedback.

```
cq de xsg xsg xsg qru~ bk cq cq cq de xsq xsg xsg qru~ bk cq cq cq de

12856.6 kHz    cw   Shanghai Radio    tst  11:30   31.3.96


16 22 mhz obs~ amver~ k cq de
wsc qsx 8 12 16 22 mhz obs~ amver~ k cq de wsc qsx 6 8 12 16 22 mhz obs~

12878 kHz    cw   Tuckerton Radio   New Jersey  USA  tst 13:10   24.3.96


cq de a9m qsx ch 4/5/6 k cq de a9m qsx ch 4/5/6 k cq de a9m qsx ch 4/5/6

12709.5 kHz    cw   Bahrain Radio   tst   13:20  24.3.96
```

Figure 87: CQ-loops in CW Morse indicated in which maritime bands, the shore station was QRV or ready for exchange. Short-wave radio communication was used by every ship on any ocean in the world to keep in contact with shore stations in their country or continent. The examples above show transmissions from Shanghai Radio in China, Tuckerton Radio in the USA, and Bahrain Radio.

Communication between naval ships and naval bases was quite similar to the described procedure. Also amateur radio used an operation mode called AMTOR which was run the same way.

Even international professional services and embassy communication operated with similar half duplex modes. Interpol as an example used AMTOR in the 1980s and later an advanced operation mode with four carrier frequencies and a more complex structure of commands and data words. Similar to the basics, but with significantly higher complexity and many more details, all this is typical for modern public cell phone networks. Cell phones and their switching centres exchange status information, and they send commands for starting a phone call or ending a call. The high complexity is a result of a variety of service information, and a lot of status information coming from the active base station and their neighbouring channels, for example. Quality of service, commands to reduce RF power, all this is exchanged while we have a call in a cell phone network. Voice traffic is only one type of data that is sent through the networks.

When we think of internet use, there is still some more data with special content to be exchanged. Authorization data and authentication data are needed to get access to defined applications. Internet also needs exchange services for transmitting pictures and emails. All this reveals an immense increase of data exchange for communication services.

To come back to the technical processes on the data exchange layers, we have already looked closer at some protocols that are needed for data exchange. But there are further standards and rules. Some of them were already mentioned in chapter one in the context of the different types of modulation.

Other ARQ Modes

Automatic repeat-requesting modes (ARQ) exist in different variants, as explained in [8]. One of them is Twinplex, working with four RF-carriers instead of two. Twinplex is working like AMTOR, but with four bits per state. Twinplex was used by Interpol on shortwave. Data exchange, as in ARQ-AMTOR, was the forerunner model. Two stations transmit alternately, while one of them is the master and the other sends only status messages as a slave.

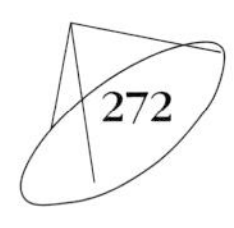

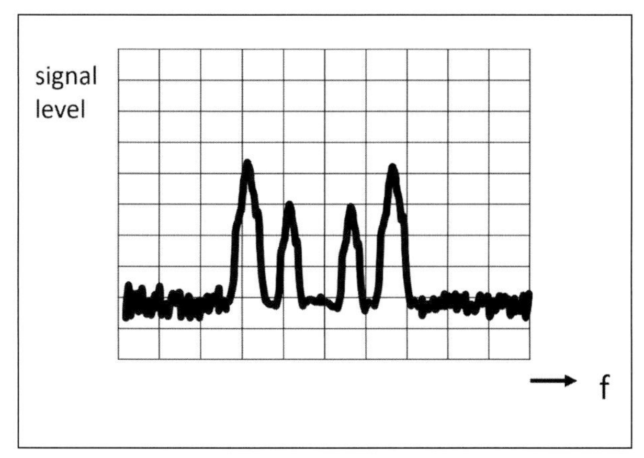

Figure 88: Typical spectrum of a Twinplex transmission. Derivatives with 8 or even 16 carriers, as in Piccolo or Coquelet, use a wider spectrum due to more carriers. Also DMR uses 4FSK as a modulation type with four carriers.

Another category of ARQ modes use continuous transmission with two carriers and two channels. That means two channels with continuous broadcasting are active. On the protocol level, these modes work similar to data exchange in ARQ-AMTOR mode, with alternating roles of both conversation partners (master/slave). These continuous transmissions in ARQ-E mode and ARQ-E3 mode were often applied in embassy communication. During monitoring they were often mixed up with FEC transmission modes due to their structure and outer appearance. ARQ-E, ARQ-N, and ARQ-E3 are examples of advanced coding technologies used by Ministries of Internal Affairs and also by embassies in foreign countries. Today most of them have no practical value, but they still have their place in history. Professional transmission modes have replaced them. They use secure transmission modes like ALE (Automatic Link Establishment) to be safe against eavesdropping and any intruder.

Error Detection and Error Correction

Although Forward Error Correction (FEC) is a general feature in information technology, this term and its abbreviation are also applied to some dedicated data transmission modes, mainly used on shortwave. When we use it in this context, FEC has a more specific meaning, denoting a special code family. Broadcast transmissions for data on shortwave without a back channel for status messages were done using FEC (FEC 96, FEC 100, FEC 144, FEC 192). In the case of unstable transmission channels which is typical for shortwave propagation, this mode can detect errors and lost data on the receiver side. And the mode is also able to correct transmission errors. First, the check of a 4:3 share of ones and zeros is executed, and when an error is detected, it can be corrected. For that reason FEC modes

518 kHz NAVTEX Rogaland NOR
SITOR-B 100
25.2.99 21:48

BUOY FLASHING QUICK WHITE HAS BEEN LAID OUT APPROXIMATELY 150 METRES NORTH OF THE POSITION OF THE LIGHTHOUSE. FURTHERMORE A YELLOW BLACK YELLOW LIGHT BUOY FLASHING 9 QUICK EVERY 15 SECONDS HAS BEEN ESTABLISHED APPROXIMATELY 150 METERS WEST OF THE POSITION OF THE LIGHTHOUSE. SHIPPING IS REQUESTED TO KEEP CLEAR OF THE POSITION OF THE LIGHTHOUSE
NNNN

ZCZC LB85
ROGALAND RADIO 990225 2010UTC
GALEWARNING NO 39
THE COAST STAD - FROEYA:
FROM EARLY FRIDAY MORNING SOUTHWEST NEAR GALE FORCE 7.
NNNN

ZCZC LB86
ROGALANDRADIO RELAYED FROM LYNGBYRADIO
990225 2130 UTC
WARNING NO 258
GALE WARNING FOR
EASTERN- AND WESTERN BALTIC, THE BELTS AND THE SOUND:
FRIDAY WEST 15 M/S.
SKAGERRAK, FORTIES:
SOUTHWEST 15.
FISHER, GERMAN BIGHT:
THIS NIGHT SOUTHWEST 15.
TAMPEN, VIKING:
SOUTHWEST VEERING WEST 15.
FAIR ISLE:
WEST 15.
THE SEA WEST OF THE HEBRIDES:
WEST 15, THIS NIGHT SOUTHWEST 18.
NNNN

Figure 89: Nautical warning for maritime areas are broadcast from European shore stations like Rogaland Radio in Norway on 518 kHz.

transmit a redundant version of the content with a slight time delay. In the case of errors, the receiver uses the redundant data during the same transmission to replace wrong or lost data words. Still today, a prominent maritime service uses FEC transmission, also called SITOR-B mode, with 100 baud for shipping. In areas of the sea that are covered by NAVTEX messages, ships can have the great benefit of reading about thunderstorms, nautical warnings about ice or the positions of dangerous wrecks, and reports from other ships. Monitoring this current maritime information can be life-saving for each ship's crew. Storm warnings, areas of dangerous mil-

16804,5 kHz unid
GMDSS
24.5.99 15:35

FORMAT SPECIFIER: Call to an individual station
ADDRESS (Coast): 219 1000
CATEGORY: Safety
SELF IDENTIFICATION (Ship): 272 875100
TELECOMMAND 1: Test
TELECOMMAND 2: No information
CALLED STATION RECEIVE FREQUENCY: No information
CALLED STATION TRANSMIT FREQUENCY: No information
ACKNOWLEDGEMENT: Call requires acknowledgement
Checksum OK

FORMAT SPECIFIER: Call to an individual station
ADDRESS (Ship): 416 236000
CATEGORY: Routine
SELF IDENTIFICATION (Ship): 416 234000
TELECOMMAND 1: J3E telephone
TELECOMMAND 2: Decoded value not valid
CALLED STATION RECEIVE FREQUENCY: 16585.0 kHz
CALLED STATION TRANSMIT FREQUENCY: 16585.0 kHz
ACKNOWLEDGEMENT: Call requires acknowledgement
Checksum not OK

Figure 90: DSC data telegram, received on one of the DSC emergency frequencies.

itary operations, and positions of explosions, all this is provided by several shore stations in all navigation areas. In European maritime areas, ships can monitor Rogaland radio, Cullercoats Radio, Coruna Radio, Niton Radio, and some more. The Ship's crews have to rely on current news and warnings as well as weather reports and forecasts. Without the internet and satellites, there is only shortwave communication left, or, as in this case, mediumwave. This successful format of transmission has been used for many decades. Accidents at sea could be avoided, and many lives could be saved by the permanent availability of up-to-date news and warnings. DSC (digital selective calling) is part of the GMDSS (global maritime distress and safety system), and in case of emergency they send short data telegrams, while emergency voice calls can also be sent in parallel. DSC has special reserved shortwave frequencies. Any ship and any shore station can receive emergency calls via DSC.

Some of the worldwide shore stations are responsible for NAVTEX transmissions. For safe operation, the maritime world is divided up into several navareas, and each transmission is done in its scheduled time slot. All these broadcasts take some minutes, while the next station will send their warnings within their own timeslot. FEC with 100 baud is really a successful mode. It has not only been used for maritime applications but also for diplomatic services in the 1970s and the 1980s. A huge number of daily transmissions of this type were in the air during the last century. Foreign diplomatic residences like embassies and consulates had their own communication channels to the Ministries of Foreign Affairs in their home country. They were used for requests in the case of hidden operations and for advice for the foreign staff in the case of diplomatic crises. Safe links with strong signals needed effective components like special broadband antennas like the logarithmic periodic antenna (see Figure 91). Modes of operation were almost well-known ones. In the early years of their application, data were even clear text (Figure 92). But many embassies started to use codes and cyphers for secure communication after some years in the 1990s.

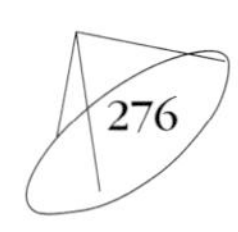

Figure 91: Embassy communication on shortwave used logarithmic periodic antennas in bands between 1.5 MHz and 30 MHz. They enabled worldwide links between diplomatic residences like embassies and consulates.

Many years ago, the internet was discovered as a transport medium to distribute breaking news to any place in the world at any time. But where did news come from in the time before the internet and where did they originate? Press agencies, domestic and international, prepared and managed news before we could read about in newspapers or view it on TV.

Some great national press agencies like DPA (Deutsche Presseagentur) and DWD (Deutscher Wirtschaftsdienst) flood their assigned RF-channels 24h a day. For domestic distribution, they used channels in the very long frequency band (VLF band) with special FSK modes operating with 4 carriers. They were special derivatives of popular FEC broadcast modes. Some of them, with four carriers, have also been used for ARQ operation with duplex channels. Interpol was one of the organisations that introduced more complex transmission methods to make the content secure against intruders by using individual codes.

```
ZCZCZC
RTI-15
FROM    PAREP RABAT
TO      FOREIGN ISLAMABAD
NO      POL-22/98  DATED 8 JUNE 1999
        S.O. AFRICA (II) FROM HOC
        REFYR FAX MESSAGE NO.MOR-5/12/98-AFR-II DATED 4 JUNE 1999
REGARDING FINALISATION OF THE MOU FOR ANNUAL BILATERAL CONSULTATIONS
BETWEEN THE MINISTRIES OF FOREIGN AFFAIRS OF PAKISTAN AND MOROCCO.
2.      WE HAVE REFERRED THE MATTER TO THE MOROCCAN FOREIGN OFFICE
AND ARE AWAITING THEIR RESPONSE.  REGARDS.
PAREP RABAT.
COL/RTI-15.
NNNN
ZCZCZCZC
RTI-10
FROM    PAREP RABAT
TO      FOREIGN ISLAMABAD
NO      ACT-7/2/98  DATED 4 JUNE 1999
        SECTION OFFICER EQ(II) FROM HOC
        REFERENCE YOUR FAX MESSAGE NO.EQ(II)-9/3/96 DATED
04 JUNE 1999 REGARDING NON RECEIPT OF OUR FAX MESSAGE NO.
ACT-7/2/98 DATED 12 MAY 1999.
2.      IT IS NOT UNDERSTOOD AS TO WHY THIS FAX, SENT THRICE,
IS NOT RECEIVED AT YOUR END WHILE ALL OTHER FAX ARE RECEIVED AT THE
OTHER END. MOREOVER, THE BAG CONFIRMATORY COPY WAS ALSO SENT,
THAT SHOULD HAVE ALSO BEEN RECEIVED. ANYWAY, WE ARE FORWARDING THE
MESSAGE AGAIN BY FAX, WIRELESS AND DIPLOMATIC BAG TO ENSURE THAT
IT REACHES YOU.  THE TEXT OF OUR ABOVE REFERRED MESSAGE IS REPRODUCED
BELOW.-
BEGINS
```

Figure 92: Diplomatic services used shortwave transmission for worldwide exchange of messages, like the following transmission from a Pakistani diplomatic service in 1999. Today, most of the data traffic of this type is encoded.

FEC as a family of error identification and error correction codes, is used in nearly every application in information technology. The objective is clear: During any transmission or data exchange, the participants need instruments to check if the transmission has been correct and complete. There are several codes and methods for checking this, like using hamming codes or making parity checks as an example. This is a large field in information technology, and the interested reader needs additional literature about the methods, efficiency, and functionality of encoding and decoding.

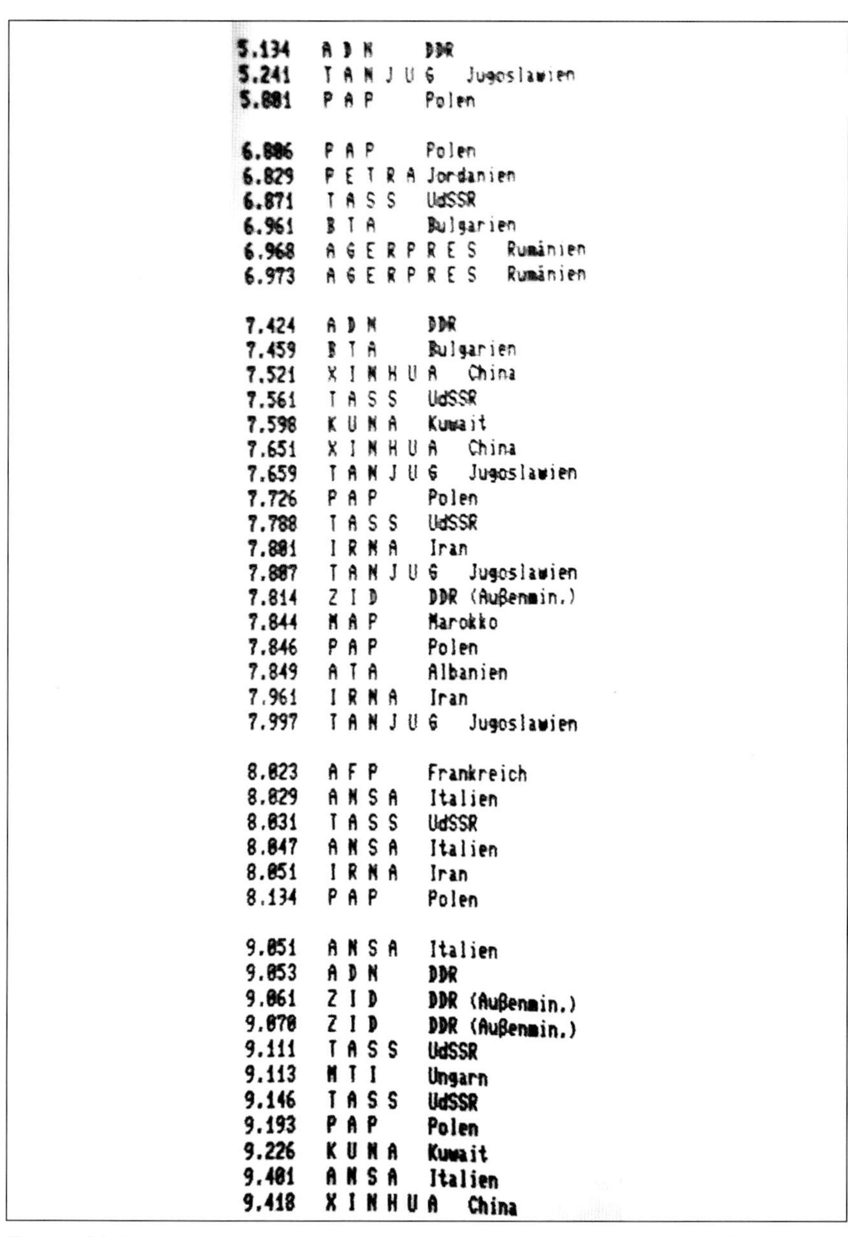

5.134	A D N	DDR
5.241	T A N J U G	Jugoslawien
5.801	P A P	Polen
6.806	P A P	Polen
6.829	P E T R A	Jordanien
6.871	T A S S	UdSSR
6.961	B T A	Bulgarien
6.968	A G E R P R E S	Rumänien
6.973	A G E R P R E S	Rumänien
7.424	A D N	DDR
7.459	B T A	Bulgarien
7.521	X I N H U A	China
7.561	T A S S	UdSSR
7.598	K U N A	Kuwait
7.651	X I N H U A	China
7.659	T A N J U G	Jugoslawien
7.726	P A P	Polen
7.788	T A S S	UdSSR
7.801	I R N A	Iran
7.807	T A N J U G	Jugoslawien
7.814	Z I D	DDR (Außenmin.)
7.844	M A P	Marokko
7.846	P A P	Polen
7.849	A T A	Albanien
7.961	I R N A	Iran
7.997	T A N J U G	Jugoslawien
8.023	A F P	Frankreich
8.029	A N S A	Italien
8.031	T A S S	UdSSR
8.047	A N S A	Italien
8.051	I R N A	Iran
8.134	P A P	Polen
9.051	A N S A	Italien
9.053	A D N	DDR
9.061	Z I D	DDR (Außenmin.)
9.070	Z I D	DDR (Außenmin.)
9.111	T A S S	UdSSR
9.113	M T I	Ungarn
9.146	T A S S	UdSSR
9.193	P A P	Polen
9.226	K U N A	Kuwait
9.401	A N S A	Italien
9.418	X I N H U A	China

Figure 93: Press agencies and state news agencies occupied many frequencies on the shortwave during the Cold War. Some examples from 1986 show channels from government-controlled agencies up to 10 MHz. They almost used RTTY with 50 Baud, whereas the use of FEC 100 was not so common.

Encoding

This chapter deals with a method for how an analogue signal gets a new representation with bits and bytes (Figure 25). The basic steps in the process of encoding and decoding are basically always the same. Digitalization is applied to analogue signals with music or voice, while the signal is changed into a data stream. One use case is a phone call where the voice is changed from its analogue representation coming from a microphone, into bits and bytes to be sent through digital networks. According to this simple transformation, no matter if voice, music, or a picture is the source, the digital representation can easily be sent over data highways. The representation in bits and bytes can be handled by digital devices like computers, CDs, servers, controllers, and other devices in information technology.

PCM (pulse code modulation) is one of the standards of information technology that is used to create digital representations. Music is one of the most popular examples, because everybody knows the good old analogue records and tapes, which have been replaced by compact discs (CD) carrying digital data. Music as a source provides analogue signals, which are encoded according to PCM as data words with bits and bytes. As it is an open standard, PCM is not about encoding to avoid eavesdropping by using encryption and keys. Features like this are part of a secure network operation, with TETRA for police as an example. TETRA used by radio amateurs, however, is also an open standard that generates digital transmission. Analogue voice is encoded without encryption, and intended to be heard by other OMs.

Encryption – from ENIGMA to TETRA

In this chapter, we have a look at encryption and its application in historical examples. We will see how important their use really is. When encoding procedures follow open standards, decoding can

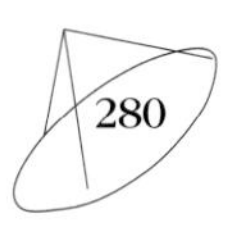

be achieved with only little effort by using just that open standard, too. It takes no big effort to reproduce data coming from any source. Even unauthorised users can do this. To avoid eavesdropping, encryption methods are applied to data streams that are intended to remain secure. We can find encryption in many areas, but mostly in environments where any unauthorised access is strictly unwanted.

Looking at WW II it was the German naval command who introduced ENIGMA, is a kind of typewriter machine that includes encryption options. Submarine fleet units should be informed about targets. When a CW code was transmitted, no enemy who was listening to those signals could understand the meaning of its content. ENIGMA was typically used on submarines where the encrypted CW Morse data were decoded easily. The naval authorities were sure for 100% to operate a system, encrypting messages so securely that no enemy would ever find out the meaning of the messages. This worked for a while. And nobody knew about the immense effort that was made to find out the secret mechanism inside ENIGMA. British experts did a great job cracking ENIGMA codes just in time to change the course of the war. The hunter became the hunted. Since messages from the German naval command were no longer hidden from British ships, it was the end of German supremacy in the Atlantic. Most of the submarines were discovered, fought and sunk. How careless to trust the promises of a machine!

Encryption is also a feature in modern messenger services and, of course, in emails. Again it is about, how to avoid eavesdropping and intruding by unauthorised persons or even authorities. Encryption is exceptionally important for organisations like police and fire departments. For a long time, they were damned to allow unauthorised listening to everybody with a little radio by slightly extending the frequency range below 88 MHz.

The state-of-the-art in professional radio for authorities with security tasks is currently TETRA (Terrestrial Trunked Radio). TETRA radio is used in many countries, and it is operated with a network structure similar to public cell phone networks like the former GSM. One important feature of TETRA is the use of secret keys for encrypted transmissions, so nobody can listen to police traffic any more. The use of encryption in TETRA requires huge effort in data processing with standard keys and encryption keys. Intruders would need immense computing effort and a lot of time. Still today, decryption of strongly protected traffic in real time is an illusion. Until the 1990s any radio enthusiast with ambitions for listening to police radio simply detuned their kitchen radio to extend the FM frequency range from 88 MHz down to 86 MHz. TETRA, however, enables secure communication through the smart use of a complex keying system. Only the operating authorities own the secret keys that are needed for the use of TETRA devices. Nobody outside has adequate means to compromise those encrypted data streams.

Encrypted transmissions also attracted attention in the Cold War. They were monitored on shortwave and included number groups, endlessly sent on frequencies from 3 MHz to 30 MHz. On these channels, the originator could easily address people in any other country. Receiving equipment was compact and could be hidden in small cases. As a message structure, almost five or three number groups were sent for at least half an hour. These voice transmissions sounded mechanically or electronically generated, and they could be monitored in AM or AM-SSB mode. Similar transmissions were done in CW Morse. Experts supposed that special departments of the Ministries of Foreign Affairs were behind that. The strange broadcasts were obviously meant for people who did not want to attract any attention. They were supposed to be persons affiliated with intelligence services, and they should remain hidden wherever they received the broadcasts. Any content behind the number groups was completely secret. According to insiders, decryption was only possible with the assigned one-time pad. Only the addressed person had this special, unique information as a decryption key on a little pad. After the transmission, the key was no longer valid and expired.

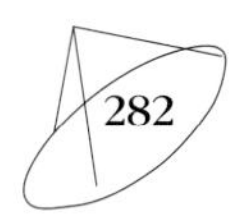

Eastern countries, such as the Warsaw Pact States, and also Western allies, like NATO countries, all broadcast an immense amount of classified information to their agents in foreign countries. Any transmission could be easily monitored by everybody with a shortwave radio, but the content remained hidden except for the recipient.

One initiative about strange transmissions was the CONET project in the 1980s. A large community was founded and activated to find answers to all these questions about number stations. Who were the originators? Where are their antenna locations? CONET generated a lot of valuable material about frequencies, schedules, and even transmitter sites. Number stations were no longer completely mysterious. But the world of spies and intelligence activities still has some secrets unsolved. Maybe just the things we don‘t know, like the secrets behind the one-time pads, will make the scene fascinating in the future. Still today, number stations continue their transmissions.

ALE Operation and Encryption

Operation methods in professional radio, like automatic link establishment (ALE) have replaced many of the recent ARQ modes [8]. Today industry provides small and very fast computing units (embedded systems) for dedicated applications. They have been designed for effective and secure transmission modes. Relevant information can be protected, and agreements between recipients and transmitting instances enable secure traffic. Those agreements can be about used frequencies, bit inversion for selected bits or for all bits, scrambling of transmission blocks, and more. This shall make a transmission secure against intruding and data corruption, and it was generally assumed to be a strong protection. But over the years, some methods turned out to be vulnerable to attacks and eavesdropping. Since then, the information has been completely encrypted by strong means. Some of the transmissions still use clear text for link establishment, but the content has been processed and transformed with strong keys to an encrypted data stream.The issue is that any intruder would need immense calculation power and fast computers, as well as a long, long time for decryption. In the end that makes no sense for private or semi-professional use. By using complex algorithms and keys, any content of secret transmissions is meanwhile safe and secure.

Frequency Hopping

The following example is about the RF level. Most transmissions use one fixed frequency for wireless applications like broadcasting. But there are reasons and options to enhance this scenario.The reason can be a compensation of amplitude dips when signal levels are getting lower by fading. The countermeasure is frequency hopping. It works similar to a switch to another frequency band with better conditions, almost like a very fast switch.Any amplitude minimum coming from interference can be compensated by switching to another wavelength.This will shift the minimum in the space domain.The second reason is to select another frequency band with better propagation conditions.

But there is still another reason, the third one, why the frequency should be changed during a transmission. It is about shaking off pursuers, which means escaping eavesdropping activities. When you change the used transmission frequency suddenly and the receiver knows the new one in advance, the transmission mostly goes on without any unauthorised monitoring.Those frequency changes can be done systematically, and they can be executed within very short intervals. Any intruder will remain on the left frequency without a signal, and he will not know where the transmission continues. Those ad hoc frequency changes simply need a proper schedule, which is known only on the receiver side. In public phone networks with TDM use (time domain multiple access), like in GSM, frequency hopping was used to improve the S/N (signal to noise relation).The logic for frequency hopping is fixed in the GSM system as a feature that can be activated, and mobile devices, and base stations know

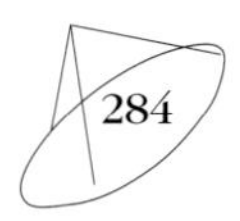

how channels are changed continuously. Using this feature, amplitude minima caused by interference can really be filled. As we see, frequency hopping is an effective feature, and it can be used in different applications with different objectives. Its use can increase the quality of service in radio traffic (2…3 dB typically), and it can also be used as a protective feature for secure data traffic and also voice traffic.

Localization

This chapter is about localization, or, in other words, methods by which persons and objects can be localized. There are very different applications on the market, and some examples are considered here. Electromagnetic waves are excellently suited to be used in some very prominent applications. Digital technology and localization methods make a perfect match. Identification and localisation belong close together in this context. Some technologies are applied for very short distances like NFC (cars, credit cards,…). RFID tags are applied for medium ranges, i.e., to identify persons in mining environments.

Other applications for localisation cover long distances, like the primary radar in air surveillance, which covers more than 100 km. While NFC is primarily used for identification and authorization procedures for access, the distances used are within a near-field coupling area, which means only a few millimetres. RFID is almost always used to identify persons or objects within a radius of a few metres. A stolen jacket at the exit of a warehouse can trigger a very unpleasant alert when the jacket is equipped with an RFID chip. The chip's response to the electromagnetic field near the exit triggers the alert. Generally, there might be a lot of applications for RFID use, like in mining, where huge machines are moving on their own and people are moving in danger zones. Similar scenarios exist in areas with radioactive radiation where people have to keep away. Some very special applications exist for tracing and tracking vehicles and machines in case of theft. Vehicles can provide their coordinates on their own when they are equipped with a SIM card for access to public phone networks. Last but not least, the same method is applied to cell phones. Most of us have already used the localisation function to show our position on a map.The number of applications is huge.

GPS

The satellite-based Global Positioning System was once developed for military use in the 1970s.Today it can be used by everybody, i. e. in cars. Since the permission for public use in the 1990s, we can all easily find out our current coordinates, which can be shown on a display with a map. Positioning information is quite precise with deviations of 3 to 5 m. This was a great booster for transport and traffic applications. Higher precision for positioning can be achieved with an additional DGPS receiver. This method can also be used for civil applications with deviations of 0.5 to 1m.

How does GPS work at all? Signals are transmitted via GPS-satellite on a primary frequency of 1575.42 MHz with a data stream of 50 bit/s. The format is periodically sent in a frame of 30 s for transmitting positioning data.The signal is processed to a special modulation according to spread spectrum technology. As a result, the signal is wider in spectrum and needs a bandwidth of 2 MHz. Spreading codes are digital patterns of 1023 bit/s and they are called chips.

They do not carry any information. They act as a kind of wrapping with a complementary code that is needed on the receiver side to reproduce the originating signal from the transmitted wide-band signal.This all happens according to CDMA (code division multiple access), which has also been used as an UMTS (universal mobile telecommunication system) in 3G public cell phone networks and also in space travel. Signal transmission on a secondary frequency (1227.60 MHz) is active for military use with higher accuracy. Each of the 24 satellites has a very precise atomic clock on board for broadcasting timing data and positioning data to earth. This is the

base for any positioning application on earth, while even the propagation delay of signals is considered while processing the data. GPS needs line of sight to 4 satellites for a precise determination of coordinates, which is done automatically by any individual GPS receiver.

Radar

Radar systems are very important for aviation and shipping. Of course, there are lots of other applications such as in space travel and for military use.The ISS gets warnings due to space junk from radar detection. Meteorological applications on earth and in space are in operation all around the world.They can be found even in consumer products like cars. Since the era of autonomous driving systems has begun, cars will be able to "see" with their own artificial eyes and process immense amounts of data in real time. Signals come from radar and laser detectors, as well as from camera systems. The time of autonomous driving with the highest level 5 comes closer. It is highly automated driving (level 5), which follows the previous variants for fully or partially automated driving. For many years, assisted driving has supported drivers in handling their vehicles with additional basic controls. Completely autonomous driving is still a vision and a real big challenge for engineers and for ethics experts. But there are still less prominent applications based on radar systems. Soil analyses in scientific archaeological projects and simple level measurements of a fluid surface, all this is done by people in many different fields of development.They all learned about the great benefits of radar systems.

The basic functionality of a passive radar system is always the same, no matter which application. Electromagnetic pulses of almost high energy from a transmitter are focused and radiated by an antenna with strong directivity.When the microwave pulses meet an object, scattered fields are reflected from the object with its individual radar signature.The backscattered fields contain information about the object based on its individual radar signature.The radar system can detect the reflected fields with their specific information. One of them is the runtime of the pulse, which is used for the determination of the distance of the identified object. As the reflected pulse has a special characteristic in the time domain, it provides additional information about the object. By runtime measurements, the resulting distance is calculated in real time, and the pulse can be monitored on a screen combined with the incident angle.

The angle is the result of the antenna's current rotation position, and it is used as the azimuth in processing routines. Area control centres in aviation use those large circular screens to observe all flight levels in the airspace to avoid collisions. In addition to the primary radar, the secondary radar provides more information needed by the air traffic controllers to give the right commands to all aircrafts in their sector. Secondary radar is an active radar, that communicates with aircraft by triggering their transponders. They answer with short data packets that contain essential information like id, airline, speed, and height above ground. Any information is routed to the display application, which shows the air traffic controller all the data which are needed for an overview of height, speed, heading, and some more to guarantee safe air traffic without issues.

Passive radar systems are always based on processing scattered signals from aircrafts, meteors, ships, and even cars (speed trap). Looking at electromagnetic waves which interfere with backscattered shares coming from an aircraft, radio enthusiasts can learn more about interfering of waves. When a signal of a strong transmitter like a radio broadcast produces a backscattered field by an aircraft above, the result will be some unexpected and exciting field interference. It is very helpful to use a waterfall diagram for monitoring. This special discipline in radio wave observation is a real challenge. Nils Schiffhauer wrote about that in [4].

For a long time, we knew radar from shipping. Each of us has already noticed the rotating bar antennas on small or big ships or towers near the shore. Most of them belong to radar systems transmitting

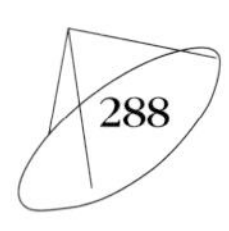

electromagnetic waves at approximately 9 GHz.Their design is comparable to waveguides or slot antennas for microwave radiation, and they are used for localising other vessels. In the azimuth, the angle in the horizontal plane, the antenna beam is narrow to achieve a good resolution in the horizontal direction on the display. The elevation angle is wide due to the small vertical dimensions of this antenna type. Height information about objects is not relevant because the application is focused on vessels.Airplanes use their own radar systems.Weather radar, for example, is installed in the nose of the aircraft and uses signals near 10 GHz to identify rainy areas and thunderstorms on their path.Weather radar stations are mostly operated by state authorities.They observe any meteorological event over a distance of 150 km.

Triangulation

Localising objects has been an exciting discipline for a long time. Electromagnetic waves provide information about the distance of objects by measuring and processing runtime information. Relevant information about the locations of objects or people is needed in mining, for example. Large machines move automatically while removing coal from a mining tunnel. Persons in the danger zone need to be identified, and the machine has to stop automatically when a person with a tag is identified within a certain distance. Public mobile networks can also locate people and objects.Triangulation as a method is very effective when base station signals are used.Any cell of a base station can use its runtime measurements.To get them, is part of the standard operation mode during daily use at any time. Distances can then be easily calculated. While combining results from three base stations measuring the signal runtime of an object, it is a proven method to draw three circles around the base stations with a radius resulting from the distance calculation.The object of interest is located where the circles are crossing. Base stations have almost three sectors or cells.They handle a lot of parameters at any time, like the identifiers for 6 to 12 further neighbouring cells.Triangulation as an instrument to locate an object is not as precise as GPS location, but every year several people or cell phones are found by this method. In some criminal cases, it was the last hope to get any information at all. Direction finding systems can also be applied as triangulation for measuring signal levels with directional antennas with a sharp beam. Long-range direction finding of short-wave transmitters is a good example.

Direction-Finding Methods

It has always been exciting to find out from where transmissions are sent. That is all the more true for radio enthusiasts and DXers who are, of course, curious about any transmission details. Professional applications and methods have been developed to find radio operators of pirate stations and anybody who interferes with professional radio. Direction-finding systems have always been very important in aviation and even in military use.

Doppler direction finders do a good job for estimating the movement of an object and the originating direction of wave incidence. Does it move towards our own site or away? Doppler direction finders reveal valuable information at a single site. When 5 to 15 vertical dipoles are placed in a circle with a reasonable spacing related to the used wavelength, the incident angle of an originating electromagnetic wave of any transmission can be found with surprising precision. The method is based on the Doppler effect, which is rather simple in theory, but it needs some processing effort to provide good results. Each antenna in the circle provides its own received signal. By querying each antenna one after the other with a fast rotation frequency, we get the relevant information. This method means a kind of circular movement of one single receiving antenna, but simulated by a virtual rotation with several antennas. Within one turn, this rotating virtual antenna moves towards the radio source and away from it within one cycle. By observing which antennas detect the movement towards the radio source, we will re-

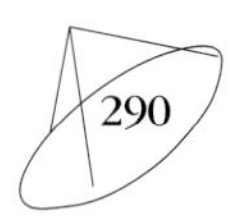

ceive signals with slight phase differences and an increase in frequency. The same happens for other antennas that simulate moving back from the source. The phase difference will then change in the other direction, with a slight and very short frequency decrease. It is similar to a slight frequency drift, which we know from the siren of an ambulance or a police car. When the radio source does not move, it will show similar frequency drifts in both cases of one turn. Moving radio sources can add additional phase shifts and frequency drifts by their own movements almost in the case of the radial movements of the radio source. Tangential movements will not change the drifts and shifts, but the affected antennas of the circle will change. So the azimuth changes, too. This means additional angular information about the object and its direction of movement.

A simple type of direction-finding antenna is the magnetic loop. We know already that its beam is ambiguous in the azimuth because it has two minima with 180-degree spacing in the horizontal plane. But an additional electric antenna, like a simple rod or pole, can compensate for the ambiguity. As a result, the combined antennas provide a very sharp and deep minimum in only one direction. Just this minimum can be used for manual direction finding.

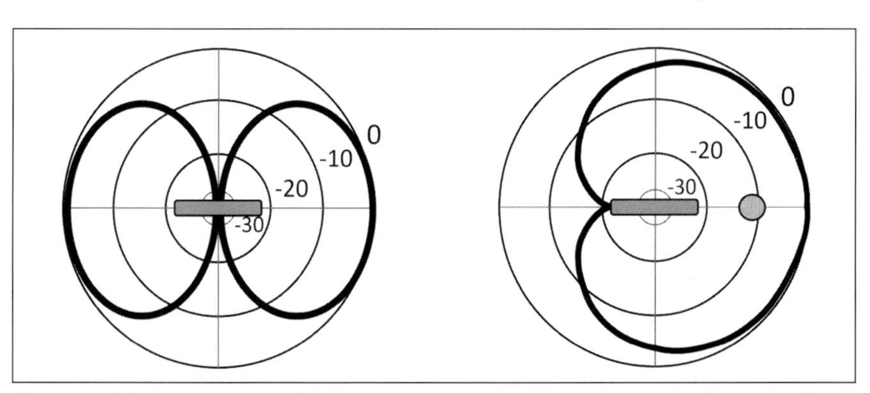

Figure 94: Diagram of a direction-finding antenna (magnetic loop), without an additional antenna (ambiguous, left side) and combined with an additional antenna (right side) to determine the angle of the minimum incident power level for clear direction finding of an incident signal. Top view, scale in dB.

The secondary antenna, almost an electric type, adds its signal to the signal of the magnetic loop antenna with a phase difference of 90 degrees. The result is really surprising. The horizontal diagram changes to a cardioid (Figure 94) and has one clear direction for its minimum.

This method was used in aircrafts in an era when jet planes were still a vision. Today, airliners and military aircrafts are still equipped with these simple elements as a secondary or backup system. On older turboprop airplanes, we could observe the loop and the stretched wire antenna as components for the radio compass.

Still today, any pilot of an aircraft or helicopter is able to find the right heading with the use of a radio compass. To be honest, today any airplane is equipped with intelligent direction-finding systems that use VOR (VHF omnidirectional range) radio beacons and MLS (microwave landing systems). They are very precise and support the landing procedure even in foggy and rainy weather with poor visibility.

Localization in Amateur Radio

Direction finding is an entertaining activity for radio enthusiasts who like to go out to search a hidden radio transmitter. Rough estimations of the incident angle of a signal can be done with a simple two- or three-element Yagi. This can even be sporty and funny when it is done in a group as a competition. When using Yagi antennas, it is usual to find the angle with the strongest signal level simply by turning the antenna in the assumed direction of the transmitter.

This method is precise for high-gain Yagi antennas, but they are almost too long and hard to handle. Using antenna types with a narrow and deep minimum in their diagrams, i.e., as in cardioid diagrams, the adjustment of the angle is more precise and has better results in finding the hidden transmitter.

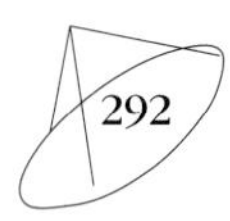

Movements in Space and the Doppler Effect

In 1986, the GRE (Giotto radio experiment) was activated, when the Giotto space probe came close to the Halley‘s comet. Scientists could observe the basic effect of wave propagation on one of the most exciting space missions. During the flyby near Halley, several experiments started their planned procedures, like the high-resolution camera and the mass spectrometer. On earth, engineers observed the encounter and used a huge parabolic antenna near Canberra in Australia. With a diameter of 64 metres the antenna monitored any detail of the encounter far away in space.

Scientists of the Giotto radio experiment (GRE) explored the mass occurrence of electrons with low energy (10 eV) and the mass expulsion of the comet‘s coma. Particles with different dimensions and weights hit the probe. As a consequence, a negative acceleration changed the speed of the probe, which could be measured in the radio signal as a deviation of phase and frequency. The responsible supervisor for this experiment was Professor Peter Edenhofer from the Ruhr Universität Bochum in Germany. The Doppler effect could really be observed and showed a frequency shift, which we also know from the siren of an ambulance car passing by. That revealed detailed information about the probe‘s movement in space, especially about its deceleration.

Sensors and Systems in Aviation

When we use the term system, it is almost the context that makes clear what is really meant. We already discussed naming and wording in an earlier chapter. Depending on the level or area being focused, the term system can mean different things. In this chapter, we look closer at methods and technologies using electromagnetic waves in aviation. Modern airliners are full of different antennas, sensors, and electronic processing units. They communicate with the ground stations even when it rains and severe weather makes landing an adventure. An airplane is a real treasure trove for those who are interested in technology, as well as for radio enthusiasts, of course.

Antennas on board

Looking at the antennas and sensors of an airliner, we will recognise that many applications and their systems are based on electromagnetic waves. Well-known examples are antennas for VHF and UHF connected to a voice transceiver inside the plane. Also RF antennas for long-range communication on shortwave are on board. Some antennas are placed under the fuselage of plane, i.e., as small shark fin antennas. Others are integrated into the leading edge of the rudder and in the wing tips. No matter where an antenna is placed, they all need free space for radiation and a minimum of other parts shadowing free radiation. ILS (instrument landing system) is implemented in airliners to enable safe landings, even in bad weather. No matter if fog reduces visibility or rain even in darkness causes a visibility of zero, ILS is a reliable system to support pilots in landing approaches.

ILS indicates if the airplane is on the glide path or if corrections in height or direction are needed. Signals from the ground-based ILS transmitter are received on board by integrated ILS antennas. The system is operated with a localizer and a glide path transmitter that broadcasts signals on frequencies between 329 MHz and 335 MHz. Antenna arrays of the localizer on the ground transmit their signals with amplitude modulation and 25…50 W RF power near the runway. Signals in a frequency range of 108 MHz to 112 MHz are radiated with information for the pilot about needed corrections. The ILS System has additional marker beacons to indicate the approach at a certain distance. These antennas radiate signals between 74.6 MHz and 75.4 MHz straight up, and they indicate the landing zone.

During a flight, it is most important to know the current position and the destination. The radio compass is a system that has been

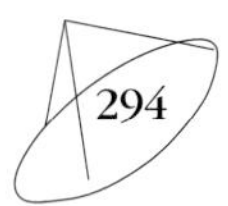

used for 100 years. The ADF is an automatic direction-finding system that uses NDBs (non-directional beacons), working in a range of 190 to 1750 kHz. These omnidirectional radio beacons radiate signals with little RF power in the medium-wave and long-wave bands. They are placed near airports and broadcast a Morse identification. Those reliable systems have done a good job, and they still do today. However, the more comfortable VORs with their large circle and group direction finders, provide navigational information to approaching aircrafts. To be honest, VORs are much more expensive and need plenty of space compared to NDBs.

But there are still other systems that provide important data to the pilot on their cockpit displays. One of them is the radio altimeter (RA), using microwave frequencies near 5 GHz. This radar system measures the real distance to the ground. 5G phone signals are suspected of interfering with the RA according to reports from the USA. Another system that supports safety during flight includes transponders for ground communication and provides their own position to other airplanes in the vicinity. By doing so, air traffic controllers have any relevant information about each flight on their screens, and they can manage all movements in the sky above to avoid collisions.

Last but not least, we have to mention the TACAN System. The Tactical Air Navigation System operates in a frequency range between 962 MHz and 1213 MHz, and it is a fixed system on military air bases and aircraft carriers. It broadcasts an identification code in CW Morse. The main functionality provides navigational information like the distance from the aircraft to the ground station. The operational range is up to 370 km. Pilots get the correct heading by using a special antenna that works similar to direction-finding systems, but in this case by transmitting a signal that is produced as a virtual rotating signal coming from the destination. The distance is processed by runtime measurement. The direction comes from the virtually rotating circular dipole group. Combined with the CW signal, the system works like an omnidirectional radio beacon, which provides additional navigational information for aircrafts to safely return to the base.

Aviation and Safety

When we are sitting in an airplane enjoying the flight, no one thinks about all the systems running to help the pilot in each phase of the flight. Starting and landing are two of the most spectacular procedures, and there are several systems using wave propagation to manage their functions without issues. Radio communication on VHF, UHF, SAT, and shortwave is needed to keep in close contact with the ground control for the whole flight. Messages about sudden thunderstorms and turbulence can be requested, and passing a critical area can be avoided. Reports about anything that might produce critical situations have to be communicated to area control immediately. In emergency cases, SOS calls have to transmitted.

To avoid this, transponder data with essential information about the flight status is provided. Air traffic control centres need these data to manage traffic and support the crew from the ground. Controllers need to know id, heading, flight level, speed, and position to manage air traffic in their sector. Other airplanes in front of the pilot are also displayed to get a certain overview of the traffic situation. Ground systems operating as primary radar scan the whole airspace over the assigned region. Other radar systems use electromagnetic waves to provide glide path signals for a safe landing even during fog, rain, and at night. In the air the pilot uses weather radar to estimate a course avoiding severe weather. There are still some more systems, and they are all available as redundant systems. On a flight, nobody needs any incident jeopardising safety on board. All systems follow strict procedures since they have been developed and tested. Safety and security standards are the backbone of safe and secure flight traffic. So everybody can really enjoy their flight without thinking of all the systems around them.

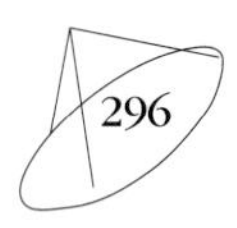

Repeater for Mobile Radio

Repeaters are bidirectional amplifiers, that are combined with two antennas to enhance the coverage of base station sectors. They are also called cell enhancers. This can be an effective and cheap improvement in signal coverage because, in hilly terrain or deep valleys, radio coverage can be very poor. As base stations are more expensive and need transport network access, the effort can be reduced by

Figure 95: Components of a repeater station: connecting antenna, repeater, antenna for coverage and cable.

using repeaters as cell enhancers. They pick up the signal of a base station from the air with a connecting antenna and radiate it after amplification over a second antenna, which generates additional radio coverage, i.e., in a shadowed area. Repeaters are operated on the channels of a dedicated base station. Broadband types amplify not only selected channels but a whole frequency band (band-selective repeaters).

One popular scenario for repeater operation is to feed a tunnel with radio signals. In long tunnels, radio signals disappear after some hundred metres, radio channels remain silent, and phone calls are dropped. Therefore, repeaters are operated by picking up the signal of the closest base station cell with its connecting antenna at the entrance of a tunnel. After amplifying the signal, the repeater radiates the signal at the same frequency into the tunnel with a high-gain antenna, that is with a narrow beam. As an alternative to the supplying antenna, the repeater can feed its signal into a slotted cable, which is radiating over a long distance along the tunnel. Slotted cables have a shielding with disruptions like slots, which couple a defined portion of electromagnetic energy to the vicinity for some metres radial distance from the cable. This is an appropriate method to provide service for phone calls, radio programmes, and professional radio like TETRA communication. Radio signals from the devices inside the tunnel run just in the opposite direction (reciprocal functionality). They are picked up from the slotted cable or the supplying antenna, amplified in the repeater, and then transmitted over the connecting antenna to the base station.

Antenna Preamplifiers

In some cases, base stations are operated with antenna preamplifiers (APA) to keep the high sensitivity of a base station high, even if a long coaxial feeder cable causes 2…4 dB loss in the receiving path. The same loss in the transmitting path can easily be compensated by increasing the transceivers‘ RF power level by some decibels.

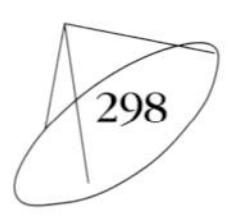

Compensating the lost sensitivity will be successful by using the antenna preamplifier close to the receiving antenna. Using the APA at the bottom and close to the base station while the receiving antenna is on a high tower with 40 m as an example means a poor S/N compared to an APA on top.The cable loss degrades the sensitivity on the receiving path and increases the noise level.

To operate a receiving system, it is always important to place the most sensitive stage close to the antenna.This can be a sensitive amplifier with a low noise figure of 0.9…1.2 dB for example.Those who use a long cable first and the sensitive amplifier behind will add the noise of the cable, which means the cable loss is added to the received noise level and reduces the S/N. In this case, amplification is applied to S and also to N and the S/N remains poor.

Converters

Some applications need a transformation of a whole frequency band to a lower frequency band, i.e., a radio signal in the operational frequency band near 11 GHz will appear 10 GHz lower. This happens in satellite receiving systems with a dish and an LNB on the roof and a receiver in the apartment. Connecting the antenna signal to 11 GHz directly with a cable to the receiver, a huge loss of the coaxial cable with 30 m length will make the signal unreadable. But the high loss can be avoided easily. The solution is a "Low Noise Converter" (LNC) or a "Low Noise Block Converter" (LNB).The converter picks up the signal in the focus of the paraboloid with a feedhorn. Then it is transformed to a lower frequency band with the help of a local oscillator, i.e., 10 GHz.At last, it is amplified and ready for descend, down into the apartment with a coax cable. What is the difference? The same signal in a lower frequency band (i.e. 1 GHz) is reduced by only a little loss in the coax cable. The loss of signals at 11 GHz in a coax cable would be enormous and in the end, unreadable. By the way, it is in the intermediate frequency (IF) range, which helps us use cheap consumer cables and adapters.

Practical Experience

This chapter includes a lot of practical aspects while operating and using radio systems, especially some aspects when technology does not work as it should or as we expect. But without a little theory, the scenarios cannot be described.

Most of us have already experienced this in the past: During a phone call, one participant can be heard without problems the other comes in with disruptions or cannot be heard at all, maybe because he is located at the edge of a cell where his signal does not reach the base station any more. An annoying imbalance. The link balance is lost, and uplink and downlink do not behave the same way. This was the diagnosis in most cases in the first decade of digital cell phone networks. Sure, there are a lot of other issues that caused bad quality and disruptions in phone call links. But let us look a little closer at the described phenomenon. Due to its frequent occurrence, there might be a systematic root cause behind it.

Issues with various causes can make users impatient. Poor quality or disruptions are scenarios that are typical for radio systems and less for landline phone networks. Issues occur on any network level and even on user devices. Diagnoses are usually difficult to make. In many cases ,the initial diagnosis is „disturbed link", a starting point that promises some effort.

Basic Properties

Radio systems and their components are generally precisely specified. Surprises or even malfunctions rarely happen. But RF technology, a field that needs to be handled with care, contains, of course, special RF properties that have to be considered. Specifications of components are helpful, but not all issues can be treated by specifications. That is especially important for scenarios, where components interact.

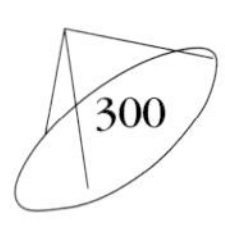

Link Balance and Imbalance

When we look at radio links and transmission lines more closely, we find issues like poor coverage or interference. But there is still another topic, which is sometimes underestimated by developers and radio planners. We observe it when pairs of links like the uplink and downlink of a duplex channel are unbalanced.

When we try to approach that phenomenon of imbalance, we can imagine a transmitter using an RF power level that is much too high, a really strong transmitter with good coverage, i.e., 50 W or more. What will happen during a phone call? Any conversation partner calling over a landline could be clearly understood by another conversation partner with a cell phone, booked in and moving into the specific base station cell. Coming closer to the edge of the cell, the participant‘s landline will still be heard clearly. But the mobile user transmits with little power, which cannot reach the base station receiver over that enhanced distance. Cell phones typically use 1 W.

To analyse this scenario, the link balance can be calculated for both directions. The first equation relates to the downlink of the base station, including RF power, antenna gain, cable loss, and propagation loss.

$$P_{S/BS} + G_{S/BS} - F_D - Kd = S_{E/Handy}$$

We have $P_{S/BS}$ as the transmitter output power. Typical GSM base stations of the 2nd generation used 38 dBm. $G_{S/BS}$ is the antenna gain of the base station antenna, which ist almost 18 dBi for standard antennas. F_D is the propagation loss and Kd is the cable loss for the connection between the base station antenna and transceiver input.

2 dB is a typical value for cable loss. F_D will be calculated later, because propagation loss is valid in downlink and uplink. $S_{E/Handy}$ is the sensitivity of the receiver in the cell phone. We assume that cell

phones and base station transceivers have a similar sensitivity. This value is also not needed because it will be calculated out when the equations are subtracted.

As a second step, the link balance from the cell phone to the base station (uplink) is set up the same way (Figure 96). When the second equation is subtracted from the first one, we can see below if the links are balanced or unbalanced. In the presented example, it is a pair of balanced links. We can recognise this by the validity of the equal sign. In this example, we assume the use of an antenna preamplifier for the base station. That means its use has to be considered in the uplink equation with an amplification of approximately 6 dB. We also assume the same sensitivity of both receivers, base station and cell phone.

$$P_{S/BS} + G_{S/BS} - F_D - Kd = S_{E/Handy}$$

$$\ominus \quad P_{S/Handy} - F_D = S_{E/BS} - G_{S/BS} - V_{AVV}$$

$$8\ \text{dB} - 2\ \text{dB} = 6\ \text{dB}$$

Figure 96: Both links, "base station to the cell phone" and "cell phone to base station" are balanced thanks to the use of an antenna preamplifier. The equal sign in the last line is then valid. When the equal sign is not valid because there are different numbers on both sides of the last equation, the links are unbalanced.

The value of 8 dB has its origin in the different transmitter power levels of mobile phones and base station. 2 dB cable loss is compensated by the antenna preamplifier. Therefore, this loss in the uplink needs not be considered. The high sensitivity of the antenna preamp is similar to the sensitivity of both receivers. When both links work with tuned parameters, as in the example, they are balanced. Without antenna preamp or simply increasing the base station power, this results in a significant imbalance. The cell phone user cannot be

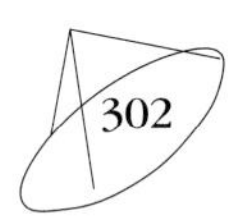

heard anymore at the edge of a base station sector. But he has a loud and clear signal transmitted by the base station. When we use an amplification of 0 dB, the equations are unbalanced as well.

Compensation of the imbalance is the topic, which has to be considered when long feeder cables are used for a base station. Except for reducing the transmission power of the base station, it is the use of antenna preamplifiers close to the base station antennas that can work against the imbalance. The preamps have two effects: increasing of the power level of the received signal in the uplink, and the second effekt is shifting the sensitivity of the base station receiver up to the antenna, as if no feeder cable would add any loss or noise to the received signal. Additional losses from duplexers are also compensated by antenna preamps. The strict objective is to maintain the high sensitivity of the receiver input.

Antenna Diversity

On many sites, we find three sector antennas for each sector, one for transmitting and two for receiving signals. The use of a second receiving antenna per sector is a standard practice for mobile radio sites. The following effect is in focus: While a vehicle is transmitting and moving through the streets, multi-path propagation causes minima and maxima, and the position of any receiving antenna on a roof cannot receive the maximum power, standing at a minimum.

A second antenna a few metres away from the receiving antenna has a good statistical probability NOT standing in a minimum of the transmitter's field while the first receiving antenna does. When both receiver inputs in the base station are combined, the result will be an increase of the average received power. The spacing between both receiving antennas is 2…5 m for GSM 1800 as an example, and the gain in urban environments can be 2…4 dB. Antenna positioning is like in Figure 98. The best effects can be achieved when antennas on a flat roof are placed at the edge or near the corners.

Figure 97: Mobile radio antennas for two frequency bands and different sectors for sectored coverage. At the bottom of the pole, there is a microwave antenna for connecting to the transport network.

Figure 98: Antenna diversity in a typical urban environment: two receiving antennas, one antenna for transmitting per sector. We can also observe that several providers share this site. Microwave antennas are also installed to connect the base stations to the assigned transport networks. Collocation of antennas is no problem, as we can see here.

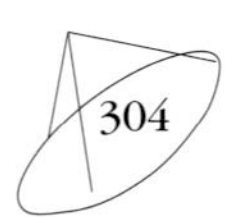

The effect of SDMA (spatial diversity multiple access) can be achieved, as shown in Figure 98 by a second receiving antenna with an appropriate spacing, which depends on the operating frequency. Inside the base station, a hardware and a software component combine the two signals from both receiving paths. As a result, the gain for the receiving path has a significant increase of 2...4 dB. The spacing should be 2...5 m for GSM depending on the used frequency band of 900 MHz and 1,800 MHz. The compensation of minima as in the receive path is not needed for the transmit path, because in case of low transmission power, the level can easily be increased by base station parameters for RF output.

When there is only little space for installing antennas for mobile radio without space for a second antenna, another scenario for antenna installation is preferred (Figure 97). It is the pole installation combined with the use of polarisation diversity. A single antenna of this special type contains a double number of dipoles. Some have a +45° orientation, others have -45°, two types for two receiving paths. Pairs of dipoles are constructed like an x inside the antenna.

This is an alternative strategy to compensate for minima in both receiving paths called polarisation diversity. As a result, the S/N of the received signal after combination is increased by 2...3 dB. In Figure 97 we can see two connectors for each path of the antenna at its lower edge, which means RF connectors for both polarisations.

Frequency Hopping

A similar effect can be achieved by using a second path that slightly differs from the major path. The difference between both receiving paths is then the result when different wavelengths are used instead of two different positions (of receiving antennas) as in one chapter earlier. Different path lengths can be provided by frequency hopping. When a frequency change is done very quickly, at the point where was a minimum before, there is probably a maximum of received energy on another wavelength. Many frequency changes in a short time frame generate a certain increase in the average received power level. Statistically, this method is proven to be successful. The process works with a schedule of agreed-upon frequency changes within a timing interval. Frequency hopping is done within milliseconds. The plan to generate additional gain in the receiving path with frequency hopping can be implemented just that way.

Frequency-hopping is also used as a strategy to prevent eavesdropping. When the schedule and the method of frequency changes remain secret to unauthorised persons and only the transmitter and receiver have that relevant information, intruders are locked out very quickly.

We also remember the reasons for frequency changes. When wave propagation changes due to solar flares, for example, and the loss gets too high, another frequency band might enable communication again. To sum up: the quality of radio links can be improved in their average power level simply by changing to different frequencies with frequency hopping as a method.

Resilience

For some systems, the importance of robustness has only a low priority, as in the fields of consumer electronics and hobbies. Professional systems, however, need more attention for stability and safety, due to their use in challenging environments.

Circuit and Implementation

When a circuit of an electronic system is still in its early development phase, mechanical robustness and other basic items can still be considered. One impressive example is when special electronic circuits are integrated into the door handles of a car. Before the final installation is done, they are almost filled with liquid resin to avoid water

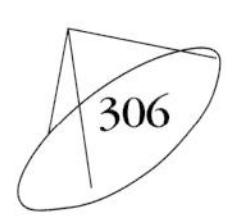

and moisture formation. Long ago, similar methods to improve the stability of elements in a circuit were used in coils and inductances. Their ferrite cores were also fixed with resin to maintain the tuned inductance. In automotive production processes, as in the case of NFC circuits, it is most important that the circuit remains dry, that the tuning of elements is fixed and safe, and that no shaking and no shock can degrade the functionality.

Modulation and Operation Mode

For radio traffic with good quality and without issues, it is important to exclude any expected or sudden interference just from the beginning. Therefore, developers consider different scenarios when they design the product. The method is to follow adequate requirements.

Which interfering signals do we expect, is the question. Are they short peaks of a millisecond? Or is it a broadband jamming what can be expected? Or is it simply the unstable propagation loss on shortwave? FEC (forward error correction) can help to implement a robust mode. FEC almost uses redundant transmissions with a certain shift in time. Other scenarios need frequency hopping or systematic frequency or band selection. Sometimes it is the modulation type that meets the requirements. Modulation types can be very effective, very resistant against interference, or suitable for extremely low power levels for transmission and receiving. By finding the right operation mode and modulation type, development and design may continue.

Bad Boys and Destroyers

The list of issues that can have a negative impact on receiving electromagnetic waves is long. When radio broadcasts were done with long vertical antennas combined with Yagi antennas for TV on the roof, the phenomenon of crackling interference was sometimes observed. The origin were electrostatic discharges when the air was charged, i.e., before and during thunderstorms. Radio amateurs with long vertical antennas observe this still today, depending on the weather conditions and charging of the air. The solution was a metallic ball on the tip of the antenna. This reduced the crackling significantly. The ball is a kind of capacity that collects charges and degrades the crackling effect.

On some sites, we can still see these antennas with a ball on them, most of them because nobody has deinstalled them in 50 years. Today, this is not relevant for consumers because radio no longer need tall verticals on a roof.

Interference-free reception and transmission with delicate components like antenna switches are sometimes impossible. Radio enthusiasts generally trust their equipment and use a switch in the antenna feeder to select between different antennas for reception. Others use splitters or similar components. They all suppose the components to be ideal elements. But in many cases, their use causes trouble. But why? When the cable had a good match at both ends before, some components cause interference when they are added.

This may have different reasons. The best measure against a bad surprise in this case is to make sure that the new added component is really specified for the used frequencies and that all gates have a good match, almost with a wave impedance of 50 Ohm and little loss. If possible, transmission loss and return loss should be measured too. Identifying interference sources can be really annoying. Some of them require a lot of effort. Others come from components we would never assume to be interferers. Official departments such as the Bundesnetzagentur in Germany can help. They have professional equipment like broadband directional antennas and spectrum analysers to fight any issue. Some interferers are presented in the following list (Table 4). They are almost man-made and cause the so-called man-made noise.

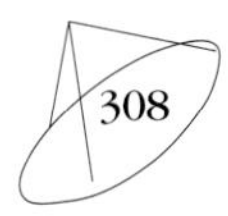

interfering sources	comment
electromagnetic pollution	Noisy interference is almost underestimated, and it has mostly a considerable level, and it is very annoying. The domestic spectral pollution is very strong inhouse, but it decreases after a few metres outside the building.
standing waves on shielding	When the shielding of the cable is getting a radiating element, interference comes like an invisible unwanted visitor into the house. Other devices are being disturbed. More noise while receiving signals.
switching power supply and charger	Widespread and hidden causes.
heating controls	This is usually thought of too late.
PLC - powerline communication	"Powerline communication" means severe interference coming from the domestic electricity distribution. Extremely undesireable for radio amateurs and radio listeners like swl.
interfering signals from the neighbourhood	i.e., from electric fences and plasma screens. Any device without or with a poor interference suppression are candidates.
interfering signals, coming from the receiver itself	It occurs in some superhet receivers with local oscillators. The intensity of interference depends on the filter quality and circuit concept.
receivers with poor quality	i.e., cheap receivers with poor signal processing. RF-filters like a band pass at the antenna input, are almost helpful.
circuits with poor quality (i.e., antenna amplifiers and distributors in SAT-systems)	attention: long-range interference due to installation under the roof. They can make trouble even in a distance of some km.
motors and generators with poor interference suppression	i.e., vacuum cleaner, drill, handmixer in the neighbourhood

Table 4: Man-made noise. Many devices turn out to be bad interferers.

Lightnings and Overvoltage

When lightning strikes, it may cause tremendous damage. They can start life-threatening fires and destroy whole existences. There are good reasons to take the force of lightning seriously. Antennas on buildings are almost the highest points of an area, and they attract lightning more than other objects. Therefore, it is a good idea to call an expert to install professional protection against lighting. Any radio amateur and enthusiast with a roof antenna should consider that.

Relevant information on lightning protection can be found in standards like DIN EN 62305, a European standard and additionally in VDE 0185 305 in Germany. Only personnel who are qualified and authorized to install lightning protection should really do that.

Certified experts can plan and do the installation, and not to forget, this is also a good reinsurance in case of damage when lightning strikes and causes immense costs. If cross-grounding is allowed at all or how lightning protection should be planned at all, each of these questions should be taken very seriously and belongs in professional hands.

Attracting lightning and leading it to the ground is nothing, which relates simply to high currents in cables.The violence of a lightning strike can be so dramatic, that the overheating of unsuitable cables may cause a fire inside.As currents are extremely high, only cables with a certain diameter shall be used because the temperatures of cables caused by high currents can increase unexpectedly.

Any careless and unprofessional installation increases the danger of fire and the risk of explosions. Grounding rods are put into the ground more than 9 m deep to dissipate high currents.As explained above, only an expert should plan adequate protection. Please ask them for support.

But why is lightning so dangerous, and what kind of damage does it cause? Lightning strikes in the vicinity can generate high peaks of voltage in the public or domestic power grid. Such overvoltage peaks can damage any device that is connected to the power grid at the moment of the lightning strike, even if it came from a certain distance. Devices can also be overheated by that and melt away, which causes high damage inside. In the worst case, this can start a fire. One very old measure against it: simply pull the plugs from the public power grid during thunderstorms and lightning. Today, we can use secure sockets at home to prevent the worst.

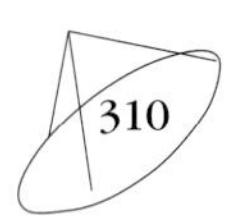

Last but not least: When a thunderstorm approaches, anybody shall immediately leave the roof. Any work on roofs during thunderstorms and lightning is strictly forbidden. It is also a bad idea to run around outside during thunderstorm. Lightning is almost always attracted by large trees. Therefore, one should keep away from trees. When there is no protective building or car to go into, it is getting dangerous. On a meadow, for example, it is helpful to squat down, feet close to each other, hands away from the ground.

With this protection, we can prevent the worst. Any voltage caused by a lightning strike in the vicinity can induce high currents inside the bodies of animals and humans who have a distance between their four feet on the ground. As the current runs through the meadow and two feet are on the ground, the level of induced current inside the body is reduced when the feet are closer together. High currents through a human body can lead to death or severe injuries such as skin burn or cardiac arrest.

Good Guys

Disturbed transmissions and bad reception sometimes spoils the joy of radio listening radically. Let us look at measures which can help us, the good guys. They give us hope that improvement is possible.

The first and most effective measure seems to be evident or even ridiculous. But it is not. Any expected interference and probable interference shall be excluded first. See the chapter "Bad Boys". In most cases, it will help to have a much better reception. But that would be too easy. In many cases, there are still sources that cause trouble. Domestic noise as an example of man-made noise is really insidious. But there are some good strategies and tricks with little effort to avoid severe and annoying interference. One basic precondition for good reception and transmission is a good matching of the antenna. Taking care of this is underestimated by many beginners. But this matching is the most important requirement, that has to be followed. When this is done, anything else will be a minor issue. But if there is no matching in the line, there is interference in the feeder. Sometimes it comes from the wrong adapter, which increases noise and disturbances and prevents clear signals.

When electromagnetic waves run through the cable and see 50 Ω anywhere from the antenna to the receiver, including adapters, then some big issues will not come up at all. This is valid for both directions of the feeder cable, transmission and reception. Any signal from the transmitter output to the antenna needs to see 50 Ω without sudden changes in the wave impedance of the line. The same is valid for received signals. When the antenna is matched properly, 100% of the transmitter power can be radiated, and received signals have almost good quality. Exception: when disturbing signals or noise run through the radio channel, i.e. transmitted from the neighbourhood.

Good Match

That means considering the correct wave impedance of all components which are connected. In most cases, the antenna socket of a transceiver is designed for 50 Ω wave impedance. So it is the best idea to use a cable with the same wave impedance. Also the antenna shall have an input of 50 Ω wave impedance for the used frequency band. Some antennas have a wave impedance of 240 Ω or even more.

Therefore, other types of feeder lines can also be a good choice. In the case of a symmetric antenna input, feeders like parallel wire lines are useful. Taking care of the right match between cable and transceiver socket impedance is most important. Matching is correct when 100% of the generated power runs from the transmitter to the antenna. Any reflection above a certain limit can destroy the transmitter. While receiving a signal, a good match means that even a weak signal is not interfered with by any other signals from the environment outside the cable, like interfering patterns of domestic man-made noise. Generally, this is a reciprocal system without a third gate when we achieve perfect matching.

But is the model of a two-port or two-gate device really the appropriate explanatory model in this context? When we extend the model to a three-port device, we might be closer to real-life circumstances. And we might understand much better some effects that could not be explained without the third port. Which ports could be meant, or which are useful for the model? Let us consider a virtual T-shaped component with three ports as a model. In addition to the end of the cable and the antenna input, there is another (invisible) port that might radiate or pick up signals, even if the antenna and cable end are connected. When we are sure that we have no matching loss i.e. in case of perfect matching (at the transition cable/antenna), then we need no explanatory model with three ports, but only with two. Just here, we need to be careful in our consideration.

In any scenario, when a guided wave meets a sudden change in wave impedance, we can expect reflections and even unwanted radiation. As a general approach, we start with a non-ideal transition between cable and antenna. The transition cable-to-socket of the transceiver, however, does not need further attention because we do not have any change in impedance at this point. In case of pl-connectors, we should know that above 30 MHz the wave impedance of a pl transition changes to undefined values with effects on the transmission properties, as shown in Figure 67.

But how can we avoid issues and interference at the input of an antenna, just where the RF cable is connected? And what happens if we are not successful? The answer is short and clear: If the antenna input matches (for a defined frequency) to the wave impedance of the cable (almost 50 Ω for coax cables), there is no leak in the RF line. We can use the explanatory model with two gates. A prominent example is a $\lambda/4$ radiator on the roof of a car, which is connected with a coax feed under the roof with 50 Ω wave impedance, while the shielding of the cable is connected with the flat roof. Similar results can be achieved by using balanced antennas with a parallel wire line as a feeder, assuming the same wave impedances for both, the antenna and feeder.

When a balanced antenna like a dipole is connected to a non-symmetric feeder line like a coax cable, the result will be a sudden change of wave impedance. In use cases like this, a transformation in the wave impedance can avoid that mismatch. Baluns (balanced/ unbalanced) or a gamma match can be applied in that case. Baluns can make an up-transformation or a down-transformation, depending on their wiring and their coils inside. Advanced literature and assembly instructions for practical use can be found in [7].

Once again, to make it clear: when electromagnetic waves are sent through a coax cable to the antenna, we can generally expect two cases.

Primary: The whole electromagnetic energy is radiated from the antenna. With a measurement, we would see a very good VSWR and a return loss of 20 dB or more.

Or secondary: A certain share of the energy to be transmitted by the antenna is reflected from the antenna input, and another significant share at the transition from the cable to the antenna is radiated from just this transition, like through a leak in any direction. This share does not reach the antenna, but it is radiated and sometimes even moves as a standing wave on the shielding of the cable. This standing wave causes a lot of trouble. It radiates, and it even comes back into the shack like a Trojan horse. Also in the second case, the measured return loss has an acceptable value like 12 dB or better and we might assume that everything is alright.

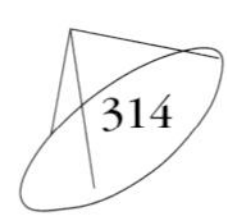

But measuring only the return loss does not show a big difference compared with the primary case. With a precise free space measurement, we could verify this because, on the transmission line, there is too little power level in the far field. The missing part interferes with leaking radiation and causes trouble, i.e., in other devices like in neighbours‘ TV or radio. The described scenario needs patience and effort to measure all shares. And who has the opportunity to make precise far-field measurements? A challenge to find creative solutions.

But how can we imagine such an unmatched transition (cable to antenna) in the case of receiving a signal (antenna-to-cable)? The explanatory model is quite simple. Imagine a cable with three coupling ports instead of two. Where the cable is connected to the antenna, an additional leaking port is assumed.

One is the port for energy coming from the antenna and going into the cable. The other port is the leak, a kind of virtual hole in the shielding, the virtual third port in our three-port model, as described earlier. This gate is really an unwanted gate because it is also the entry for man-made noise in the near field and for any other interference from disruptive emissions such as generators, neon lights, and vacuum cleaners.

But there is an obvious indication of a leak or a poor transition: When the VSWR of the antenna is measured to be moderate or even good and, at the same time, received signals have poor S/N, less than we expected, then the existence of a third leaking port is probable. The result is an increasing noise level.

The N in S/N increases. That‘s a big annoyance for all who are damned to operate from a poor site, sometimes even with non-ideal antennas and feeder lines. Interference is coming in from the antenna (the RF-channel) and also over the shielding of the cable when the matching is poor, two sources of noise.

Figure 99: The design of the marked area (antenna base) is most relevant for a good match between the two opposite arrows (ports). The third arrow indicates a probable virtual port, which is responsible for receiving man-made noise and interference from the vicinity (near field) in cases of poor antenna matching.

Looking at a portable receiver with a plug-on antenna, we can apply what has been explained before. The two different antennas in Figure 99 can be seen in Table 5, marked as antenna 4 and antenna 6. The arrows focus on the critical zone, which can already radiate energy before it stimulates the antenna in the case of transmission. When signals are received, this can be a leak for the ingress of near field interference and noise.

But how can this be? Recommended and purchased antennas generally have a good matching. Their connectors are designed for 50 Ω antenna base impedance. What might be the issue? The antenna on the right side has indeed a good matching for two bands, 2 m and 70 cm, according to the specification. On these frequencies, interference and noise levels are really low. This is what we want. One little exception should be mentioned: when a noisy transmitter is received or the source is a noisy broadcast. Of course, on receiver side any received noisy signal from the far field cannot be compensated anyway.

When the matching is really good, then interference cannot intrude from outside through the leak, which means from the vicinity (near field). While using broadband receivers, we will often find frequency bands or sections with poor matching and bad reception with a poor S/N. Just in this case, the third leaking port is "open" (third arrow next to the opposite arrows in Figure 99).

When we consider the arrows as the three gates of a three-port element, where energy gets lost while transmitting and where ingress causes interference while receiving, then the scenario can be explained as follows:

The marked area in Figure 99 is generally a transition with a good match between the socket of the receiver with 50 Ω and the antenna base with 50 Ω as well. But the world is not as ideal as we might assume. We can use a model to explain the real properties of the marked transition.

According to the explanatory model, we have a transition with parasite elements as additional capacitors, inductances, and resistors, some as a series resonant circuit, some as a parallel resonant circuit, connected in series or connected to the shielding. All those combinations will define a special individual behaviour of your antenna. Altogether, it is an equivalent circuit, which stands for the characteristics of the real mechanical implementation of the transition area.

This area has a decisive influence on the behaviour of waves passing here. In addition, this area and its equivalent circuit have frequency-dependant characteristics. We know this frequency dependence from many antennas, like the RH660 (right side in Figure 99).

The RH660 is optimised for operation in some special frequency bands. But in some others, the antenna only works with moderate or poor results. A closer look at the return loss of an antenna reveals,

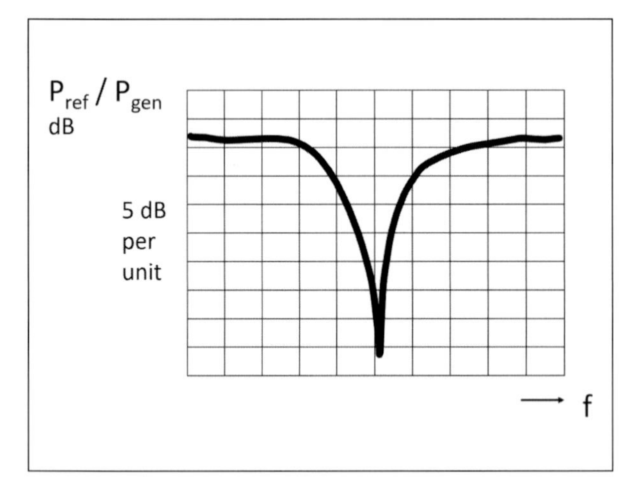

Figure 100: The ideal resonance of an antenna (represented here by measuring the return loss at the antenna base). Unfortunately, this cannot be observed for every antenna.

if it is really suitable for operation in the relevant radio bands, i.e. when operating a broadband receiver. Generally, we expect that an antenna will show good results for reception in the specified frequency bands, as they are optimised for them. This is at the same time the weakness of most antennas for broadband reception. Of course, they are deaf in some other radio bands, for which they are not specified.

Consequently, when antennas are tuned to preferred frequency bands, they will work really efficiently here. The matched range is sometimes 8 % or less of the complete range.

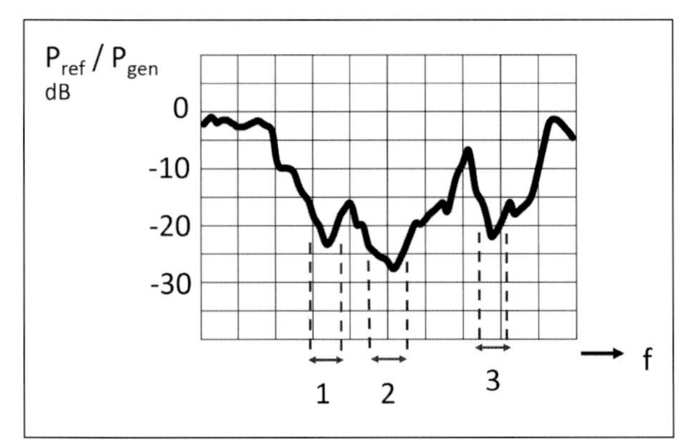

Figure 101: Broadband antennas have their preferred radio bands for good radiation and reception (section 1, 2, and 3). Between and beyond these frequency bands, the antenna shows poor results.

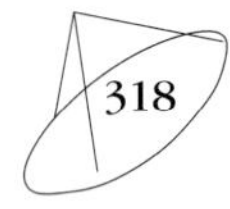

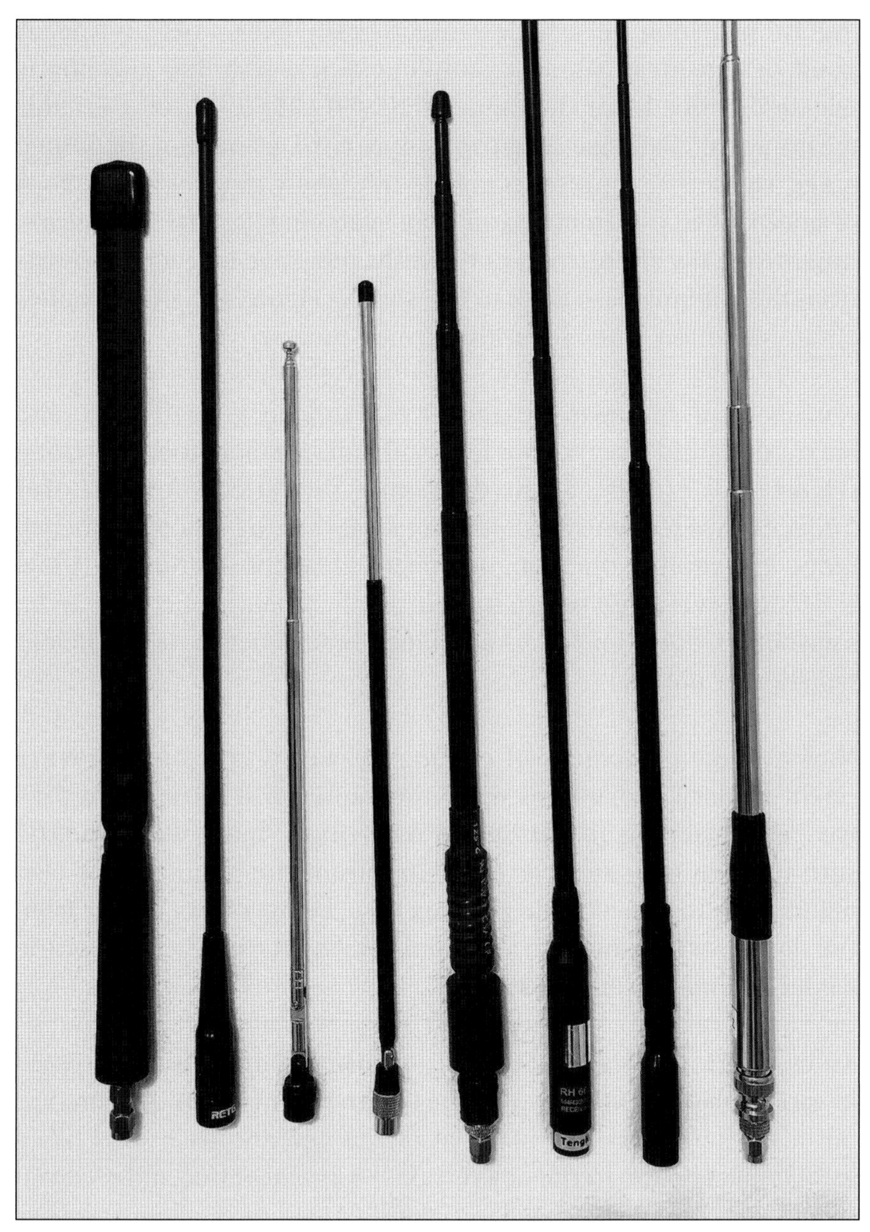

Figure 102: Rod antennas and telescopic antennas show surprising results. Some are broadband types; others provide strong signals or are deaf, depending on the frequency bands. Despite similar outer appearance and construction, they have completely different electrical characteristics (Table 5).

Precise matching for a wide range is really hard work and needs good design ideas. As an example: when a stripline antenna is matched perfectly for the range of 10.950 GHz to 11.700 GHz, this means a real challenge. And this is only a 6.4% share of the whole range. Figure 101 represents a wide frequency range with sections of good antenna matching and poor matching. Return loss means, how strong the attenuation of the reflected wave is when it returns from the antenna base back to the source. Figure 101 shows values about reflection in the logarithmic measure.

Narrow-band antennas are almost resonant in a small range and have a deep and sharp curve for good radiation and good matching (Figure 100). Looking at the diagram of the reflection factor over frequency, we recognise that it is not ideal over the whole frequency range. And another phenomenon needs to be considered too. A low reflection level (high return loss) of an electromagnetic wave to be transmitted, does not tell us anything about whether it is efficient radiation or not.

Maybe a high percentage of energy is changed to a standing wave on the feeder. This happens when the transition at the antenna base has poor matching. In the case of receiving a signal, we cannot be sure that we only receive the signal from the antenna, but also ingress into the area of the antenna base. Using a coax feeder from the device to the antenna, a standing wave on the shielding of the feeder line might cause trouble.

Antennas for hand-held devices are constructed and manufactured in various designs. It is not only the length that makes the difference, but also the design and material of the antenna base. This has a great impact on the frequency response for transmitted and received signals. Of course, the mechanic stability is important for the user, too. When we look at the different models, it is quite noticeable that their frequency responses are obviously different, even if the outer appearance of some models is very similar. This is even more a sur-

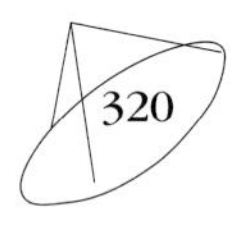

prise when we compare telescopic antennas. One of them can master the short wave with high efficiency; the other one is totally deaf up to 30 MHz. And that difference does not result from their difference in length. Some antennas are suitable for 70 cm with the best results, others are the champions at 2 m. These properties cannot be easily recognised by their outer appearance. Also length is proven not to be the main criterion. What really helps is an experimental analysis to find out the strengths and weaknesses of each antenna. And it is really worth doing. Only one way gives certainty about the real properties, and this is by operating different antennas and receiving signals at different frequencies.

Figure 102 shows various products with their individual designs of the antenna base. This is the area that shapes the frequency response for receiving and transmitting signals, in combination with the length of the antenna. The consequences of the mechanical design of this area are almost underestimated. Of course, the length of the antenna has a certain impact. But it depends more on the overall antenna design. In some cases, the variation in length can even optimise the matching for a selected band, and it is worth trying.

No.	antenna	length	characteristic	strength	weakness
1	Bingfu Military Look	33 cm	wide range 100-500 MHz	good or vy good in the specified range	without special advantages
2	Retevis	36 cm	144/430 MHz	good performance in both ham bands	in other bands poor performance
3	original-ant.	70 cm	wide range 100 kHz - 3 GHz	moderate or good performance in the specified range	only moderate performance
4	Ant-500	84 cm	wide range, good performance	on VHF and UHF good or vy good, also on shortwave	thread stiff
5	Nagoya NA-767	92 cm	SMA adapter needed	starting at 100 MHz, VHF very good	UHF weak, no shortwave
6	Tengko RH 660 S	107 cm	long and unwieldy	vy good > 100 MHz, VHF / UHF vy strong	no shortwave
7	unknown supplier	122 cm	long and unwieldy	> 3 MHz good, VHF vy good, UHF moderate	vy long, UHF moderate or poor
8	Twayradio	130 cm	pure sw-antenna, vy long	vy good performance on sw	only sw, rather long

Table 5: Rod antennas with different receiving properties. Telescopic antennas can even be varied in length, sometimes with a slight tuning effect. But length is not the master criterion for any conclusion about the frequency response. See also Figure 102.

But how can we find out more about the properties of an antenna in case of receiving and transmitting? In the end we can check each antenna by looking at what goes in and what goes out in the case of transmitting. For reception, it is the best check to find out if even weak signals from afar can also be received in the considered frequency band. These answers might not be satisfying, but comparing an antenna with a reference set can really give the best and clearest answers. Antenna measurement chambers, RF measurement stations, and free field measuring stations are generally needed for industrial development and are almost not available to consumers.

When we allow the three port considerations in this chapter, then we draw some conclusions: A passive three-port device cannot be lossless, symmetric, and reciprocal at the same time (physical law).

Our model antenna with three gates is indeed not symmetric. But it can be lossless and reciprocal. Ohmic losses are negligible at the antenna base. Reciprocal means that the behaviour in the case of transmitting is similar to the behaviour for reception. Reciprocity for our model also means that an antenna with poor matching will interfere with the vicinity during transmission. While receiving with poor matching, any ingress from outside, like man-made noise from domestic appliances, will interfere and reduce the S/N much more than with good matching. When the antenna base can provide a wave impedance of approximately 50 Ω to a 50 Ω cable, we can expect good matching and interference-free operation. And of course, the model is lossless when we assume that no ohmic share will transform electromagnetic energy into thermal energy.

It is most important to operate your radio system with good matching when you strive for good links and good results. Even if there is not sufficient space around the shack for installing large antennas, there are still alternatives for radio links over a long distance. With the right ideas it will work. Wire antennas, some metres long, are popular options. Combined with an antenna matchbox or the right balun (or unun like unbalanced to unbalanced), wires will almost always have good results. This can be achieved immediately. Even worldwide contacts seem surprisingly easy. End-fed antennas can also be matched by a "Fuchs circuit" as an example. This practicable option can easily be implemented. It is described in [7].

Balancing

Closed dipoles are balanced antennas, which can be fed with a balanced feeder line like a ribbon cable with a 240 Ω wave impedance as a supply line. Their input impedance is also 240 Ω and a good match can be expected. Feed lines and antenna base impedances shall have the same wave impedance. Impedance matching does not only work with a balanced "chicken ladder" as a feed line. Even unbalanced feed lines, such as the coaxial cable with a 50 Ω wave impedance, can supply a dipole. But attention, take care!

The antenna base impedance of a dipole must match that of the supply line. And without any additional measures, this is not the case. However, there are tricks to overcome this. Also, a balanced supply line with a 240 Ω wave impedance needs an additional adaptation when the base impedance of the antenna is significantly higher or lower.

How to realise these transformations between unbalanced and balanced components or how to manage any significant deviation of impedance between two components, all this is part of the knowledge of any radio amateur. Most of them manage the described use cases with a lot of practical experience. Literature like [7] is recommended. Many use cases can be solved with π-filters or Collins-filters. They include two variable capacitors at the input and output connected to ground and an inductance (also variable if possible) in series between both variable capacitors. Those who prefer the comfortable way can use an antenna match box that adapts any antenna or metallic device, like a gutter, as a radiator.

Standing Wave Barrier

In the case of a justified suspicion that high-frequency energy is on the shielding of a coax cable, it is almost a standing wave. And this type of wave should not be there at all. Standing wave barriers can avoid this scenario, which causes a lot of trouble. Standing wave barriers are available as a ferrite around the coaxial feeder with a clip. There are different basic types on the market.

Some standing wave barriers are coupled inductances on a ferrite core in a box close to the antenna to avoid any coupling of the antenna with the shield of the feed line. Standing waves on a cable are really an annoying matter because they transport trouble directly into your shack.A typical use case that generates standing waves on the feeder is poor matching of the antenna to the connected supply line. We remember:Any sudden change of wave impedance on the way of a guided wave, as in unsuitable adapters or transitions, may generate reflections, spurious emissions, or cause ingress. All this means loss, interference, even trouble with the neighbour, and, in the worst case, a damaged RF power amplifier.

When you connect a ground plane antenna (Figure 27) to a 50-Ohm-coax cable without any additional measures, you do not need to worry. Connecting a 240 Ohm ribbon cable with parallel wires directly to the closed dipole is also a good match. Scenarios resulting from poor matches will almost have severe consequences, which should be avoided. As an example, it can happen, that neighbours hear your voice on their radios or on their TV when you transmit. They will soon find out whose voice it is. Some will say it‘s just what I predicted: a radio amateur means trouble, and therefore antennas are forbidden in the future.

Generally, devices are not a source of interference and trouble. Developers of RF-components and circuit designers take good care of spurious emissions and ingress. Especially in automotive production any new component or system will be checked according to cleary defined procedures with ambitious limits. Experts in accredited testing laboratories use professional equipment to detect any unwanted behaviour of circuits and devices.

Strict standards for spurious emissions, ingress, and some other scenarios with high electrical charges are checked in endless tests until products get their approval. Consumer products are also checked, but according to less demanding test procedures.

Second Antenna to Compensate Interference

Eliminating interference is always a time-consuming topic in the wireless world. A smart approach is, to beat enemies with their own weapons. That means, in detail: Get the interfering signal with a second antenna, shift the phase by 180° and add this to the mixture of the primary signal, i.e., a radio broadcast signal plus the interfering signal, (i.e., man-made noise).

When the addition is done with the right amplitudes and phases, there will be only the primary broadcast signal as a result, without interference, or at least with significantly less interference.

Doesn't that sound much too easy? How shall noise get reduced? But it is not a magic trick, and it even works, at least for man-made noise coming from the environment. To avoid a reduction of the primary signal (the radio broadcast), the signal of the secondary antenna should include only a small share of the primary signal.

This can be achieved by placing the secondary antenna just in the domestic noise and the primary antenna at any place with good reception of the radio broadcast. Two physical quantities have to be balanced for good results: the phase shift of the secondary antenna and the amplitude of the interfering signal. Then the interference will be much less annoying.

Good Antenna Placing

Many issues while listening to distant or weak station will not occur at all, when the right place for the antenna is selected right from the beginning. But which place is advantageous and which is not?

The basic rules are easy. Antennas need air. That means that a roof or a pole is a good position for any antenna. Please don't forget lightning protection, planned and done by an expert. The antenna shall stand high, in relation to the environment. But it should not be too high over the ground because overreach signals from abroad might interfere with the primary wanted signals. This may happen when the frequency reuse coordination of regulatory authorities does not consider sufficient spacing between sites in two neighbouring countries operating on the same channel. One bad example for placing an antenna is within the domestic man-made noise.

Some radio amateurs and shortwave listeners do not have the opportunity to use the roof of their house. In some cases, it is a pity that neighbours spoil the pleasure of radio enthusiasts, and so radio listeners have to look for an appropriate antenna to place in their own protection area, on the balcony, on the window sill, or even on the living room table. But even for these use cases, we have good news. Reception in the VHF bands can be easily improved in some cases by varying the exact place of the antenna simply by shifting the antenna or the radio by 20…50 cm to an alternative position. Just try.

This effect originates from near-field reflections and the superposition of waves in the environment. Inside a house, several reflections interfere with a complex field distribution with minima and maxima. Positions with a minimum will have poor results. But looking for a better position inside a house or an apartment may improve the S/N by 20 dB or more. Positions of nearly zero level on one hand and a maximum of signal strength, on the other hand, are sometimes close together with a spacing of 12 cm or less. The numbers result from the wavelengths used and can vary with operating frequencies. But it is worth checking it out anyway. Some dB as a reward will always be achieved.

In very bad scenarios without any signal reception, however, the same curiosity and creativity are really useful and will be rewarded in most cases. When no signal at all is received, there might be an unexpected or unknown interferer with a strong and wide spectrum close to the receiver.

Wide noise floors, sometimes with a certain spectral pattern, can shut down a whole radio band. That means any spacing from potential interferers should be increased. To put it in other words: keep away from any switching power supply, heating controls, engines, and generators, and take precautions against neon lights.

What remains open is the question if antenna collocation with several radiators on a roof may cause trouble by mutual interference. The good news is that a shared use of one roof with antennas from different providers operating in different frequency bands, will not interfere in most cases. And in a whole country, there are a huge number of stations that share one roof. With this large experience from many thousands of scenarios, we don't need to worry. In 2001, the provider of a European TETRA-network started an analysis. The focus was to find out if the recent analogue trunked radio stations (still in operation) and the new digital public TETRA stations would influence each other anyway. The result: Even if antennas have a close spacing of a few metres on shared roofs, engineers could not observe any negative impact during the operational mode of both analogue and digital. It is most probable that different providers executed similar analyses. The results are supposed to be "all clear, no problem." This is also proven by the close neighbourhood and peaceful sharing scenarios, as in Figure 29, Figure 57, Figure 73 and Figure 98.

Voice Quality

Good intelligibility in a voice transmission is essential. But some operators don't have this in mind at all. To be honest, most of us do not worry, because any new radio set with a dedicated microphone is expected to be a good combination. But looking at the results in real life and also by listening, it can be a disappointing surprise.

We observe that voices in transmissions sound dull in some cases or very quiet in others. Sometimes the signal appears unpleasant, like amplified with overdrive like electric guitars in the 1970s. They seem to be much too loud. Such an overdrive can modulate so strongly that the frequency deviation of FM is too large for the channel, and the squelch of a receiver reacts with short cutoffs in the signal. Noise of wind in the microphone, driving noise, and hiss when pronouncing consonants, all this spoils transmission quality. Consequently, in some cases, it needs an analysis from the base when voice transmissions sound disastrous.

But what is typical for good voice and good audio quality at all, and how can that be achieved? There are typically two areas that we shall take care of independently to get voice quality to a certain level. One is the technical channel itself, which needs to provide an appropriate bandwidth the other is the sound generation with a microphone and its capsule, and probably any additional pre-amplifier. Both aspects, in combination, need much care. The objective is to tune the audio level as a function of frequency to a strong and natural sound, all within a frequency range of 300 Hz to 3000 Hz as an example. Equalisers can help to achieve this.

But as a first pre-condition, the transmission channel of the radio set has to be checked. The channel shall primarily provide a sufficient maximum of modulation to a signal. When the channel offers a low maximum modulation and, secondly, a narrow characteristic to the microphone or even cuts off frequencies above 2500 Hz, even the best equaliser cannot help. A good dynamic microphone or a good electret microphone or even an amplifier microphone will produce the natural sound of the original voice. Many do this even if they come as supplementary components off the shelf. But take care; this does not help if the channel has poor characteristics. As well, a few microphones have a poor characteristic and generate a dull sound, or they add resonant frequencies.

The audio may sound sharp and unpleasant. Here we have a good chance to improve the station and its transmission properties. Electronic fine-tuning by digital audio processing can match the audio to the transmission channel. When the channel offers good bandwidth and amplification, the result should be a strong and natural voice sound.

In some applications, voice sound is processed by a digital system to emphasise special frequencies to make the sound sharp and intrusive. This can be observed in the voice communication of military systems, where background noise levels are pretty high, like in a cockpit. When compression is applied to voice signals, they can also dominate in a noisy environment, as in the tactical aviation communication of military aircraft.

Those who use an amplifier microphone like the legendary „Turner plus 3" should take care of the amplification level when using it in a large room with bare walls. Reverb from the empty room is added to the voice sound and an amplifier microphone often increases the reverb effect during transmission. In the end, a completely overdriven voice signal becomes unreadable to the receiver. Microphones like the Turner are so sensitive that their amplification control should be carefully tuned to optimised sound quality. The more, the better is not true in any case. Sometimes it is just the opposite. Those who like it loud with the highest amplification factors may use that modulation type. QSO partners will always give an honest report about the quality of a transmission when they are asked.

But which items beyond have an impact on modulation? Narrow-band FM and SSB are mostly operated with 2.7 kHz bandwidth (sometimes more) and provide good quality in any transmission. To use the bandwidth efficiently, the first audio stages should provide a good audio level to the modulator and RF driver stages. When this audio level inside the amplifier circuit is limited to a low level, then any legendary microphone and equaliser can't help from outside. The audio signal remains low, and the voice sounds weak, no matter how strong the level of the transmitted RF signal will be.

But the opposite constellation, with high levels, can produce a disastrous signal. When signals for the modulator are wide and have a higher level than specified, the FM deviation increases, and the audio received in any other shack scratches and hisses. Listening gets very tiring.

Even the squelch may cut off the audio. For a precise adjustment of voice and audio levels, you need good hearing and judgement of details. Also, patience and a certain technical know-how are helpful. The field of optimising audio is really complex; maybe that's one reason why many radio amateurs don't care. Those who purchase a radio station with good modulation and optimised voice quality right from the beginning, should be happy.

Voice sound in digital radio systems depends additionally on the voice codec used. This might be one reason why voice quality in different systems like DMR, D-STAR, NXDN, P25, TETRA is perceived differently.

Depending on the integrated speech codec, the sound characteristics and their different perception, get a real background. But some of the above-explained details about audio quality are also valid in digital systems. Overdrive or weak modulation decrease the audio quality of transmissions significantly. The characteristics of the codecs come additionally on top.

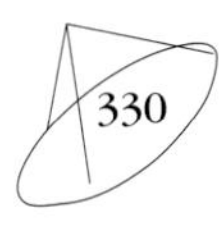

Coverage and Long Distances

How far can electromagnetic waves propagate? To answer this question, we will look at different scenarios over long distances, and later we will look closer at radio wave propagation for radio communication close to the ground.

In radio astronomy, we find relations, where it is useful to calculate with light years as a unit for distance. That means the distance travelled by light in one year. Some observed galaxies are thousands of lightyears away. Scientists receive waves in the optical range that were transmitted many thousand years ago. Light coming from our sun to earth needs 8 minutes, only a small share of a lightyear. That distance is called one astronomic unit (1 AU) and means approximately 150 million kilometres. These relation reveal how far electromagnetic waves can travel at all.

In practice, we do not only talk about waves in the optical range but also about radio communication in space. When the Pioneer probes were expected to end their mission, they still continued their work and even left our solar system. Far away from our solar system with all its wonderful planets, the probes have still been in operative mode and transmitted signals to earth.

But how far can radio coverage be on earth, and at what distance can waves travel here? Radio waves with frequencies above 30 MHz and in the GHz range can basically travel just to their optical horizon. This is basically the limit, but there is still an area behind, which is also covered by effects like diffraction. The expansion of this area is dependent on the used wave length and lies on average about 15 % behind the pure optical horizon, and can be added. This is the quasi-optical range, and it is valid for wavelengths of 0.5 m to 10 m. Above 10 m any shortwave characteristics will dominate propagation according to their own rules. Below 0.5 m wavelength, propagation characteristics are more and more similar to those of optical light.

With an easy approximate formula, we can calculate the range of a transmitter without too much mathematics. It is a simple tool for good estimation. The key parameter is height over the ground. Any other impact coming from buildings, trees, hills, obstacles, or shadowing is not considered in this simple model. By varying only the height of a transmitter, the following equation provides a good estimate.

$$S_O = \sqrt{H_S \cdot 2R_0/1000}$$

R_O is the radius of the earth. We use it for the impact of the spherical surface of the earth as a limiting factor. S_O is the (optical) range in km. H_S is the height of the transmitter over the ground. To include the quasi-optical range, we multiply the radius of the earth by a factor of 4/3. Then the use case with a certain area behind the optical horizon was also considered. R_Q is the radius of the earth with a virtual extension, to consider the quasi-optical range.

The term is $(4/3) \cdot R_0$. By pulling fixed values before the root and rounding, then we get simplified terms as:

$$S_O = 3{,}6 \cdot \sqrt{H_S} \quad \text{und} \quad S_Q = 4{,}1 \cdot \sqrt{H_S}$$

S_Q is the range of the quasi-optical sight. When transmitter AND receiver have a significant height over ground, then we can use the following extension of the equation before we estimate the range.

$$S_O = \sqrt{(H_S + H_E) \cdot 2R_0/1000}$$

When we want to consider the quasi-optical use case, too, for calculating S_Q we draw as much as possible before the root and round this. The results for pure optical view S_Q and for a quasi-optical view S_Q are:

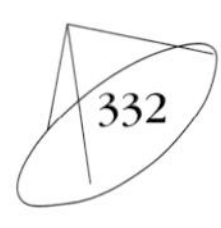

$$S_O = 3{,}6 \cdot \sqrt{H_S + H_E} \quad \text{und} \quad S_Q = 4{,}1 \cdot \sqrt{H_S + H_E}$$

S_Q relates to use cases with operating wavelengths of 0.5 m to 10 m. H_S indicates the height of the receiver antenna over the ground. As we see, it is also beneficial for range when the receiver has a certain height such as on a tall building or on the top of a hill. Based upon the equation for quasi-optical range, we can see some selected examples in the following Table 6:

height over ground m	optical line of sight km	quasi optical los km
1.5	4.4	5
3	6.2	7
5	8	9
10	11	13
20	16	18
50	25	29
100	36	41
500	80	92
2000	161	183
5000	255	290
10000	360	410

Table 6: The range of the optical and quasioptical horizons depends on the height of the receiver (or transmitter).

To explain the characteristics of wave propagation in radio communication in more detail, we look at some selected use cases. They are based on observations, which might deviate from theory at first. The reasons are buildings or height deviations as well as shadowing effects. Even trees and forest areas enhance the theory before. They almost mean an increased propagation loss, especially in summer with much more leaves. All the factors mentioned can lead to very different results, as shown before. Of course, there are a lot of other examples. But as a first overview, this line-up covers many scenarios

transmitter site	environment	coverage
two handhelds (pedestrians)	dense urban, inhouse	1-2 km
two handhelds (pedestrians)	dense urban	2-4 km
two handhelds (pedestrians)	rural	3-15 km
two vehicles (direct mode)	country	6-10 km
two vehicles (direct mode)	suburban	3-6 km
two vehicles (direct mode)	dense urban	2-4 km
central radio station (20 m above ground) – vehicle	country	10-15 km
central radio station (20 m above ground) – vehicle	suburban	5-12 km
central radio station (20 m above ground) – vehicle	dense urban	3-6 km
handheld (balcony, 5 m above ground) – vehicle	country, hilly	> 5 km
handheld (balcony, 5 m above ground) – vehicle	dense urban	> 3 km
handheld (balcony, 5 m height) – station (15 m above gr.)	hilly	> 20 km
handheld (balcony, 5 m height) – station (100 m above gr.)	little shadowing	> 100 km
air radio (tower 15 m height) – aircraft (3,000 m height)	free line of sight	> 100 km
air radio (tower 15 m height) – aircraft (11,000 m height)	free line of sight	> 400 km
walkie-talkie (100 mW RF-power, 20m above ground) – fixed station	suburban	2-4 km
2 walkie-talkies (both inhouse, 2nd or 3rd floor)	dense urban	1-2 km

Table 7: Radio communication ranges are mostly dependent on the terrain and the transmission site.

from real life. The approximate formulas can probably be helpful to enhance the list for individual use.

Quantification and Helpful Quantities

Let us also consider the quantities around electromagnetic waves. Many of us are interested in parameters like field strength and RF power, others want to know more about antenna gain. You get quickly to the point where terms and quantities lead to confusion and misunderstandings. Terms cause trouble, like the unit for the wave impedance of a feed line, which cannot be easily measured with a multimetre. There are some important relations to consider:

dBi is the logarithmic quantity for antenna gain as an example. That gain, relates to an isotropic radiator as a reference radiator. It is not a unit.

Also, dBd is used in this sense, but the term relates to the real dipole. The difference to dBi is 2.15 dB coming from the definition.

Sometimes dB appears without indication of a reference quantity. Then dB means almost the difference between two levels of a quantity like power, current, loss, and more.

Using dBm is always power as a logarithmic quantity related to 1 mW (1 milliwatt) as a reference. Consequently, 0 dBm means exactly 1 mW. 10 mW means 10 log (10) = 10 dBm and 1 W means 10 log (1000) = 30 dBm.

But watch out. The difference between two power levels is always dB (without m or anything else).

Electric voltage has the unit V (Volt). As a logarithmic quantity, it is used as dBμV as an example. Consequently: 100 μV means 20 log (100) = 40 dBμV.

Electrical field strength is V/m (Volt per metre) as a unit. The corresponding logarithmic term is dBV/m or dBµV/m .

Example: 10 µV/m means 20 log (10) = 20 dBµV/m

< 30 MHz LW/MW/SW				**>30 MHz VHF/UHF/SHF/..**		
s-value	**power level**	**el. field strength**	**el. field strength level**	**power level**	**el. field strength**	**el. field strength level**
step	**dBm**	V/m	dBV/m	**dBm**	V/m	dBV/m
0	**-127**	1,41 mV/m	-57,0 dBmV/m	**-147**	141 µV/m	-77,0 dBµV/m
1	**-121**	2,83 mV/m	-51,0 dBmV/m	**-141**	283 µV/m	-71,0 dBµV/m
2	**-115**	5,66 mV/m	-44,9 dBmV/m	**-135**	566 µV/m	-64,9 dBµV/m
3	**-109**	11,3 mV/m	-38,9 dBmV/m	**-129**	1,13 mV/m	-58,9 dBmV/m
4	**-103**	22,4 mV/m	-33,0 dBmV/m	**-123**	2,24 mV/m	-53,0 dBmV/m
5	**-97**	44,7 mV/m	-27,0 dBmV/m	**-117**	4,47 mV/m	-47,0 dBmV/m
6	**-91**	89,4 mV/m	-21,0 dBmV/m	**-111**	8,94 mV/m	-41,0 dBmV/m
7	**-85**	179 mV/m	-14,9 dBmV/m	**-105**	17,9 mV/m	-34,9 dBmV/m
8	**-79**	358 mV/m	-8,9 dBmV/m	**-99**	35,8 mV/m	-28,9 dBmV/m
9	**-73**	707 mV/m	-3,0 dBmV/m	**-93**	70,7 mV/m	-23,0 dBmV/m
9+10	**-63**	2,24 V/m	7,0 dBV/m	**-83**	224 mV/m	13,0 dBmV/m
9+20	**-53**	7,07 V/m	17,0 dBV/m	**-73**	707 mV/m	-3,0 dBmV/m
9+30	**-43**	22,4 V/m	27,0 dBV/m	**-63**	2,24 V/m	7,0 dBV/m

Table 8: Power levels, S-value and electric field strength have a close relationship. The values for voltage in this table refer to a wave impedance of 50 Ω.

At first glance, it seems to go haywire when we try to use units and logarithmic terms. But when you make some calculations on your own, they turn out not to be as complex as assumed.

Some of the terms have a close relationship and can be expressed otherwise. Received power levels and electric field strength are suitable examples. When we consider a received signal with a certain power level at a defined location, expressed as a logarithmic quantity (dBm), we can conclude the electric field strength dBmV/m at the same place. We only need to know the offset between both representations. In this case, it is an offset of 70 dB. We can find this in table 8, too. Another example is when we look for the voltage considered in a system with a 50 Ω wave impedance. Assuming that we measure or receive a certain power level (dBm) and we want to

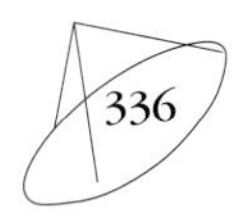

know the level of the assigned voltage expressed as dBμV, we only need to add an offset of 107 dB to know just this voltage level. Some professional receivers can even toggle between a display with dBμV and dBm as a feature and using the S-metre and the receiver as a precise measurement instrument. The relation from dBm to dBμV is:

with x dBm + 107 dB = y dBμV.

Consumer Products in the Course of Time

In history, only 100 years are sufficient to get a comprehensive impression of products coming from innovations in communication technologies. Even after this relatively short time in history, we can now enjoy the whole fascination of mature products and look closer to them. Astronomy, as a counterexample, is a field that has fascinated people around the world and in history for thousands of years.

Many of us know radio and TV sets from our childhood. Today, even children know modern smartphones as an important part of their lives. Some flops, however, are almost forgotten because they were not successful as consumer products. Not new high-tech product is automatically appreciated on the market. But this will be considered later with some examples.

Radio and Television

It was always exciting when the tubes of a radio or TV started to glow and heat up. They got a pale magic orange colour thirty or sixty seconds after switching the devices on, until finally an entertaining sound and pictures filled the room. In the 1970s the number of technical devices and media consumption were still moderate. Channels with three TV programmes and four local radio programs were available. They came through a cable from a roof antenna or from an indoor antenna like Hirschmann‘s “Libelle” as the first extension and as a second channel in the era with only one programme. Small pocket radios and portable radios were state-of-the-art to get music and news into any room, sometimes even into the kitchen and bath room.

Suddenly, TV programmes were no longer black and white but full of colours. Just suitable for this time when Glamrock with new fantastic bands made the radio and music scene also colourful. Legendary pirate stations like Radio Caroline started their innovative programmes, by transmitting from a ship in the north sea. But they were pursued by state authorities because private transmissions were not allowed, both on land and at sea. Therefore, a ship as a transmission site was more useful. Radio Luxemburg was also a source of new and fresh radio entertainment.

Decades of innovation in all areas have made the world exciting. Computers were still far away in the 1970s, but CB radio as a new hobby brought an exciting technology. CB provided speaking and listening to anybody just in their individual manner. It was the first technology that revealed the great need to communicate with any other CB fan in the neighbourhood, no matter if they were strange people or not. Communicating over the air was the new number one.

With the new century starting in 2000, things changed dramatically. New technologies like the internet and circuits with immense speed opened new doors to new spaces, some of them only virtual. Mobile radio networks provided wireless contact with a new virtual world from any location. Where only mobile phone service was possible, networks were enhanced, and the internet brought the new world of social media with all its features and consequences. Everybody had their personal phone and their personal access to any application, anywhere. In the course of time, just this has had a big impact on us and has brought many changes, which even modify our personalities and our behaviour.

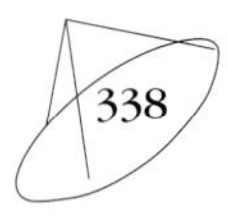

The era of long-play records and CDs is slowly expiring in the new century. Today we stream, anywhere, with any device like a cell phone, tablet, iPod, PC, TV, and so on. Radios are now called SDR and can get additional functions through a software update. The time of expensive hardware is over; hardware has become quite cheap. Big and heavy devices combining and including radio, cassette deck and turntable are no longer useful, except for nostalgic people who still pay a lot of money for their hobby. Software, streaming, and fast internet make the scene and make money for providers. Of course, LCD and plasma screens still remind one of a time when TV sets got the best place in the living room of hard-working people. Long play records have their revival, but on a lower level. Also, tube amplifiers are in use, and some of them are even state-of-the-art, like the T8000. This Japanese model is probably the best headphone amplifier in the world. It is the figurehead of a technology that has often been described as dinosaurs or even declared dead. But good quality survives, as we see. The good remains, and there is always something new to add. We live in a time of greatest variety and have the privilege, simply to enjoy… music, TV programmes, and games.

Mobile Devices

Radio devices and RF sets changed in the course of time. Radios have almost disappeared, or they are integrated behind displays, as in modern cars. In some kitchens and living rooms, we can still find radios, but mostly they are integrated into other devices like mobile phones or TV sets. Flat but still large, still today TV sets have their prominent places. But TV can also be consumed with smartphones, mostly from streaming databases. Radio communication sets did not change their outer appearance. Handhelds are still what they have always been in the last century. But now they are digital, smaller and lighter in weight, current-saving, small and robust.

Once, there were also those special devices called pagers. One very early provider was the Eurosignal service. It was usable in 1974 to

1994 via transmissions at the lower end of the FM radio band. On 87.5 MHz a mysterious transmitter sent out strange multitone sequences in intervals of 5 or more seconds. What first seemed mysterious was a method to signalise something to assigned receivers. The user should call a pre-defined phone number when receiving the signal intended for him. It was displayed in a kind of electronic box with a simple display. Another attempt to establish signalising over pagers started in the 1990ies, when small pocket devices with a display should conquer the market. But the average consumer had no need. And this was not a surprise because the first cell phones have been developed and are ready for sale. The small pagers, however, were able to receive a signal with short strings from one special receiver. But they were designed only for unidirectional communication. They were also called beepers, as a nick name in Germany. Their real use case was just another. Emergency alerting was the professional purpose, i.e., for doctors and volunteer firefighters. As a mass product, they were more or less a flop. That was obvious when a provider with the product QUIX entered the stage and pushed his product, small pagers with a display for a short text. The provider used a country-wide radio network with base stations according to the POCSAG standard (Post Office Code Standard Advisory Group). Maybe the use was too uncomfortable because any sender had to call a service number and tell the dispatcher any short message. The dispatcher routed the message to the base station network, broadcasting it with a special id to the receiver. For consumers, the use of cell phones surpassed this application immediately because they provided bidirectional communication without any fee for SMS. But unidirectional communication by pagers has survived and found its place in professional alerting. For decades penetrating beepers started thousands of rescue operations and are still an important part of rescue chains in many countries.

Today, mobile phones and professional radio are the dominat technologies, one for consumers and the other for professional use by police, fire brigades, ambulances, and other governmental security services. It is remarkable that the use of SMS was underestimated for a long time. Sending strings of letters and numbers as a short message was only a free by-product without any meaning for the providers.Thanks to the developers, they considered SMS an important feature and pre-installed it in the GSM system right from the beginning.The useful feature got more and more attention, and after some years of operation, providers recognised SMS as a money-making feature and part of the business model.

In Germany this happened in the mid 1990s. At the beginning, this useful data transmission tool was limited to 160 characters. But even with this limitation creative developers used this as a data service, i.e., for remote control of network elements, means sending commands and polling to get status information.

Recent and Modern Receiver Technologies

Receiver technology is worth looking at, especially how it has changed within the last 100 years. Starting with mysterious crystal detectors, engineers have created amazing technologies and devices like SDR, even for consumers. The whole story could fill an extra book, but here we can at least give a short overview. As a start, we ask ourselves what we expect from a good receiver.

Parameters of Receivers

So what makes a good receiver or a good radio? Of course, the frequency range is the first criterion. Radio bands shall match the needs of listeners, including their special requests. The frequencies of a typical portable radio from the 1980s are shown in Figure 1. But there are still some more inner values that play important roles.

Sensitivity. Typical receivers are designed for sensitivities of 2…7 μV for AM on shortwave and 0.32 μV (12 dB SINAD) for FM in the higher bands like VHF, UHF, and higher. It seems paradoxical, but the sensitivity of a receiver is almost not a limiting quantity. SINAD means signal to interference ratio, including noise and distortion. Any type of distortion, whether "man-made noise" or natural sources, can be the real limitation for receiving scenarios. Just in the shortwave range, the increase in sensitivity is unnecessary. Natural sources produce noise. This will limit the possibilities for reception. In the lower shortwave bands, as an example, the limitation comes from atmospheric noise.

Strong signal behaviour. It seems paradoxical that strong signals spoil the fun, i.e., when DXers listen to their favourite stations with weak signals from afar. Those who climb hills for better radiation and reception hope to catch stations over long distances and get better results on top of a hill than in their own shack. But reception on a hill needs special attention for pre-selection. The sensitive first RF stage of a receiver needs to stay protected from strong signals outside the used band. People who ignore this might be surprised on the hill. They can be heard far away, but they receive only a noisy mixture. Strong signals from anywhere can fill the receiver stage in such a way that nothing but noise and distorted signal fragments are the result. It is the consequence of intermodulation products, which are caused by non-linearities of electronic elements and circuit stages.

They are produced by the receiver itself and can be eliminated by a pre-selection filter or by using the RF-attenuator. Ghost signals as an interfering product result from low-cost circuit concepts with a low intercept point in the second and third orders. Low-cost receivers can produce a lot of signals that do not belong in the spectrum. Receivers with a good intercept point in second or third order promise much better results. With an IP2 of 30 dB or even 40 dB, the receiver has properties like professional types. Some cheap devices from Asian mass production lines with highly integrated cheap RF-stages can spoil the fun. No real signal will be received, but noise, crackling, and distorted voices come out of the speaker, a terrible mixture, nothing that can be used anyway.

Cross-modulation effects can also be annoying. They do not originate from your own receiver with a poor filtering concept, but they come from other almost strong transmitters, located mostly not very far away. They imprint their modulation with their strong emission somewhere in the spectrum where we do not expect it anyway. Radio listeners discovered this effect many years ago and called it the Luxembourg effect.

When listening to a station on medium wave on its regular frequency, an additional signal from another medium wave transmitter in the neighbourhood could be heard at the same time. This special type of interference was not produced in the receiver, but it came as an unwanted signal and decreased reception quality. This effect can take unexpected proportions, like the excitation of the ionosphere in a certain area. As a result, the modulation of a strange transmitter can be heard as unwanted interference on the listener's tuned channel. Signals are mixed up outside the receiver and produce annoying modulation shares.

Selectivity. When many broadcasting stations are active in one band, it is good to operate a receiver with high selectivity, especially while listening to weak signals side by side with strong signals. Radios with poor selectivity can play two stations simultaneously through the speaker. Selectivity is created by the intermediate frequency filters (IF filters). On shortwave the AM filter bandwidth is almost 5 kHz. Higher values sometimes mean a parallel reception of neighbouring channels. For ssb it is 2.7 or 3 kHz.

Operating the NRD-545 or the Perseus SDR, the needed signal in the spectrum can easily be cut out of the spectrum. Even a complete sideband can be removed completely with surgical accuracy. This is the way to get clear signals even from transmissions in the tropical bands coming from Central or South Africa. Also, weak signals from New Zealand suddenly sound clear, and you can understand any audio detail like never before. Religious stations from the United States like WYFR Family Radio or once HCJB, the Voice of the Andes from Quito sounded clear like local stations. For the first time, listening to local Chinese stations or Caribbean music broadcast, is getting possible. But even without stepless IF filtering, the efficiency of products with fixed IF filters also showed good results. It was always dependent on how much effort the developer put in to create effective filtering in a receiver. As a first step, switching from a 5 kHz filter to a 3 kHz can be an alternative. The result was: a significant increase in selectivity.

2000 memories. For some users, it is most important to store all their important receiving channels, including additional data like operation mode, bandwidth, and individual text. The intention behind this is, to have quick access. The use of memory operation, is needed especially for broadband monitoring, where even 2000 memories seem not to be enough. The organisation of stations is supported excellently by most receivers.

Search and scan. Quickly searching in different ranges of frequencies with the right step is an important criterion for broadband receivers. Radio monitoring while searching with a speed of 200 channels per second is no problem today. Once, without fast circuits, it was always a time-consuming job to search in several wide radio bands, and the results were almost a few random hits.

Early Concepts

This seems lightyears away when the first attempts started with demodulating radio signals. It was approximately 100 years ago, in Germany, that the first transmissions were produced. Rectifying radio signals was already possible, even without semiconductor diodes. A metal tip pushed with a balanced pressure on materials like pyrite or galena, is connected to an electrode. This was the prototype of a crystal diode. The 1950s were too early for transistors as a mass product. Tube receivers appeared and became a standard, but they were expensive and heavy. Detector receivers were suitable do-it-yourself projects for people in the first years after WW II. The only precondition was a strong local broadcasting station. The concept was simple, and the assembly included a crystal, or later a semiconductor diode, a resonant circuit, and an earphone. The strong transmitter was needed because energy was taken from the electromagnetic field; in some cases, it was the only power supply because batteries were rare. These were the first detector receivers. A few years later, receivers with one or two amplifier stages appeared. They were called audions and were realised with one or more stages operated with tubes.

With a good performance and even with one stage, they could do different tasks like demodulation and amplifying. Some types worked with a feedback loop in one stage. A certain share of energy from the resonant circuit was fed back to the input with a sensitive control for manual use. This had an enormous increase in amplification and even selectivity. But the concept also had a weakness. The feedback loop oscillated when too much energy came back to the input of the resonant circuit.

As a result, a sharp whistling noise disrupted the audio. Therefore, the ease of use was also miserable because any user needed two hands for tuning, one for the frequency and one for the feedback control, which had to be re-tuned after each frequency change.

A little later, radio technology with tubes conquered the living rooms. Large radio sets with good speakers provided an amazing sound at home. Similar to PCs, they needed 30…60 seconds to start because the tubes heated first before any station or noise came from the speaker.

Before the first luxury tube receivers were available, people used detector circuits and simple audion stages for reception. They had an acceptable performance and an increased selection. Receivers with a feedback loop brought more innovation and better selectivity. Even CW signals could be detected when the feedback loop was used. When the coupling share was adjusted sensitively, good selection and CW mode were the rewards. Those who tuned with a rough hand had almost to accept the overdrive whistling noise. We know the effect of audio systems on a stage when the microphone comes too close to the speaker.

Figure 103: Nordmende Elektra from the 1950s was the attraction in many living rooms after WW II.

Tube receivers were also improved in the 1960s and they came as so-called superhet receivers (superheterodyne). They used a local oscillator (LO) with a signal for mixing the received RF signal down to an intermediate frequency. Between antenna input and the first stage a filter blocked any unneeded RF share.

After mixing with the LO signal, the RF signal was available in a much lower frequency range. For those low ranges, filters with good selectivity could be easily integrated. For higher frequencies, that means that without a mixing stage, filters would have been more complex and expensive. But then the transistor was ready for the market and conquered many applications. It must have been in the 1960s. Small portable radios were popular, and later even integrated circuits became part of any receiver and TV set. The superheterodyne principle convinced consumers and engineers, and still today it is one of the most successful technologies, both for consumer products and for professional radio.

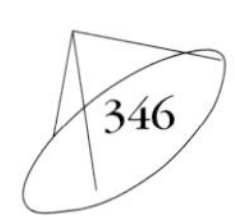

Figure 104: Radio scale and tuning knob of the Elektra radio. A large speaker provided a full sound.

Proven Concepts

For many decades, the superhet concept was the backbone of receiver technology. There is no need to observe the fine tuning of the feedback energy, no whistling noise, sufficient amplification, and good selectivity, and a full sound were the arguments why detector receivers and audion receivers disappeared as fast as they came. The mixing concept of a superhet, however, had its little weakness, too. While mixing with the LO frequency, an ambiguity arose. In a few cases, it was annoying during monitoring in a neighbouring frequency range. A station from another frequency appeared as a ghost station, where it did not belong, at all. This can be explained mathematically: While mixing two frequencies physically, there are always two outcomes. Two results, represented as two frequencies, satisfy the equation:

$$f_{ZF} = | f_{Empf} - f_{Osz} |$$

$$f_{ZF} = | f_{Osz} - f_{Empf} |$$

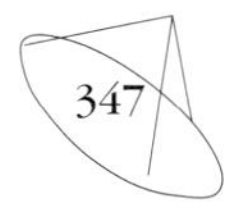

In a good receiver, one of the frequencies was always filtered out, and the other one was used. Efficient filter circuits could manage that ambiguity and ghost stations (on mirror frequencies) at the wrong place in the spectrum were strongly attenuated. High-quality receivers did this without any leftovers.

In the 1980s of the 20th century, receivers with improved performance and advanced technologies have been developed, world-receivers or in German: Weltempfänger. They came on the market as cheap radio sets. World receivers like Grundig's Satellite 600 or Sony's ICF-2001 got their deserved place in history.

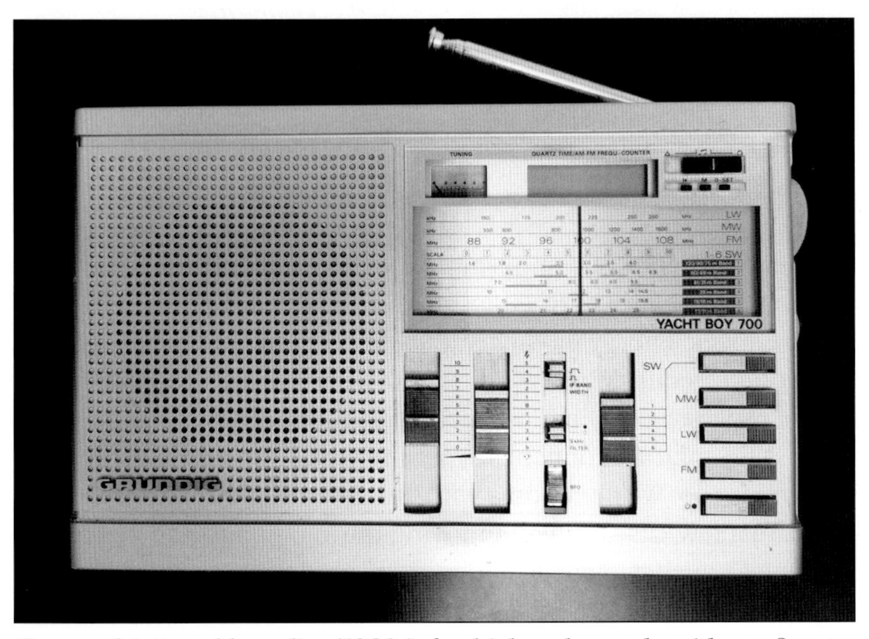

Figure 105: Portable radio (1980s) for higher demands with an S-metre and BFO for SSB signals.

For all curious people, asking themselves what kind of signals were between the radio broadcast bands, an exciting era started. Since radios with a BFO and product detector were available, new transmission types with new operation modes could be used. Transmissions

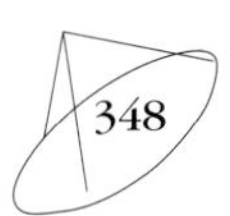

in CW Morse and radio teletype ITA 2 (Telex) have been interesting to shortwave listeners for a long time, but now they could be detected. With a digital frequency display, it was even possible to locate them on the radio scale very precisely.

While CW Morse could be learned by everybody who was patient and ambitious to decode transmissions, RTTY transmissions needed an electronic decoder to transform the two-tone bit stream into letters and numbers on a display. New products with a high performance, developed with fast semiconductors in their circuits revealed a lot about wireless data communication when suddenly any decoded text ran over the screen or through the display.

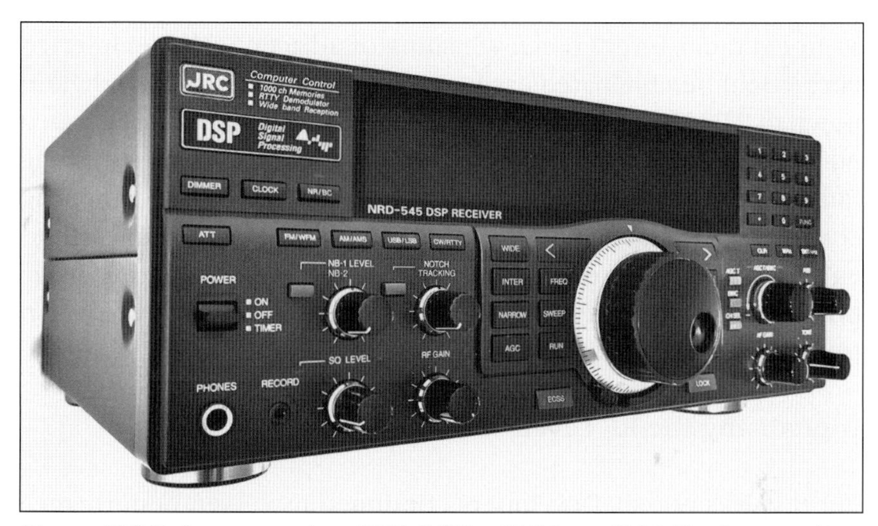

Figure 106: Reference receiver NRD-545 by JRC from 1998. The intermediate bandwidth could be tuned with a control, nearly stepless. Most types of interference could be cut off or faded out completely. Communication receivers were the new class of radios.

When new digital circuits appeared, a new type of circuit entered receiver technology. PLL (phase locked loop), which means control loops for the phase of a signal, were used to generate oscillation on defined frequencies. Even in the early car radios of the 1970s these PLLs had already been in use. Frequency stabilisation was done by feeding back a share of the signal at the output of a PLL back to its input. This technology was used in stabilised receiver circuits, for frequency synthesis, in mixing stages, and in FM demodulators.

Over time, developers created receivers with more features and improved performance, called communication receivers. It was towards the end of the 1990s, when JRC's NRD-545 came to market as probably the best communication receiver in the consumer segment. Still today, this radio is our first choice when we talk about top performance in short-wave listening. One feature was completely new: the stepless adjustment of IF bandwidth with a control. For higher demands, users could finally select extremely narrow or also very wide IF bandwidth to cut out a CW signal or to enjoy music quality as an FM radio. Interference whistling or adjacent channel interference could be eliminated easily, i.e., by reducing the IF bandwidth down to 2.7 kHz or less. Also, an interfered sideband could be faded out completely, and the other one was used. While listening to strong music stations without interference coming from neighbouring channels, it was always a pleasure to increase the RF bandwidth to enjoy the complete audio spectrum. Of course, automatic pre-selection close to the receiver input was essential to avoid distortion from other bands while listening to very weak signals.

Digitalising the RF Level

But this was not all. Developers continued increased performance and, added more features. There was still space for improvement. But how can we improve this class of high-end receivers? When communication receivers seemed to be nearly perfect, like the NRD-545, another leap in technology was made by developers.

This time, the origin was in Italy. Thanks to microprocessors with immense speed and other extremely performant elements for digital signal processing, a completely new paradigm prevailed. The whole RF range of the shortwave spectrum could be scanned and directly

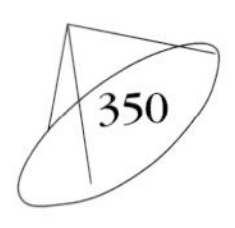

processed, without mixing stages and complex filtering as in a superheterodyne receiver. The era of SDR (software-defined radios) dawned. When Nico Palermo's "Perseus" entered the stage, experts were surprised and amazed. A small hardware box without switches and controls appeared on the market. Any manual operation had to be done with a software screen, which could be used with an app on any personal computer or notebook.

Figure 107: SDR Perseus from Microtelecom has little hardware but unsurpassed inner values.

SDR (software-defined radio) provided completely new features. What was essential: any new software with new options for the user interface and also new firmware, software that is active to control hardware functions, all this can be updated. Amazing for any user: The waterfall display, which allows you to overview complete frequency bands with all their signals is, of course, a colourful presentation. This two-dimensional display of any activity in the selected range is a great benefit, i.e., to find sudden transmissions in the band while listening only to one fixed frequency. Of course, the intermediate frequency range (baseband) can be adjusted without steps. A channel can be selected as a symmetric range, but it is also possible to shift the IF range of one channel to one sideband while the other sideband is cut off. The baseband of an AM signal can be focused on the undisturbed range within one channel. It works like a magnifying glass that can be shifted. This is an important and beneficial tool to eliminate interference. When weak signals have one disturbed sideband, music from Caribbean stations becomes clearer than ever before while using this tool. Even in cases of extreme interference and disturbances, that can help make a station readable for the first time. Of course, the NRD-545 has a similar option, but Perseus reveals always an efficiency that is some percent superior compared to the NRD-545. All in all, the communication receiver cannot keep up with that. Finally, the times are over when mixing products and mirror frequencies or other ghost frequencies, like those of a superhet receiver, appeared or caused noise and confusion. The frequency processing that Perseus applied was an essential innovation in consumer electronics, and it has been implemented in the most effective way.

But how can SDR do all these jobs with a small metal box? The answer that a powerful pc adds its performance, would be too easy. The developer, Nico Palermo, combined the principles of RF technology with new concepts of digital data processing. That means, in detail: After the antenna input, it starts with an effective filtering process, using automatically switched band passes for different shortwave bands and a low-pass filter that keeps out strong signals from higher communication bands. The core of the new concept is very interesting, because it transfers the needed RF spectrum (shortwave) to a baseband that consists of a digital representation of all signals. It seems to be similar to the superhet radios, but it is not.

The classic mixing can be omitted here. This job is done by digital processing, first by an a/d converter (in this case, a LTC2206) with a speed of 80 MHz and then with a digital direct conversion (DDC).

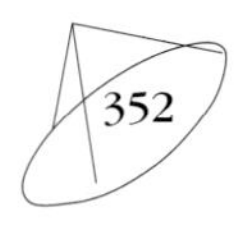

It is 80 MHz, because the Nyquist-Theorem has to be considered. The signal spectrum of shortwave with 30 MHz has to be scanned with the double frequency at least. The organisation of the data stream and the processing for a baseband representation are done by an FPGA (field programmable gate array) labelled as XC35250E. This is the unit that provides the baseband signal, which is then sent over a highspeed-USB-controller to the PC, but only as a reduced stream of 500 kB/s. Any manual operation is done on the PC as a user interface. The screen has a useful structure with controls for each feature and, of course a professional S-metre and a big, coloured waterfall display for all signals in the selected frequency range.

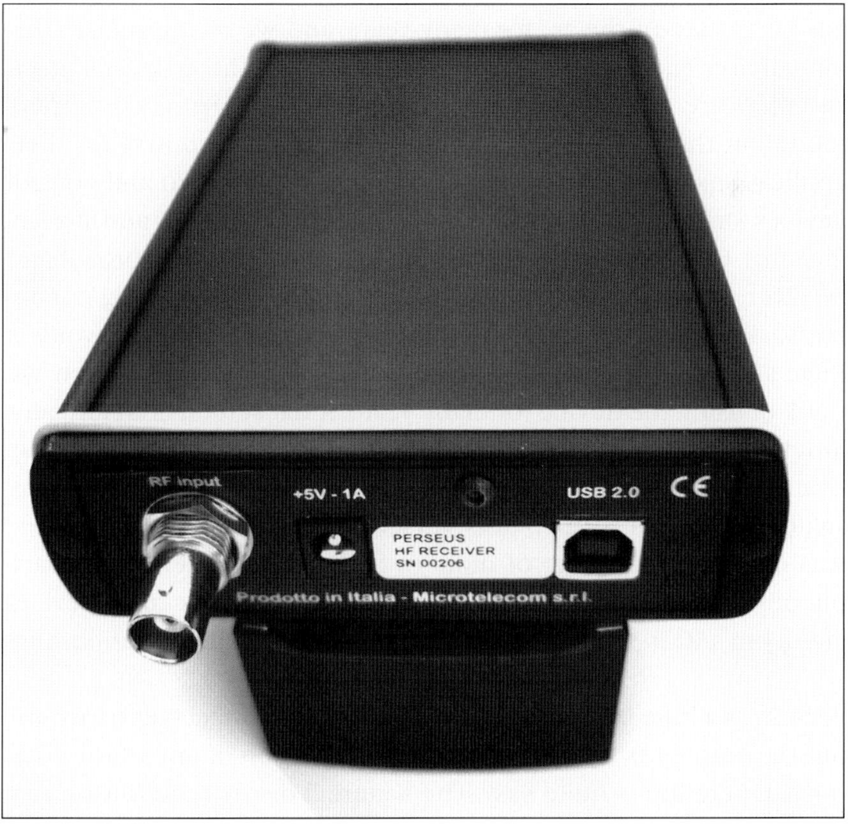

Figure 108: The back of Perseus with an antenna socket, a power supply socket, and the USB-interface for the PC.

Similar to the general development of automotive displays, the philosophy of manual operation changes dramatically. Many controls, switches, and LEDs have been characteristic of any high-tech product with the highest comfort. Today we observe a change to flat user interfaces and integrated controls and information on flat screens. Modern cars are designed with large displays covering the whole area below the front window. In the past, the dominant elements were knobs, controls, and switches just in this area. Now, the driver gets any status information and can activate functions with a flat screen.

Large displays and touchscreens are used in many areas of technology such as of radios and other handheld devices. The use of smartphones has been state of the art for many years, and we all appreciate this kind of user interface. Select any apps by wiping through catalogues, and there are, of course, a lot of new applications. Some developers found out that smartphones are the ideal remote control for their applications. They are the most flexible interactive i/o and control devices for different applications like network receivers and streaming apps. One special use case is the remote control of a broadband receiver. Pressing buttons, scrolling up and down, checking parameters, all this can be done by wiping and touching. This avoids a huge amount of wear and tear on the control elements when we think of using the device for many years. A flat screen is an efficient input element for commands, letters, and numbers, as well as for activating functions. No more ageing and mechanical use of knobs, switches, and control elements. Perseus with a PC as a user interface was only the beginning of modern and flexible operation; smartphones and tablets are the consequent evolution of operating devices like Icom's IC R-30 (see Figure 111) even for remote operation.

Perseus as a little box is, of course, only a piece of receiver hardware, but the highlights come with the control screen of any PC or notebook. Operation is quite easy. The waterfall display and buttons for different functions are presented with good structure and placement. And the results are amazing.

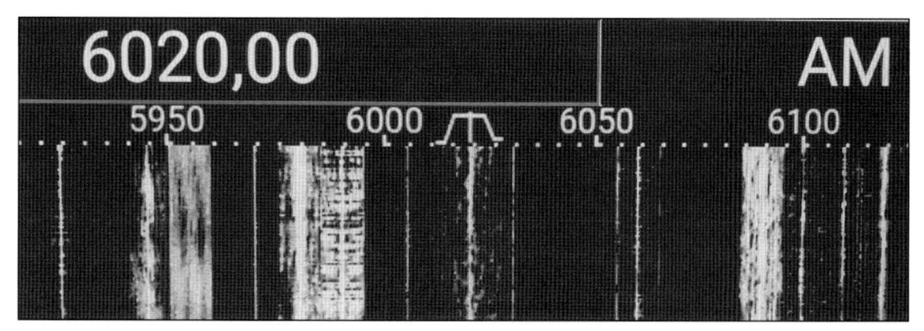

Figure 109: Waterfall displays are also provided by remote-controlled web radios. We can see the stronger carrier of an AM transmission in the middle of the signal, like above at 6020 kHz. Both sidebands carry less energy which can be recognised even visually.

Another everlasting category of receivers that can detect any kind of signal, even at higher frequencies, are broadband receivers. They also cover the long wave range up to the GHz bands while providing and managing different operation modes and their parameters, like bandwidth and channel spacing. Modern broadband radios can also be used for D-STAR, NXDN, or DMR. They can be programmed in such a way that they scan certain groups of channels or a whole frequency band with all channels and user-defined edges.

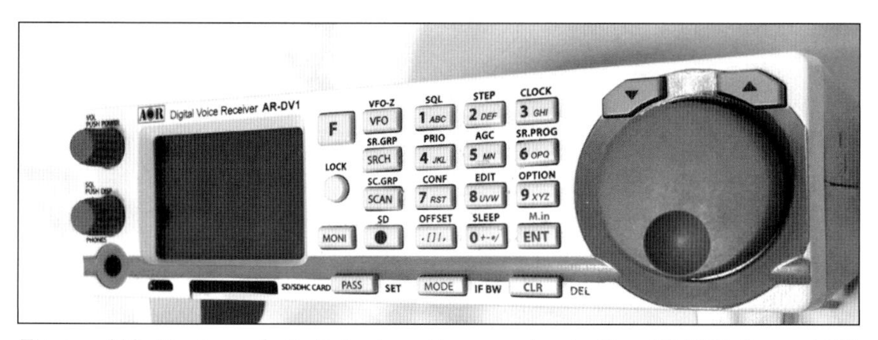

Figure 110: Master of digital operation modes is the AR-DV1 from AOR. Of course, analogue signals can be professionally handled as well.

Modern broadband receivers like AOR's AR-DV1 provide DMR, D-STAR, NXDN, APCO P25, TETRA, and of course they are also able to satisfy any demand for receiving analogue signals. Even TETRA is available, but only without a decryption mode. Transmission from state or police authorities remain hidden from public eavesdropping. The processing effort is much too huge for a consumer system.

Those who like to listen to any programmes and stations when they are on the way, mobile or outside, can find very interesting devices, too. One of them is Icom's IC-R30, which is also a master of all important use cases of a monitoring device like the previous model, too.The IC-R30 provides digital operation modes like D-STAR, NXDN, APCO P25, and some more. DMR is not part of the list, but all the excellent properties, like scanning with 200 ch/sec and 2000 memory channels make you quickly forget the missing DMR mode.

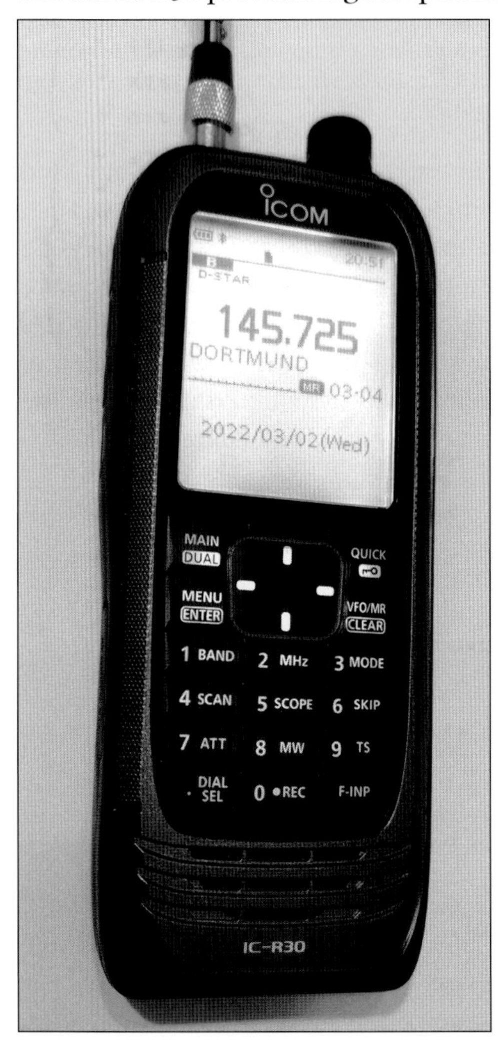

Figure 111: The IC-R30 is a modern broadband receiver with a frequency range of 3 GHz. Using the right antenna, even shortwave signals can be monitored with good sensitivity and sufficient selection.

Developers and manufacturers have now fixed the weaknesses of previous broadband monitoring receivers, and performance has significantly increased. Sensitivity has been improved by using a smart filtering concept; scan and search can be used with an enormous speed of

200 channels per second. Even ghost signals have mostly disappeared and large signal behaviour is much better now. Strong stations in the neighbourhood are no longer annoying. But if a permanent silent carrier on any channel does not leave, it can be marked as a skip memory and will be ignored for any scan and search operation.

Operational concepts have become more user-friendly. Even remote operation is possible and the receiver can be operated with a smartphone using a blue tooth connection. Bluetooth can simultaneously be used for headset use. That means a lot of comfort which previous devices did not provide.

Today, the operation of a broadband receiver has many advantages. You can operate it per remote, and the use of a flat screen is more efficient and means less erosion of the control elements on the receiver. Using a cell phone to conrol the monitor receiver gives you more free moving space, because it can be annoying while moving with a device and its long antenna.

When you are familiar with the new user interface, all the advantages become obvious, especially the avoidance of abrasion and the use of a wireless headset.

Figure 112:
Smartphone user interface of the IC-R30 to operate per remote. Even audio will be sent to a headset via bluetooth as well.

Professional Systems

At first glance, any professional system or component, like a professional receiver, does not perform better than a good consumer product. And they are more expensive, too. So how can they have advantages? And who should buy a professional receiver? Looking closer at specifications and use cases makes it clear.

The reasons for the declaration as a professional device come from the product's development according to strict standards and requirements. Also, the target group is not the same as for consumer products.

MIL-standards as a criterion for quality and approval for use in critical environments, all this is characteristic. Robustness and shock resistance while using it in a monitoring vehicle driving on cobblestone and resilience in the case of immense electromagnetic pulses and discharging are a few examples.

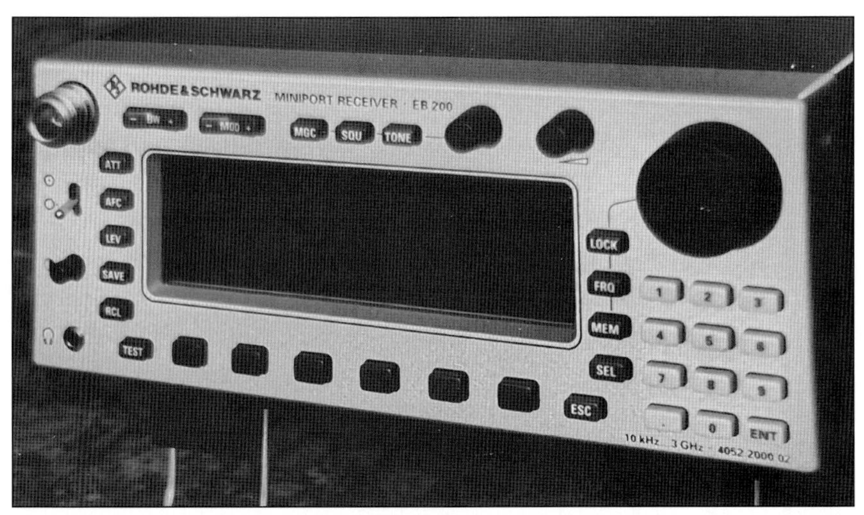

Figure 113: The Rohde & Schwarz EB 200 is a professional receiver. Its behaviour in the case of very strong signals and in critical environments is amazing, due to strict specifications that even consider strong electric pulses as well as shock and vibration resistance.

A striking example is the S-metre display. Most receivers on the consumer market have only poor displays for rough estimations in the best case. The display of the EB 200 shows very exactly the received power, and you can even toggle between dBμV and dBm. Engineers can use it as a precise measurement instrument for electrical fields and transmission levels.

One of the very important parameters is the intercept points of 2nd and 3rd order, which represent the resistance against distortion caused by extremely strong receiving signals. With an IP2 of 40 dB, the receiver belongs to the leaders in this field.

Distortions and overdrive effects in the vicinity of very strong transmitters. These intermodulation products can be excluded anyway. The difference becomes obvious when using radios on a hill, where a lot of strong signals need to be handled and separated. Cheap modern radios are not able to receive anything while their input stage is completely filled with signals.

They often use cheap processors designed to provide any function that is needed for receiving signals. But they have mostly no preselection and they fail in critical situations, such as on a hill with strong signal reception.

Those who believe in the cheap solution against better knowledge will hear noisy fragments, crackles, and not a single reasonable signal, even if they are in the air. The goal of catching some weak stations from afar cannot be reached. Without a certain effort in a receiver, like efficient filters, mixing stages, and demodulators well thought out, results will remain poor or simply noisy.

Other parameters that make a difference between professional receivers and comparable consumer products are the temperature ranges. Operation must be guaranteed even far below zero and also in a very warm environment.

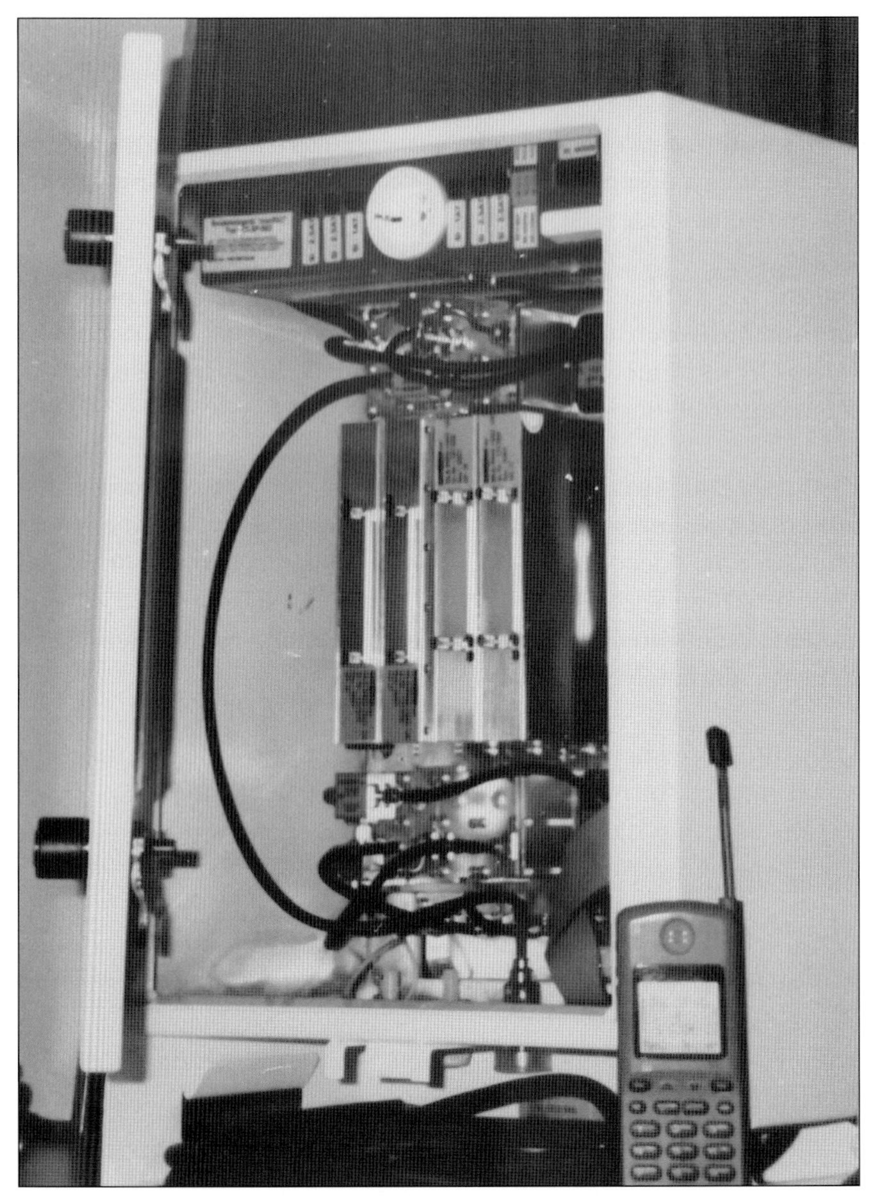

Figure 114: Repeaters enhance radio cells, and have to endure wind and weather and are easy to maintain. In operational mode, they transmit and receive radio signals from cell phones and base stations simultaneously. High quality filters are used for good selection in these bidirectional amplifiers.

Algorithms Meet Electromagnetics

Long before computers conquered the world, RF technology played the biggest role in communication systems. Most transmission scenarios over a certain distance included radio systems for communication. Today, both technologies are available and reveal completely new options in product development and applications. Since the 1990s performance of computer systems has increased steadily.

Expensive computers with enormous speeds could be used in development departments. Some applications seemed to be completely out of reach before. The most prominent example is the cell phone, which everybody can easily take wherever they want. Before the implementation of digital phone networks, nobody would have taken a prophecy about small cell phones for one's pocket seriously.

One example of complex applications are the functions of mobile switching centres in a mobile network. They work behind the level of base stations, which communicate with our cell phones all the time. The switching centres, however, provide the communication lines of a call to the cell phones when we are dialling or when we end the call. Pushing a button starts always a complex process, which most of us would never think of.

Another example where fast computers and efficient algorithms are used, is in the planning process and the prediction of radio coverage for a mobile radio network. Sites and sectors, antennas and frequencies; all these elements are part of a large and complex network design with thousands of transmission sites in a country.

Simulation for Radio Network Planning

The planning of a cell phone network could have been done by checking every potential site with a test transmitter. Many thousand installations and measurements are needed to find out if the antenna site is a proper one, would be an immense effort for many years. But it was done much easier with significantly less effort. Since the 1990s the whole design of cell phone networks was done with predictions, calculated with very fast workstations and complex software programmes and databases with detailed topographical data, i.e., height above ground. Also categories like urban, rural, forest, metropolitan, and more were part of the input for the coverage estimation.

This method of using predictions calculated by work stations saved a huge amount of investment. Installations with test transmitters would cause endless measurements and extreme costs and investments. Also, the time of implementation would have been exceeded by many years. But also effort and invest for work stations, databases, and qualified engineers had immense dimensions.

Still today I remember some fellows of neighbouring specialist groups when I was leader of the expert group for radio systems. One of them was responsible for making mathematical models mostly reliable when using them for predictions. The other one was responsible for the use of databases with precise topographical data. Some years later, both have aligned their professorship according to their experience in planning with scientific methods. Today, each of them holds a significant job at famous international universities.

Numeric Calculations of Electromagnetic Fields

Maxwell's equations are most essential when calculations of electromagnetic scenarios are done with a computer.

Maxwell's equations can be used as a base for setting up equation systems. The equations represent special use cases and scenarios, including electric and magnetic field quantities. That means antenna structures and fields are expressed in numerical terms to enable computers to make realistic simulations. We have seen that in an example with back scattering elements that were resonant for certain

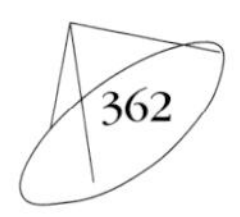

frequencies when excited by an electromagnetic wave at the end of chapter one.

But there are still some more applications of that technology, like the development of waveguide filters. They can also be calculated and precisely optimised by setting up a linear equation system based on Maxwell's equations. When the needed coefficients are calculated, any mode in the waveguide can be considered, and the whole filter characteristic, like transmission and reflection parameters for a defined structure is received as a result. This is mostly cost-efficient and a smarter method compared to manufacturing a lot of different samples and throwing away most of them.

Signal Decoding

Starting with the RF signal of a data transmission, it takes some steps to obtain clear text on the receiver side. Demodulating a signal can be done with different concepts.

Some digital signals can be decoded easily. For radio enthusiasts, this can be a challenging, do-it-yourself project. Professional decoders like the famous W4010 of Christian Kesselring, who died far too early, are a legendary product.

This excellent decoder had incredible sensitivity while decoding signals within a noise floor. When the human ear struggled for any reasonable-sounding fragment, the decoder still did a good job. Also, its selectivity while different FDM signals were received, was phantastic. The W4010 used an early concept of updating and enhancing functions with the help of eproms. At this time, the internet was not available for software updates. Followers of the W4010 in the 1980s included several demodulators and understood a whole sequence of alphabets. Monitoring data traffic was a pleasure using all those different modes, even for analogue signals as facsimile transmissions (FAX) and SSTV.

Operational modes					
ASCII (IRA) (ITA5)	Coquelet-13	O FEC-A	ITAS (IRA) (ASCII)	Piccolo-6	SITOR-Auto
BATIS	Coquelet-80	Feld-Hell	Meteosat-Weatherfax	Piccolo-12	SKYPER (POCSAG)
AUTOSPEC	OCW (Morse)	FMS-BOS	Morse (CW)	OPOCSAG	SPREAD-11
Baudot (ITA2)	DSC	G-TOR	MPT1327/1343	POL-ARQ	SPREAD-21
BR6028	DTMF	U GMDSS	NATEL	Press-Fax	SPREAD-51
BULG-IRA	DUP-ARQ	Golay	NOAA-Weatherfax	Radiofax	SSTV all modes
CALSEL	DUP-ARQ-2	HC-ARO	Packet Radio	RUM-FEC	SWED-ARQ
U CIS-11	U DUP-FEC-2	3 HNG-FEC	PACTOR-1-1	SI-ARQ	TWINPLEX SITOR-ARQ
CIS-14	EEA	ICAO Annex 10	PACTOR-1-2	SI-FEC	UNO-PACTOR
CIS-36 MFSK	DEIA	ICRC-PACTOR	PACTOR-1-3	SI-Auto	VDEW
CODAN	ERMES	IRA (ITA) (ASCII)	PACTOR-1-4	SITOR-A (ARQ)	ZVEI
Coquelet-8	D EURO	ITA2 (Baudot)	DPACTOR-1-5	SITOR-B (FEC)	

Table 9: Operational modes of a high-end radio teletype decoder of the 1980s.

Alphabets
Arabic ITA2 5-bit
ARQ1a 7-bit (used in ARQ-E, ARQ-N, FEC-A)
DATU-80 Arabic ITA2 5-bit
Bauer 10-bit (AUTOSPEC, SPREAD)
BCD (ACARS, ATIS, ERMES, FMS-BOS, VDEW-SELCAL, ZVEI-SELCALL)
binary (MPT1327/1343)
Bulgarian IRA (ITAS) (ASCII) 7-bit
Cyrillic Morse
IRA (ITA5) (ASCII) 7-bit (BR6028, GOLAY, Packet Radio, Piccolo-12, POCSAG)
IRA (ITA5) (ASCII) 7-bit with block coding (DUP-ARQ-2, DUP-FEC-2, G-TOR, PACTOR,
Packet Radio)
ITAI 5-bit
ITA2 5-bit (BR6028, Baudot, Piccolo-6)
ITA2 5-bit with block coding (ALIS, DUP-ARO, HC-ARQ)
ITA2 with 10 redundancy bits (HNG-FEC)
ITA3 7-bit (ARQ-E3, ARQ-M2, SI-ARQ. SI-FEC)
Cyrillic and third-shift Cyrillic M2 (CIS-14, CIS-36 MFSK)
Cyrillic and third-shift Cyrillic M2 with 2 identification bits and 4 parity bits (CIS-11)
SITOR 7-bit (ARQ6, POL-ARQ, SITOR, SWED-ARQ, TWINPLEX SITOR)
Standard Morse
Third-shift Greek ITA2 5-bit
Transparent
8-bit binary alphabet (CALSEL)
16-bit alphabet (RUM-FEC)
16-bit modulo 10 alphabet (RUM-FEC)
80S (Coquelet-82)
82S (Coquelet-82)
401 (Coquelet-13)
402 (Coquelet-13)
403 (Coquelet-8)

Many of these transmission modes became less important. A big share of data traffic is not sent on shortwave or longwave any more, but over the internet. On shortwave, new protocols and operational modes, like ALE, are state of the art, because they have options for encryption, which cannot be easily removed. As the

Table 10: Alphabets of a high-end radio teletype decoder of the 1980s.

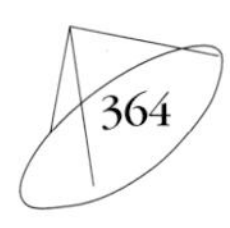

frames for transmissions use almost clear text without encryption the identification of some stations is still possible.

Some of the transmission modes, however, have been popular over decades, and they are still in use. Some have been extended to new, special applications. Baudot, PACTOR, and CW have their place in history and are still part of resilient transmission systems, as in maritime applications.

The use of many different operational modes and several alphabets like Cyrillic, Arab, Hebrew, and more does not make decoding easier. Automatic routines help the user find the right mode, and they even propose parameters for the right settings, which gives almost correct results. Modern decoding systems are quite complex and expensive today, too. High investment and a wide use of encrypted traffic make them unattractive, except for special intelligence organisations.

Chapter 4: Networks

The use of the term "network" has increased dramatically within the last 50 years. One reason is the implementation of modern technologies such as internet and digital radio networks, but the importance of social networks gives this term a new dimension. Humans have a great need to communicate with each other, and everybody is building their personal networks. Technologies invented and implemented within the last 100 years help us satisfy this need in several areas. The differentiated consideration as follows is worthwhile: Technical networks and social networks. It is really amazing how fast the importance of both has increased.

Terrestrial Networks and Satellite Networks

For many decades, we have used a number of networks that provide good service to us. In the 1960s, power networks as an essential standard for all people in a country, were taken for granted. Countrywide telephone networks were implemented soon after. Even TV programmes were transmitted to any household over a countrywide network of broadcasting towers, with only one programme at the beginning. Transmission systems for phone networks and TV networks were located on several high towers to achieve a certain signal coverage (TV) and to connect different regions and nodes by microwave links (phone). But there were also fields of new radio technologies that had lower acceptance.

Trunked radio networks as an improved technology for professional radio communication found not the greatest acceptance in Germany. The idea, to make the use of professional radios more efficient and to provide better radio coverage, seemed like a logical evolutionary step. The German Telekom introduced a service called "chekker," based on the MPT 1327 standard, and implemented transmission stations in several regions. That was really new. Trunked radio uses frequency resources in the 70 cm band as an example and organises

the use of frequencies automatically. Dynamic assignment of channels over control channels should make trunked radio more efficient compared to the use of fixed channels as before.

This concept was even extended to a public digital trunked network, as for bus traffic, taxis, private ambulances, and more. The implementation of the first public TETRA network started in 2001. It was planned as part of a pan-European TETRA network. At the same time, its provider, Dolphin Telecom, already operated a TETRA network in England, where the implementation was faster. But much too soon, the business model was considered fragile and the financial base for finalising the implementation in Germany was withdrawn. The pan-European operator Dolphin Telecom stopped all activities halfway, while in London, several taxis have already been equipped with TETRA devices that sounded really good. The transmission quality was extraordinary, and the voice over the speaker in the taxi was understandable as never before.

In 2001, I was able to assure myself about that. Approximately 20 years later, TETRA is also used in Germany, but not in a public TETRA network. Police authorities, fire departments, and other governmental security organisations introduced TETRA as a professional radio system; not as a new technology, because the life cycle of digital technologies will expire within a few decades and new technologies will come up, but as a technology that at least avoids eavesdropping by unauthorised persons. During the big flooding of 2021 in the Ahrtal in Germany, one of the most dramatic floodings with nearly 200 dead, TETRA was no great help because the flooding of controller stations and base stations had obviously not been foreseen in the planning.

Today we have a lot of standards available for many different use cases in wireless transmissions, like ZigBee networks since 2012 or Z-Wave networks, which are a good choice for local applications such as home automation or electronic sensor networks. They use

ISM bands (2.4 GHz, 868 MHz, 915 MHz), and their implementation and maintenance is not rocket science, thanks to pre-manufactured units available on the market. But first, we will look closer at the large public networks.

Telephone

It was after 1970 when telephone service started getting popular more and more. In the 1960s many people had to ask friendly neighbours to allow a phone call over their phone device. Some corner shops, like Grocer Jack already, had a telephone, and for a good customer, it was, of course, allowed to use the phone now and again. There were only a few yellow phone booths in German cities and only a few red ones in Great Britain.

Rotary phones and reduced local calls for 20 cents and counted time intervals for long distance calls with a fee, depending on the distance, were the standard service for a long time. Button phones needed a long time to come to market. But in the 1980s they came with colourful cases and brought colour into the grey of everyday life. At the beginning, they also used pulse dialling according to the older rotary phones because the old mechanical switching technology had also to be replaced. Rotary selectors and rattling relays disappeared within a short time, and modern dial-tone technology made phone calls more comfortable. Some years later, modern digital technology was implemented in the switching centres, and finally, IP-phone service could be provided to consumers. Today, we all make a great profit from new generations of digital switching technology and fibre optic networks, even for internet service and television.

Networks for Radio and TV

It was December in 1920s, when the first radio transmission in Germany was launched. This brand-new service was transmitted from

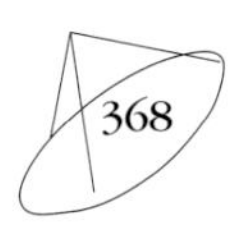

the famous transmission site in Königswusterhausen, near Berlin. From this prominent location with its large antennas, a Christmas concert was transmitted as the first broadcast in Germany. The frequency was 130 kHz, a little below today's public long wave range. European countries used frequency ranges on LW, MW, and SW in the first 30 years. Since 1960, many radio transmission sites have provided diverse programmes in different languages, a proper European or even worldwide broadcast network. Since 1949, a countrywide network of FM stations has been implemented with broadband FM signals and improved quality for listeners. TV experts made their first test transmissions in the 1930s.

But soon after WW II, the first TV broadcasting network was implemented for regular operation. The second TV channel followed and filled the channels countrywide. TV programmes grew in popularity year by year. Some decades later, new technologies also entered the TV networks. According to digital networks for phone service and internet, radio and TV should be digitalized as well.

In the 1990s transmission quality and efficiency were improved by new digital standards, and the capacity of channels and quality of service were significantly improved in any field of broadcasting. Transformations like this need some time for planning and implementation. Also getting the acceptance of the audience was sometimes a hard job.

Today's public radio and TV networks in Germany and other countries are digital single-frequency networks or single-wave networks. That means one frequency for a whole region. Radio has been in DAB+ mode since 2011. DAB+ provides better audio quality than the former DAB and additional text services.

The used frequency bands are 174...240 MHz and 1452...1490 MHz, one of the former TV bands and the L-band. Modulation is now QPSK and QAM. Radio programs are packed or multiplexed by

COFDM with a total data rate of 1.136 Mbit/s. This is OFDM with an additional FEC (forward error correction) to ensure good signal quality. Single stations use data streams of 24 to 144 kbit/s (typically 96 to 104 kbit/s).

Digital terrestrial TV uses the DVB-T2 in Germany. After DVB-T1 from 2001, this improved standard provided a good signal quality. Channels are multiplexed by COFDM in bundles with a total data rate of 21.600 Mbit/s (net 13.271 Mbit/s) and is transmitted in the former TV band of 470 to 690 MHz. The used modulation types are QPSK and QAM16/64/256.

Digital radio and TV programmes come also on satellite and cable as DVB-S and DVB-C. Since 1995, the popular ASTRA satellites have provided digital TV, using QPSK and OFDM signals. High data rates with 10,000 kSym/s ensure the best quality for sound and pictures. DVB-C has been coming through cable networks to our home since 2019. QAM and QPSK modulation is applied for signal transmission.

Transport Networks for Mobile Radio

In the last century, it was microwave technology that provided a big share of countrywide backbones for phone services, with a lot of nodes on high towers. Microwave transmission lines connected analogue local switching nodes and regional long-distance nodes. They were added to the line grids of phone operators, and they were necessary for phone calls in every region.

Later in the new century, the great microwave antennas of the older networks were removed more and more when new digital networks were implemented with fibre optic lines for highest capacities. Some of the drum-shaped and disc-shaped objects on the platforms of high telecommunication towers are still left (Figure 29). Even new microwave antennas and nodes were added. They are used in the transport networks of mobile radio operators or energy suppliers.

Figure 115: Mobile radio tower with sector antennas for different frequency bands (4G, 5G). On the left hand below the panel antennas, a small microwave dish antenna is placed. Further down on the right, antenna preamps are installed.

Fast internet needs fibre optic lines and shall provide high-capacity data highways. We all use it, and we know the huge demand for capacity and high-speed applications. Some of the new fibre optic lines also carry data traffic to connect network elements like base stations or base station controllers (BSC). They can be used as a transport network for different services. In the past, a lot of new microwave transport lines on 30 GHz and 60 GHz were installed to connect network elements of mobile radio operators.

The demand for broadband services has steadily increased. In the last 30 years, the use of optical fibre for long-distance backbone lines or for regional transport has almost satisfied the high demands. With efficient operation modes and modulation types like OFDM as an example, the transformation of the networks has become a continuous process that never seems to end. Cable TV providers also use transport networks. Some of them operate their own grids of lines, local or regional; other providers lease lines, depending on their current demand and their business model.

Cable TV providers use their own DOCSIS standard and operate their own "local area networks" (LAN) for providing services like almost all phone, internet, and TV. For long-distance routing, they operate core networks to be independent from competitors. Some operators rent lines for transport networks but also provide their own products at the same time. The core networks have to carry a lot of data traffic. Gaming has caused an enormous increase in traffic within the last 20 years. Once, network engineers wondered why nightly data traffic seemed to grow. By analysing the root cause, it was a certain surprise when gaming turned out to be the originating application.

Networks for Amateur Radio

In the last century, we would have hardly expected that amateur radio would implement and use appropriate networks, too. But more

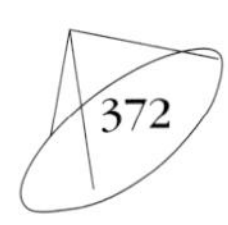

and more operation modes in amateur radio are based on networking. Once, amateur radio QSOs between one or more OMs were used as direct connections. That was the standard. To provide better coverage for longer distances, repeater stations have been implemented. They can be found in higher locations, and they are a good option for radio amateurs living in valleys or between high buildings. Repeater stations provide good connections, even to those with poor locations. On the uplink channel, the repeater can be called, and it will transmit on the downlink channel with wide coverage.

FM repeaters. Many repeaters with pure FM narrow modulation have been implemented since the last century. Still today, this modulation type is a popular mode in the 2 m and 70 cm HAM bands. At the beginning, they were stand-alone devices, and the transmission signals of each repeater covered their own regional area, which was characteristic of each repeater. Some of them are "stand-alone" still today.

Echolink. With the help of line-based services, some repeaters are connected by Echolink. This is a special technology to access network elements like repeaters by wire. This can be helpful in the case of maintenance. Even some small groups of repeaters in different regions are connected for parallel operation to provide enhanced coverage. This can be seen as a simple type of networking.

Radio amateurs can also use Echolink as remote access to many repeaters; that is, without a radio link, over the internet. When the repeater has this option, it is a good alternative without a direct radio link to the repeater. When several repeaters are operated in parallel and connected by Echolink, they cover a huge area, and many OMs can be reached. Talking from one city to another over a distance of some hundred kilometres with the best FM quality can really be exciting. But not all OMs are excited about it. Some feel deprived of their intimacy with their local QSOs. As is so often the case, there are different opinions about evolution.

APRS is another suitable example for networking within amateur radio. Vehicles can transmit their coordinates over a packet radio link to special nodes or gateways that are connected to assigned servers. They are connected to the internet and provide positions of other mobile stations on the APRS platform. This can all be operated automatically, and the locations will be displayed for authorised users. Also, for CB users, implementations of gateways are getting more and more popular. Users can be contacted over a long range. Also on Freenet, this technology based on the internet, provides signals for access even over a longer distance.

Out at sea, radio links and digital modes can be useful to keep in contact with other ships or with stations on the coast. Far from the shore, the signals of public cell phone networks are getting weaker, and after a certain distance, cell phones can't acquire any base station signals. Systems for satellite communication have mostly been expensive options. PACTOR is a good alternative when amateur transceivers are available on board. Even email traffic is possible with this technology. Assigned network transitions for email enable skippers to send an email on sea without a public cell phone. The ship station is not completely offline and can use some of the applications as usual.

WinMail and **Sailmail** are applications with a radio level and a network level that are quite similar to amateur radio applications. People who want to avoid satellite technology can use one of the alternative options. They use their own services to transmit from ship to shore, similar to amateur radio transmissions. Emails and other texts can be sent by sea.

PACTOR is a combination derived from Packet Radio and AMTOR. It has been continuously developed by several steps (PACTOR-I, PACTOR-II, and more). In the 1980s and 1990s PACTOR was also used by professional organisations that were operating worldwide. Especially the higher releases of PACTOR or even derivations for

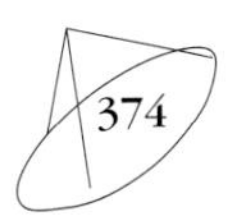

use in special services had some benefits, like better resilience and efficiency. Error correction methods have been improved, and modulation types have been optimised. In the early phases, PACTOR used AFSK (audio frequency shift keying), later also QAM (quadrature amplitude modulation), and QPSK (quadrature phase shift keying).

Network operation. Digital modes are much more based on network operation (just by their concept) than analogue radio. Digital radio operation reminds one of cell phone networks and how they are operated. The system includes a grid with relay stations that are operated according to technical standards and provide a set of features for users and for maintenance.

The audience is a community that is authorised for access. Implementation and maintenance is done by experts and sysops who are part of the community. Repeaters can also be used as groups of repeaters because connections can be selected and combined to form talk groups according to the standard used.

Talk groups, conference rooms, and reflectors are part of the concept of grouping within each digital standard. The structures of digital networks and the architecture of levels and features are not easy to overview at first glance. Of course, the complexity can push newcomers and experienced users to their limits. But everybody can benefit from the HAM spirit, which promises help from other OMs whenever they are asked.

DMR is a good example of network operation and digital transmission. It is used by professionals and also on amateur radio networks. Connections between two stand-alone stations in direct mode are also possible, but they are rare. Network use has become the most popular operation mode. Of course, each network provides its own features, but generally, some typical features are available anyway. When we look closer at DMR, there are three networks available in Germany. Two of them are more famous and more popular than the third one: the Brandmeister Net and the DMRplus-Net (DMR+). The DMR DL Net, the German derivative of the American DMR MARC Net, is not as popular as the others.

When digital networks like D-STAR, C4FM, NXDN, DMR, APCO P25, and TETRA are implemented, infrastructure like transport networks has to be considered too. Servers, connections between elements, protocols, access, and authorisation; all this is an enhancement of pure radio systems, which were almost independent of the internet so far. When we look at the variety of operation modes, concepts of grouping stations, and a lot of configuration options, there is a new category of amateur radio available. An outstanding example of network operation in amateur radio is the Brandmeisternetz, which uses DMR. It is a decentralised digital network with a high number of included repeaters. Currently, we can see the variety in technologies, which automatically splits the community into smaller user groups, and on the other hand, we observe a decrease in the number of OMs. How this will influence amateur radio, we will see in the future. The new technologies are ten or twenty years old, and of course, the younger users appreciate the world of controllers and digital systems. Maybe the digitalization of radio technology is a magnet for younger generations. They are familiar with streaming and gaming, but also with configurations, ports, and mac addresses, at least more than OMs of the generation emitter/base/collector. But there are some more interesting networks.

HAMNET is an IP network for radio amateurs that uses the internet protocol. HAMNET is used to connect repeaters, for example. Another application of HAMNET is to replace the older packet radio networks one after the other.

The use of transport networks is essential for digital communication networks. Also, while operating data networks as a LAN, sooner or later the data traffic needs to be routed in higher network levels. Nodes and controllers are added to concentrate data traffic on dedi-

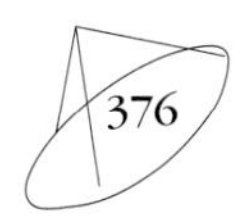

cated servers, that will use data highways for long-distance routing. Of course, backbone networks and local networks need administrators for operation and maintenance. Backbone networks can be found in the infrastructure of operators and providers of phone, internet, and TV services to ensure countrywide availability and quality of service over long distances.

Digital radio has its own characteristics, and of course, some new issues have been discovered. In some networks, such as DMR it is not easy to operate devices in a car while driving. Configuration changes to find repeaters in any areas on the way or to change to the right reflector when you are mobile are not an easy exercise. Especially configurations of devices are mostly for persons with advanced experience, and it may happen that an experienced OM with high CW expertise will resign and ask for support.

It is more complex to get an overview in fields like the structure of talk groups, reflectors, and conference rooms with their own configuration in each operation mode for digital voice. But it is at least a new challenge for those who are interested in new technologies. And if somebody does not find the right access to network technology, there are still operation modes like CW, FM, and AM to have a lot of fun in amateur radio, DXing, and monitoring.

Criticism of modern operation modes says that access technologies like the internet or applications like EchoLink lead to communication types without radio frequencies. People can talk to each other with any device, but without radio technology. This ain't the status of a radio amateur, says the critic. This is, of course, a possible use case, but it is almost not the preferred one among radio amateurs.

It is always a question of appreciation for values and what every person wants to do. Enhancements of current applications and introducing new technologies can make amateur radio more exciting and interesting. Local radio access, national and international links, even without shortwaves or satellites, can be seen as a result of the variety and evolution of technologies. Talking on VHF and UHF channels from one continent to the other means a big challenge. In the end, the user will surely be rewarded with some international contacts.

But it is also true that the increase in complexity needs a certain know-how for server and communication line technologies. And this is not the preferred field for each OM. Disruptions within a transmission with good quality can happen much more in digital systems compared to analogue ones. To analyse the origin of issues in digital systems, we also need additional know-how at some other network levels, like transport networks.

New root causes for disruption or quality issues need attention, as well as configurations and parameters for digital settings in handheld devices. All this is nothing extraordinary for people working with new technologies. With a little support and patience, any issue can be solved. And if nobody and nothing can help, look at the good old operation modes and modulation types. A lot of OMs anywhere in the world are waiting for new contacts.

Satellite Networks for Navigational Use

Currently, there are approximately 5500 satellites in earth orbit, some as small size like a basketball, others as big as a refrigerator or even a truck. One special kind of satellite seems to be positioned at a fixed point in the sky. That's very practical because receiving stations on earth don't need any readjustments of elevation and azimuth, except for an initial installation, of course. These satellites are located over the equator in geostationary orbit positions with a height of approximately 36,000 km above ground. Service can be provided 24 hours a day, like TV satellites do. Many satellites, however, orbit the earth at different lower heights and in polar orbits. At a low altitude with 400 km as an example, they also fly through the ionosphere. Short distances to the receiver mean stronger signals for reception. Another advantage is the significantly increased resolution of optical systems for earth observation. One disadvantage is the dynamic adjustment of the antenna (tracking). One of the most famous satellite networks is the GPS (global positioning system), which is installed in nearly every vehicle. Devices receive coordinates and show the current position on a screen. Combined with topographic data, the best routes to any destination can be calculated and displayed to the driver. GPS uses a broadcasting mode with unidirectional transmissions to the receiver.

There are also other satellite networks we can use. They provide global telephone service even in areas where no terrestrial infrastructure enables phone calls. GPS has also competing networks like Glonass, a Russian network with 24 satellites orbiting at a height of 20,000 km. 30 European Galileo satellites use orbits at 23,000 km and supply users with navigational data, too. In detail, it works as follows: 30 Galileo satellites are operated on 3 different orbits around the earth, approximately 10 satellites on each of the 3 orbits. 9 of them are always busy, and one is kept as a reserve. To calculate the exact position, four satellites are needed.

According to the triangulation method, three of them transmit positioning data, and the fourth transmits the precise timing base. Only with the exact timing data can the coordinates be calculated with an appropriate accuracy of some meters. Operation of GPS satellites is similar.

It was Einstein's theory of relativity that enabled that accuracy at all. Without this knowledge, the shifts in timing caused by satellite movements would not be considered, and so the needed accuracy of coordinates on earth could never be achieved. With a resolution of several kilometres the results would be much too coarse for most applications. So we all benefit from Einstein's enormous, unique contribution to scientific work as a base for amazing products some decades after his death in 1955.

Satellite Networks for Phone and Internet

There is another network with orbiting satellites at a height of approximately 780 km. The famous Iridium system with 66 orbiting satellites provides phone service at almost every place on our planet.

They have good coverage due to the high number of transmitters in space. Inmarsat, an international organisation for maritime satellite services, also operates a phone network for voice services. They need only 9 satellites for operation in the geostationary orbit, approximately 36,000 km over the equator. But from here, the polar caps cannot be seen by satellites. That means the line of sight is missing, and signals in the GHz range propagate always straight forward. As a result, the north pole and south pole have no Inmarsat service, as long as they use the geostationary orbit. Starlink is another network with a lot of orbiting satellites using heights in a range of 540 km and 570 km, and Starlink provides internet service around the globe. In the first implementation phase, the network has 1584 satellites in low orbit and satellites can be used all over the world.

VSAT

In regions without terrestrial infrastructure for data services, satellite links have always been a good option to connect business facilities with their central offices and data centers. Small terminal stations can be operated on continents, in the desert and also in the ocean. It was in the early 1990s when sat terminals were developed to connect shops and offices, because internet was not available in all regions or in rural areas. So shops were equipped and got ready for payment transactions from credit cards. Gas stations got access to the billing infrastructure of petroleum companies for example (Figure 116). VSAT systems (VSAT: very small aperture terminal) with a receiving LNB and a transmitter were used for data traffic over satellite links on 14 GHz. Today, access can be realized over landlines and internet applications in the most regions.

Figure 116: VSAT outdoor unit in the 1990s for billing system access of a petroleum company.

Public Mobile Radio and Landline Infrastructure

The idea of a mobile telephone was born long before any radio networks were available. In the 1950s the idea of public mobile radio was realised in Germany. The so-called A-net did not use a dense grid of infrastructure, and calls were switched and connected manually. A few FM channels on VHF were available. Since 1972, the B-net has gotten some attention from business people, and mobile phones in cars were an amazing innovation for very few people. Both networks were operated on frequencies above the FM radio band. 158 base stations for the B-net covered the whole country with FM channels using a channel spacing of 12.5 kHz and 25 kHz. Transceiver technology was heavy and expensive, but now calls could be started in both directions, also from a car. The era of manual switching and connecting was over. From 1985 until 2000, mobile services in Germany were enhanced by the C-net of Deutsche Telekom. Coverage was improved, and eavesdropping on FM channels was made nearly impossible by a sort of scrambling of the voice audio. Pairs of frequencies were assigned in the 70 cm band near 450 MHz. But only a short time was needed to develop monitoring receivers that were able to descramble the voice signals. However, all the forerunners of digital mobile radio had a widespread weakness. Neighbouring countries used different technologies, and roaming was not possible. This finally ended in the early 1990s with the introduction of GSM in many European countries. Handheld devices could still be used after leaving an airliner in a different country. Roaming was even international, and GSM devices could be used in most of the countries.

With the introduction of digital technology in public mobile networks, several new features were available, like SMS, the short message service. At last, phone tariffs became attractive for a large part of the population with an average income. Handheld devices were no longer like bricks, and it did not take much time when everybody carried them in their pockets. Radio coverage had strong signals and the cell phones could be operated everywhere, even for inhouse use in many urban areas. The time of dinosaurs, phone devices with some kilos and a label with some thousand dollars on it was finally over in the early 1990s. Data transmission was a feature with an increasing importance. Even if it started slowly, the attention of operators on data services indicated a quality feature for phone services. The internet could only be used from a PC at home or in the office, and it needed those loud and noisy modems with relatively slow data rates of 256 kbit/s. Today, no one would ever think about the internet as a feature. The internet is available everywhere and on every device. Today, data capacity and transmission speed are amazing.

When GSM was still young, new data transmission technologies were introduced since in 1990s. They were called GPRS (general pa-

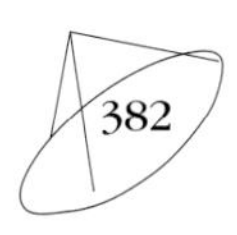

cket radio service) and HSCSD (high speed circuit switched data) with a speed of 50…100 kbit/s. This was enough for short emails, but without space for long attachments or pictures. GPRS was a packet-based data service that used capacity only during the transmission time of the data packets. HSCSD was the line-based variant, which was on for the whole time of the connection. Both data modes used interconnected timeslots of GSM voice channels to achieve their data capacities, which seem very poor compared to channels and capacities today. Data rates remained low when EDGE (enhanced data rates for GSM evolution) was introduced with 150…200 kbit/s. EDGE could improve its in data transmission by using modulation types with 8 states and 3 Bits. EDGE was operated with 8 PSK.

Fast data transmission. When UMTS (universal mobile telecommunication system) and HSDPA/HSUPA (high-speed download/ upload packet access) were implemented, data transmission became faster, safer, and more comfortable. Sending out word documents with several pages no longer took minutes or hours. UMTS increased data transmission to 376 kbit/s and with HSDPA, a kind of booster even 7.2 to 42.2 Mbit/s were achieved. Really comfortable data service finally came with LTE (long-term evolution, 4G) and 5G. Streaming of music and videos without stop-and-go, parallel use of different apps, high-speed downloads, this is really satisfying. With data rates of 20 to 500 Mbit/s LTE has become a powerful standard. 5G, the fifth generation, promises even 150 Mbit/s up to 10 Gbit/s. It is like "all applications in real time". What more do we want?

But the implementation of digital cellular networks needed an immense effort and a huge amount of money. Exact official numbers about installed transmitters come from the Bundesnetzagentur in Germany. According to estimations for Germany, there could be more than 100,000 base stations in all regions. For operators of these networks, any station needs access to the system. Therefore, access networks or transport networks had to be installed, too. Links and nodes, controller sites (base station controllers, BSC) and microwave traffic collectors are needed to connect the radio units or base stations with the switching centres (mobile switching centres, MSC) and billing centers of the operators. Access networks get better resilience when they are implemented as a ring structure, with two gates for traffic. When one link is disrupted, they can route the traffic through the other gate. Access subnetworks with star-shaped link structures do not benefit from this advantage. Disruptions or any degradation of service quality are not accepted by the users. Robustness and redundancy are important for every network region. For subnetworks, it is clever link planning, which can reduce outages. Some big data centres with extraordinary importance are operated at two different locations as redundant systems. These special control centres can toggle between each other in event of severe damage or an outage. And of course, power supply has an additional battery system for base stations and even large generators for central control stations. All these measures will prevent total failure in event of a disaster.

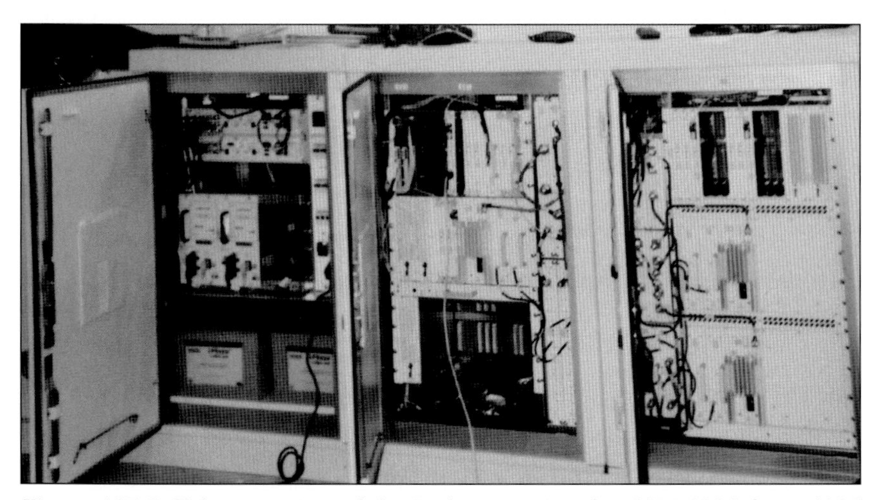

Figure 117: DCS base station of the 2nd generation for GSM 1800 from 1995 with an additional battery pack for 2 hours of emergency operation. Data access was realised by microwave links or leased lines with a capacity of 2 Mbit at least. The investment for base station technology was 150,000 DM or more.

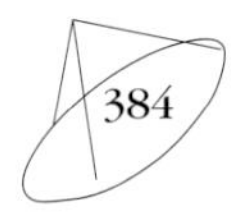

Base stations and other network elements are always embedded in a whole network by regional access networks. Today, the size of a base station is much smaller compared to recent generations (Figure 117). Smart base stations are as small as a suitcase or even a shoebox. Data traffic was once routed over lines with 2 Mbit/s as a standard capacity. Of course, they were designed only for pure voice traffic with cell phones as a narrow-band data application with no internet, or streaming service. Capacities cannot be compared to those of modern data highways, realised with fibre optic technology. The local and regional networks of the providers are now designed for real-time applications, gaming, streaming, and business applications like video conferences. Data capacities have always had to be adapted to new demands, which means a steady increase. Most providers operate their own networks, not only local, but also for long-range traffic over backbone networks and core routers for the highest data speed, too.

TV cable operators use their own technology to supply customers at home. With DOCSIS, they achieve similar results in speed and capacity as the big telecommunications providers. DOCSIS as a technology has also been improved over several releases and generations, and it is optimised for TV services, combined with the internet and telephone, of course.

With their own technology, operators are more independent of access to leased lines, which all belonged to Deutsche Telekom at the beginning of TV network implementation in the last century. Even for the so-called last mile, DOCSIS is used and increases competition between providers as a benefit for consumers. TV network operators use their own networks and are a real alternative to other established phone companies. This alternative can be important in regions with poor data rates on traditional phone networks. Meanwhile, the local access technology is based on LAN. Backbone networks with-high tech and high-speed core routers promise a huge amount of data traffic, almost without interference or outage.

The recent providers of phone service, who operated their networks after WW II are now providers of broadband services for all customers, business customers, and consumers. Today, the challenge is to provide fast internet, even in rural areas. Industrial users with offices located in the country, all need the same state-of-the-art services as anywhere in the world. Too many years have passed since providers avoided investing in rural areas with only a few customers. Operators and providers always worried about urban areas and metropolitan areas with dense population. This strategy based on economic arguments was quite good for their own business model as an operator, but it was a disaster for many companies in industrial areas in the countryside. Office staff had to listen to the sound of the good old modems of the last century for sending out simple documents. Bigger schedules and detailed technical drawings needed some hours to be transmitted with the old modems at low data rates. But step by step, the network rollouts of the big network operators cover rural areas, too. The telecommunications industry seems to remember its promises for better digitalization.

Internet

When we talk about the internet today, we mean almost all applications that are based on the internet infrastructureA lot of applications or apps allow access to services like email, knowledge data-bases, shopping platforms, and many others like social media and music stores. The list of applications and services is endless.

Before all these services could be used, the appropriate network infrastructure had to be planned and implemented worldwide. The very early years were not spectacular. The concept originated in the military area, and it was the idea of a decentralised network that could not be destroyed easily by attacking some servers and nodes or even a region. Significant functions and links should be available even in the case of a disaster, and operations have to be resilient with robust links and systems distributed over a wide area. One of the very first services was email. Still today, it is one of the most important services for business and private correspondence worldwide. Network infrastructure includes a lot of distributed servers and nodes, connected by large data highways for high-speed transmission. The use of fibre optic technology is one of the keys to very fast data transmission between servers all over the world. Some of the early checkers used the internet as a money-making machine, while others were cheated by them. The checkers were active in doubtful fields and used gaps in the legal system. They provided download services for little money with pirated copies of music. The record industry and many artists were cheated by these doubtful services. Some of the criminals became millionaires within a few years. But not for a long time.

Access to the World Wide Web is done via local networks. With the help of the internet protocols http/https (hypertext protocol/secure) and ftp (file transfer protocol) the transitions between wide area networks and local networks are easily managed and web pages and files seem to fly with high speed from the sender to the recipient. Methods for authentication and authorisation have become extremely important. These mechanisms make sure that messages stay unchanged and that intruders get no access. And the only correct recipient shall get the message, of course. Many technological details enable the use of this amazing internet. When web technology and wireless communication were mixed, some immense changes came on the agenda. Some were related to technology; other changes came slowly and were related to social life and individual behaviour. Today, we live with all these changes.

Most impressive is the amount of data exchanged at any time and at every place. And in the new century, all this can be done even with small mobile devices in everybody's pockets. Also satellite networks have been implemented to provide voice and data services over the internet at any place in the world, even without line infrastructure. Starlink is a good example. The satellite system has been activated by TESLA's owner, and it is operated by SpaceX. Decisions about service coverage, even in crisis or war areas, are taken here. Decisions about who is allowed to use the internet, and who is not.

Impact on Society

The meaning of web platforms and applications for consumers has a special history. In the 1980s and the 1990s early users of computer technology were mostly interested in hardware and some special functions. Algorithms and programmes (today we call them apps) were created by interested computer users on their own. Some programs were reproduced from books and other sources.

Other users were not impressed by the new technology, but they liked computers due to programs for writing letters or documents, combined with one of the early printers, almost dot matrix printers.

They only wanted to use special programmes running on a computer, like games and special hobby programmes as for drawing. The programs were the real user magnets. Hardware and software industry learned that. Today programms are simply called apps (applications). The use of apps has always been an easy exercise, because most programmes were designed with practical user interfaces.

The average user does not start compilers, and nobody has to link modules to get a program running. Apps have become the personal interface to the whole internet with all its pros and cons. We are uncompromisingly networked.

The apps are the new access tools of any individual, like a socket to the amazing fields of the World Wide Web, but also to any dark abyss waiting there. We are connected, maybe with a personal chip implanted in everybody's neck, but at least with projection glasses to stream new realities to our brains. Biological interfaces like this give much room for thinking about our future as a society.

Benefit

Without doubts every user of a cell phone will confirm the whole portfolio of benefits of a mobile device. In critical situations, they have saved many lives. When a large fire started in the airport in Düsseldorf in 1996, 17 people lost their lives. But there were several people who could be saved, some of them trapped in an elevator. With their cell phones, they could send an emergency call, which saved their lives in the end. Many similar examples could be described all over the world.

To better understand the meaning of cell phones for our society, we should divide the time spent on mobile phones into two phases. One is the era of mobile phones without internet, because it was simply not implemented for mobile use. The second one is the era of smartphones and all their functions. In the first phase, mobile phones got their own countrywide network to start calls anywhere in the republic, only as a phone service. Outside on the street, while driving in a car, in trains, in the forest, and in shopping malls. Silently and almost without attention, new generations of systems and technologies were developed and implemented. Until one great leap. The evolution continued silently, but a completely new type of cell phone appeared. The era of smartphones started in the 21st century. When smartphones entered the stage, the functionality of cell phones was enhanced dramatically. Being online all the time, looking at any web page, posting photos and chats whenever and wherever you want.

Visions, Concerns und Fears

Technological progress in our society was always evident when visions and visionaries pushed new ideas and developments. Many examples in history prove this. Every time a new technology was published and propagated, a group of people came together to reject and refuse that progress. Maybe the various arguments against a new technology is some kind of human reflex that is triggered by fears. Also, the implementation of cell phone networks has been rejected in some areas and by some groups of people.

When fear arises, the origin often lies in anything that is unknown and new. From all that we do not know, most of us keep a certain distance. And sometimes it scares us, like many of those radio towers of cell phone networks with their numerous antennas at the top. In some areas, residents got a headache only from seeing those new components of communication systems.

Sometimes this happens before a station is on air. According to some reports, the headache began when the steel and the pole for antenna mounting were installed, even without any antenna or cable and without signals.These symptoms also need to be taken seriously, even if some people find them ridiculous.

Today, we know much more about mobile radio networks, and everybody uses their own device, wich is a great benefit for all. The vision of total communication, that everybody can communicate at any time and everywhere over a small device in their pockets and bags has become true. But what are the next steps?

What we can observe today is new behaviour and some new individual characteristics of people, even physical changes in smartphone users. A crouched way of walking and hanging heads are typical when we observe people with a certain addiction to mobile phones.

Another affliction is an increasing deafness when we look at people who walk outside. It can happen when you talk to a person and ask for their way, and the answer can be a very loud "HÄÄ?". It sounds unfriendly, but it isn't. Many people are totally isolated from their environment, and they are frightened or surprised when another person tries to communicate with them in real life and not over an app.

Social Networks

With the introduction of the appropriate technology for total networking on any level, things happened that were never seen before. Even during the implementation of modern networks, many new appearances have been observed for the first time because they were completely unknown before.

It was again a time of visionaries, a little bit like a new gold rush in the new millennium. We all know them, and many of us use them; they are called Google, Facebook, Amazon, Netflix, YouTube, TikTok and so on. But let us discuss it step by step.

We could observe the rapid ascent of social platforms and networks with amazing acceptance in the whole audience. Engineers and other scientists in the 1990s did not expect all the impacts on society from the implementation. Technical networks revealed some extraordinary side effects. Mobile phone calls at any place and in every country are for many people, the comfortable variant of the recent use of a telephone. Now the phone could be taken anywhere, even on vacation trips and in holiday residences. In the last century, broadband and mobile internet, coloured displays, and functionality like a computer and even emails were once far away. In the 1990s most people would have smiled at you when you were talking about such ideas as an omnipresent cell phone in future. "Coloured display? Paaah... such nonsense. Nobody needs that." That was a typical answer often used in the last century. In some countries, this type of reflex is characteristic when new ideas are presented in the context of new technologies.

Some older people remember that it was similar when colour TV was introduced. Once, the world in TV news and reports was black and white. The rapid development and enormous increase in computing speed and storage capacity of the smallest microprocessors and memory chips opened the door to a new world after 2000. Ubiquitous smartphones and services, the end of Nokia's Symbian operating system, and the missing success of Nokia's MDA with its big and heavy multifunctional feature phone, all this is an essential. Manufacturers of feature phones lost market share. Phone calls, email, and a coloured display with a dot matrix were nice in 2006, but these features could not help the old hardware platforms to survive. The age of iPhones and Android-Phones has dawned. And this happened with immense power.

After a few years in the new millennium, we observed this immense technological leap for consumer devices and services, with some large movements and changes in the markets. The success of smartphone technology could not be stopped, and everybody was affected, some more others less. Looking back at this time, mobile phones were only wireless telephones, similar to those at home but with a countrywide range of operation.

Put into a pocket, a cell phone could provide access to emails everywhere. Even with coloured displays, a mobile phone was not the consumer's favourite hardware platform for chats and browsing. First, cell phones did not have the memory stacks or the performance for applications. That was the personal computer. But the performance of microprocessors and storage components like hard discs got better with each passing year.

But where did the popularity of social networks come from? When we go back to the 1970s, we meet a unique phenomenon. People who were interested in technology suddenly revealed an increased interest in communicating with each other. CB has been launched, and it has connected thousands of people in talks during the day and also at night. With a wink, we might assert that CB radio was the forerunner of the internet, not for data transmission but for voice. But at least it was the forerunner of chatting. Radio enthusiasts and electronic hobbyists talked on 12 channels, like in 12 chatrooms, about everything. With a radius of 2…5 km, at any corner, new

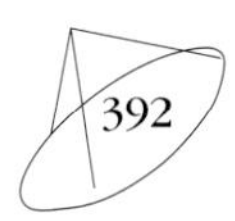

people could be found who liked to talk about what happened in their life. Young people, boys and girls, strange people of any age, and friends, of course, got euphoric as never before. All night long exciting talks and exchange of radio experience, as well as contacts to stations in other countries, everything was of great interest. CB radio even worked for long distances at the beginning because it was the time of a sunspot maximum. Sometimes QSL cards were exchanged to confirm the radio contact over some hundred kilometres. In urban areas, CB stations could be found every hundred metres wherever the radio was turned on. The devices had enormous market penetration, like cell phones in the 1990s as an example. Radio enthusiasts used the new CB radios combined with a roof antenna like the λ/2-radiator or the ground-plane antenna. Also, mobile radios were used in cars with antennas of 50 cm or more than 1 m in length like the famous DV 27 "long" with 2.5 m, most efficient.

The brand variety of CB radios was immense: Grundig, Zodiac, Albrecht, Universum, Stabo, DNT, Lehnert and more. It was a gigantic market. Manufacturers supplied shops with a colourful variety of CB radios with different designs. Young people preferred handhelds for little money, like the Universum 3 channel or Stabo's P3/P6. We can assume that it was the first indication of interpersonal communication, which should have a rapid rise when the proper technology is available for everybody. And there is still another aspect additional to communication devices when we look at our modern society: the 24h availability of consumers who are available for their devices, that is for other users at any time. Phenomena like the Tamagotchi, which came on the market 1997, predicted how fast and how severe people can be addicted to a small piece of technology if they get inspiration to use it. The submission of the individual to a duty revealed addictive and dubious demands that were suddenly much more important compared to everyday duties in family life. Other really important activities and tasks were neglected. Did Tamagotchi predict how smartphones will be used, and did Tamagotchi shape younger generations and reveal their temptations and weaknesses? When the appropriate technology for total communication was finally available, like PCs and smartphones, it was only a matter of a few years before their popularity and penetration would grow beyond earthly proportions.The time has come for new generations of applications and their consumers. Social platforms, social networks, shopping platforms, and music streaming should get a prominent position compared to other applications.Talking and writing all the time, with friends and strange people, everywhere for many hours, this was the new way of life, or at least the temptation.

Health Aspects

It has always been quite natural to look at health aspects when we talk about electromagnetic waves and their technologies. Most evident are the sources that have a great impact on all of us and which we all should be aware of. Ultraviolet radiation from the sun or artificially generated for our beauty has to be observed and enjoyed with the utmost caution, even if it gives our skin a beautiful brown colour. UVA rays make our skin brown, but the skin will also age faster with too much exposure to the rays. More severe and threatening are UVB rays for a longer time. First, they will cause sunburn, and the skin will get more and more red.

Later, it will burn more and deeper until the skin peels off and the deeper layers are destroyed. In the end, too much ultraviolet radiation may cause skin cancer.This is not beautiful, but sometimes even deadly. Malignant melanoma, also called black skin cancer in Germany, appears many years after the exposure to strong radiation in a solarium or on the beach. People who don't take this seriously can be confronted with a bad prognosis for their lifetime. But now the good news: not any spot on the skin, even if it is dark, means skin cancer. What is most important in cases of doubt? Keep calm, and if it is possible, ask a dermatologist.They are experts in the right diagnosis.

The direct impact of radio waves on people's health has been explored since the 1990s under high pressure. In addition to the direct exposition, there are other topics like social and psychosocial consequences. They are obvious when wireless technology becomes a personal interface to the virtual world of social media, especially when it is used all day. We will discuss this later. But first there was the fear that the exposition of electromagnetic radiation was a reason to worry. Still today, we observe that people are concerned.

Havanna Syndrome. Some years ago, new reports about electromagnetic waves attracted attention. That means it was supposed that the reports were about electromagnetic waves and health aspects. The reports came from people working at US Embassies in different countries. Since 2016, a new phenomenon has been observed, and they called it Havanna syndrome.

The name of the Cuban capital was used because the symptoms were confirmed at the US Embassy of Cuba as a first location. For the first time, several symptoms were observed with an increased concentration there. It was all about health impairment, which made some people worry. Unspecific symptoms like dizziness, headache, and nausea were observed, and a connection to people working near or even in the embassy seemed obvious.

What was the origin? There was no idea at the beginning of the upcoming discussions. Could it be an impact coming from technical systems close to the embassy? Could microwaves be the source, and nobody suspected this? That was the last update at the beginning of 2022. After several examinations in different US embassies, the suspicion arose again. Could this be a technical system transmitting microwaves, that had a sharp beam to focus their electromagnetic radiation on the embassies? And if this were really true, who should be the operator of such a system? Is the exposure of the embassy staff to dangerous radiation man-made? And where does it come from? The last reports did not give clear answers to all this. But we suppose that the impact comes from outside the embassies. There is room for speculation and conjecture.

Of course, health is most important for all of us. We should never forget this. But new technologies have the ability to polarise people dramatically. It was not long ago when the rapid development of the Corona vaccine caused a lot of trouble. People could have been happy about the early availability, but as the communication of the authorities was disastrous, we could see an unexpected phenomenon.

The use of the vaccine has been rejected by many people, and they refused to protect themselves. This happened between 2019 and 2022.

Once, when the first railway was installed and people came to watch the new invention, a surprising behaviour was observed. The speed of 35 km/h or 20 mph worried many visitors. People recognised that as witchcraft. And still, at the end of the last century, we could watch similar scenes. Even when mobile networks were implemented, some strange stories were told. Construction workers who wanted to build up new radio towers were driven out with clubs.

EMVU/EMCE. The German term and a derived English term as an abbreviation mean electromagnetic compatibility with the environment. Many studies about that already exist. They should answer questions about how dangerous electromagnetic waves are for us. Especially in the field of mobile radio networks, it has always been of great interest, and the studies should help to understand the dimension of the danger.

In 2022, the results of the "MOBI Kids Study" were published. The reassuring news: Mobile radio signals do not increase the risk of cancer for children and younger people. In the early 1990s network operators and providers faced accusations of starting one large experiment with microwaves without knowing what microwaves

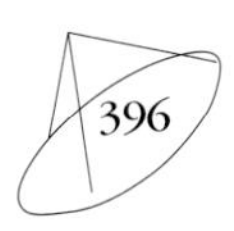

(GSM 900 and DCS 1800, the later GSM 1800) would do to us and our health. But only the very pessimistic ones talked this way. Even before some studies about radio waves had been conducted, in 2022, current results from another new study, which has been conducted over 20 years, were published. It was the "UK Million Women Study". Brain tumours were in the focus, as were women, and if there was an increased probability of occurrence. But in the end, there was an all-clear, fortunately.

Field strength limits for radio systems, especially for base stations, are defined by authorities within standards and by governmental regulation. In Germany, it is the 26th BIMSCHV (Bundesimmissionsschutzverordnung), a governmental regulation paper. Of course, any base station is operated on the basis of that. For handheld devices there is another regulation. Commissioning regulations for electronic devices, are the frame of regulations in Germany.

When we look at people's opinions about the harmfulness of electromagnetic waves, there are a lot of voices and attitudes in mobile radio scene. We can look at measurement aspects as an objective approach. When we talk about EMVU/EMCE (electromagnetic compatibility with the environment), the reference is almost the field strength and not power levels or voltage terms. We have already seen that they can be transformed easily by an offset.

For a long time, people assumed that the base stations were the worst threat. Mobile devices have almost never been considered a real danger in most discussions. Maybe it is due to the size. A huge steel structure on a house with at least six antennas seems indeed more dangerous compared to a small handheld device.

But is this assumption really justified? We can make a brief calculation for one sector of a base station, covering mostly 120 degrees in the horizontal plane. When we compare the base station and the cell phone by their field strengths, we get the following result:

The signal level of a base station at a distance of 30 m is at least 10 dB lower than a cell phone signal at a distance of 0.5 m from the user's head, even in the main beam direction of the antenna. On the ground, the base station signal is supposed to be even 5 to 10 dB lower, due to the antenna diagram. The conclusion is that base stations radiate 15 to 20 dB less energy at the head of a user compared to cell phone use at a distance of 0.5 m. The base station signal is ten times to a hundred times weaker.

These are unexpected results for most of us. The calculation was done with the following assumptions: Maximum power of the cell phone is 1 watt, maximum power of the base station is 10 watts, 2 dB cable loss, distance phone to head of the user 0.5 m. A huge tower with a lot of steel installed looks much more dangerous than a small handheld.

Media usage time. But there is still another area from which danger is threatening humans. This new category of health problems is not yet clearly visible today because it comes slowly along and is almost hidden. It is the use of a smartphone for many hours and every day. The neck is brought into a special tense posture, as are the arms and hands. And effects like addiction increase and may end in a vicious circle. We can suppose that both are getting explored more and more. And it is evident that already today, a strong impact of many hours of wireless media usage happens some million times, every day and everywhere in the whole republic. This will influence our personal health, and children and young people are exposed to an extraordinary high risk of becoming stunted.

Beautiful new technology for beauty. We should not look away from some obvious fields that really matter. Beauty is important for almost all young people, boys and girls. Consequently, they want to look perfect in photos, especially when they are posted. Instagram provides even more filters for optimising the look. Some kilos fade away with the right adjustment in the beauty app, and even fat rolls

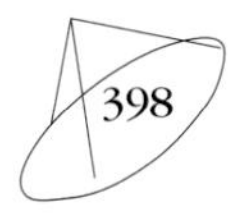

on the belly can be retouched. Smooth skin and slender figures, anything is possible, and the body looks as it should. So far, it's okay. But what will that do with us?

The answer is easy. Just young people search for their ideal look. Which is my perfect look that the others will accept? Just here, we need to look closer at the consequences: What happens if a young person is different from that ideal, different to all the others who present themselves as amazingly beautiful? The answer is alarming. With a blemish, you feel left out. You are different from the ideal. Affected girls and boys suddenly go through a process of change, which is not always good for them.

They want to be different from the person they really are, as it is good for them. Parents of girls suddenly face the consequences of eating disorder, like anorexia and bulimia. That should be taken very seriously, because it is an enormous threat to the personalities of young people. Ideal beauty can also polarise boys, and they might feel excluded. It could all be so easy: simply showing each other that they are amazing just as they are. Each one has their own strength, their own personality, and their own beauty. Sometimes it has yet to be discovered.

Hate and Mobbing. When we assume and observe how too many hours of using a smartphone can change the personality of an individuum, the intensity of discussions is extremely low. Carelessness seems to be common today, even if we can see deficiency symptoms and other ailments when we look around.

Where are those people who consider new technologies like cell phones dangerous and have a negative attitude, but who accept all the new diseases coming from misuse or a wrong understanding? These new threats, like addiction to cell phone use and the degeneration of physical and mental abilities, are quite real. We can see that every day at any place. Some of us don't even need to leave our houses to observe it. Another type of effect of excessive consumption seems to be even more dangerous: the psychosocial consequences. We can see the results when we look at real life. They are even discussed on TV channels in entertainment programs. Hate on the world wide web shows a tremendous increase, and it is one of the most important challenges in social life.

The internet is an anonymous field where offenders are hiding wherever they originate and to which social class they belong. They come from any social environment of the whole society in a country and from other countries, too. But what kind of people are they?

They belong to the whole cross-section of a population. Offenders can be attorneys or technicians. Sometimes it is the dear uncle in the neighbourhood or in the allotment club. They are unobtrusive employees, stalkers, and idlers who like to attack other people. Some of them have nothing better to do than insult and abuse their victims. Some offenders drive their victims to suicide.

We also see mobbing on social media, and we should think about the consequences. The exclusion of children and young people doesn't happen in the schoolyard alone, but much more in a secret area where offenders can hide behind accounts. Attempted suicide and traumatised children are not the preferred topics of people who praise social media use in the highest tones.

Politics is just at the beginning of recognising and developing instruments against that. We hope that our society becomes much more conscious of the psychosocial consequences of excessive media use. Parents, teachers, and people from the environment of mobbed and excluded children need a good power of observation and a lot of understanding for the affected children, as well as adults. When we consider the influence of new technologies that are based on electromagnetic waves, it is difficult to draw a simple conclusion. But we can sum up the conclusions with several accents.

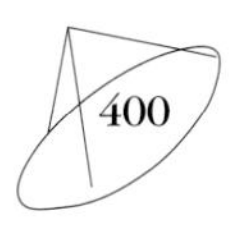

On one hand, the use of technologies based on electromagnetic waves, like computer tomography, means immense progress in the development of medical applications. Imaging methods like MRT (Magnetic Resonance Imaging) and CT (Computed Tomography) and contactless imaging with such amazing results are a boon for medicine and people.

Worldwide communication every day at any place in the world is very convenient. Starting emergency calls with a little box in our pocket has already saved many lives. Mobile radio networks are a curse and a blessing at the same time.

Generally, it is more of a blessing. But combined with the unrestrained and careless use of social networks 24/7 at any place, the personalites of people can change. And this means new challenges for society.

Excessive use of wireless technologies has an impact on personal health in the worst case. Permanent gaming or dreaming while listening to music and crossing a road at the same time can be deadly, especially when the traffic lights are red. Games and typing on the cell phone often need too much attention.

Gaming all night long, without limits, is exploiting the body and mind, and it will have pathological traits when it is not regulated early enough. Hate and mobbing on social platforms are much more threatening than most of us might assume. Parents and institutions like teachers and authorities in politics and also the police they all face a big challenge here concerning prevention and help.

When we look out at natural phenomena in the universe, we are almost fascinated and overwhelmed. Despite all scientific effort throughout the centuries and despite all our technologies, nature can stop anything on earth very soon without warning. We hope that phenomena like x-ray flashes from strange galaxies will never strike us on earth. But as scientific appearances show, they are welcome and we will look into the sky with the same fascination as people did hundreds or even thousands of years ago.

The force and its dark side. Electromagnetic waves are a special kind of energy. Humans made it on their own. Dominating this energy means to use force. We think of information coming to us from broadcasts from state authorities. Information with an opinion-forming character, which is transmitted to us night and day, shall influence us at home and abroad. We should also think of our fight against diseases like tumours, which can be identified much better and much earlier with modern imaging methods. But we should also think of the dark side. The darknet, with all its gloomy and criminal offers, is only one way to access an evil world full of threats and crime.

We can face the destruction of satellites, terrorist attacks on submarine cables, earth quakes, and solar flares. But this is not all we have to fear. There are also lies and hate. Information wars due to the truth are also part of the dark side. Not to forget: abuse of systems for theft and fraud and intrusion into the private sphere. There is a long list. But most essential are infrastructures and their ubiquity, which makes us addicted and vulnerable. We can feel it most when we are disrupted from the usual daily channels of information. One simple question makes it clear: What will be left without media and internet when we switch it off?

To come to a positive end with closing words, we go back to electromagnetic waves. The air is literally full of radio signals. This statement is still true and as valid as it ever was. Electromagnetic waves have their origin in nature. But we have learned to use this kind of energy for communication and other applications. For worldwide and local communication, electromagnetic waves as a form of energy are a blessing. People are connected, borders are overcome, and this little piece of freedom makes the world a little better for most of us.

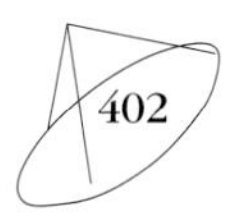

List of Illustrations

Figure 1 Tuning scale with long-, medium-, and shortwave bands plus FM. SW bands have been the preferred bands for worldwide listening to radio programmes 18

Figure 2 Ground stations are connected to the whole world by geostationary satellites. Phone calls, TV programmes, and data links are transmitted from several orbit positions 25

Figure 3 Electric and magnetic fields in their proper oscillation plane and their direction of propagation, a wave with linear polarisation. In the far field, both components are in phase, and in the near field, they have a phase difference of 90° 28

Figure 4 Coaxial cable for installations in cell phone networks with little loss up to GHz frequencies. The inner conductor is tubular. The bending radius should be greater than 200 mm. Typical diameters are 1/2, 7/8, and 1 ¼ inch 36

Figure 5 Microstrip line over a metal plate. In applications, wave impedances can be designed by the size of the line and the distance of the plate. 50…100 Ohm are typical. 39

Figure 6 Microstrip line with a patch antenna. Two variants are usual: galvanic coupling or capacitive coupling 39

Figure 7 Microstrip line on a foil between two metal plates, lying on a substrate made of foam 40

Figure 8 Feed system with a horn antenna, OMT, microwave filter, and 2 LNBs (transmission lower one/receiving upper one) 41

Figure 9 Rectangular waveguide with the electric field of the H_{10} mode (left). The electrical field strength in the middle is higher than towards the edges. Right: magnetic field lines 43

Figure 10 λ/2 vertical radiator with a typical distribution of current strength (arrows with a current maximum in the middle). The electric field has a parallel orientation to the rod. The magnetic field consists of concentric circles around the rod. Arrows: current intensity . . . 48

Figure 11 Imagination of the transition from a resonant circle to a radiating antenna when the circuit is more and more spread. The electric field lines can be seen between the plates of the capacitor (left). The energy is fed by a generator 49

Figure 12 The electric field lines leave the antenna rod when the charges oscillate. Then the antenna radiates energy, which we can receive as a radio signal 55

Figure 13 Magnetic field lines. Left: Vertical linear radiator. Horizontal beam characteristic: omnidirectional. Right: loop antenna with magnetic fieldlines in the centre of the loop antenna. Beam: maxima to the right and to the left 51

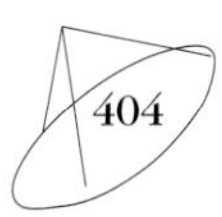

Figure 14 The ionosphere changes the direction of the wave propagation by refracting and reflecting waves. They can propagate over a distance of thousands of kilometres. The ionosphere is located at a height of 150 – 600 km in the atmosphere. 57

Figure 15 The vertical radiation characteristic of a dipole changes with its height over the ground. Left: Height over ground is 5/8 λ. Right: below 1/4 λ 57

Figure 16 The conductivity of the ground can change the propagation radius of the ground wave. a) very good conductivity, b) medium conductivity, c) poor conductivity. The vertical axis represents the elevation angle of the beam 64

Figure 17 Microwave links (i.e. on 30 and 60 GHz or more) need a free Fresnel zone without obstacles. 66

Figure 18 Multipath propagation: With the right digital signal processing, the receiver will combine different signals from the same source into one resulting signal. The reflections are added by eliminating the time difference caused by different path lengths of reflected signals. The interference is now eliminated. In GSM systems, the Rake-Receiver applies an algorithm to do this 68

Figure 19 Channel spacing for LW, MW, and SW ranges 70

Figure 20 AM signal: carrier with two side bands 78

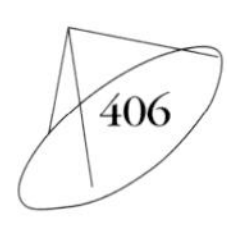

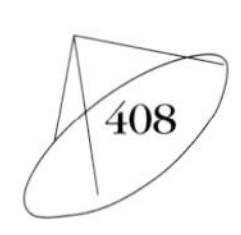

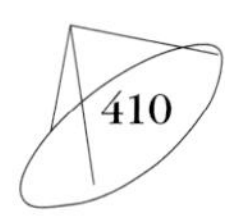

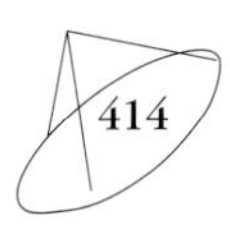

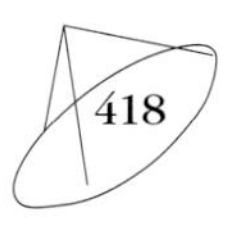

Table Directory

Bibliography

[1] HF in Funk und Radar, H.-G. Unger, Teubner-Studienskripte

[2] Conversion power/voltage, "Funkamateur" Ausgabe 02/2018

[3] Conversion VSWR, return loss, "Funkamateur" Ausgabe 02/2019

[4] Airliner-Scatter, "Funkamateur" Ausgabe 05/2022

[5] Untersuchungen zur Frequenzselektivität quasi-periodischer Schleifengitter, Thomas Kaczmarek, Diplom-Arbeit, Ruhr-Universität Bochum 1990

[6] Piratensender, Wolf-Dieter Roth, Siebel-Verlag

[7] Rothammels Antennenbuch, DARC-Verlag

[8] Radio Code Manual, Shortwave Frequency Guide, Klingenfuss-Verlag

Glossary

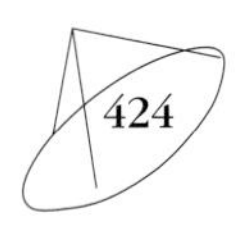

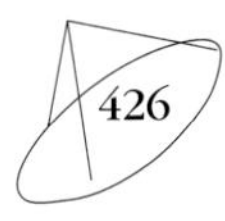

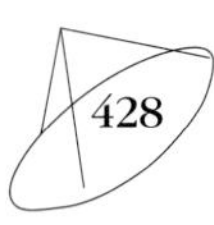

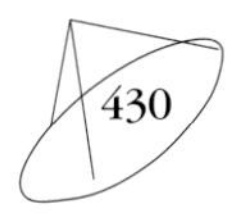

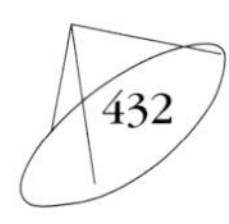

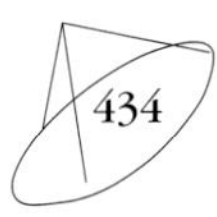